Frontiers of Targeted Therapy and Predictors of Treatment Response in Systemic Sclerosis

Frontiers of Targeted Therapy and Predictors of Treatment Response in Systemic Sclerosis

Editor

Michal Tomcik

MDPI • Basel • Beijing • Wuhan • Barcelona • Belgrade • Manchester • Tokyo • Cluj • Tianjin

Editor
Michal Tomcik
Institute of Rheumatology in Prague
Czech Republic

Editorial Office
MDPI
St. Alban-Anlage 66
4052 Basel, Switzerland

This is a reprint of articles from the Special Issue published online in the open access journal *Biomedicines* (ISSN 2227-9059) (available at: https://www.mdpi.com/journal/biomedicines/special_issues/Therapy_Systemic_Sclerosis).

For citation purposes, cite each article independently as indicated on the article page online and as indicated below:

LastName, A.A.; LastName, B.B.; LastName, C.C. Article Title. *Journal Name* **Year**, *Volume Number*, Page Range.

ISBN 978-3-0365-6121-9 (Hbk)
ISBN 978-3-0365-6122-6 (PDF)

Cover image courtesy of Michal Tomcik

Contents

About the Editor

Michal Tomcik

Michal Tomcik, MD, PhD, is an associate professor at the First Faculty of Medicine of Charles University, teaching both undergraduate and PhD students, and is a board-certified rheumatologist at the Institute of Rheumatology in Prague, Czech Republic. He is a treasurer of the Czech Society for Rheumatology, a member of the European Scleroderma Trials and Research Group, a board member of several Czech research committees and grant agencies, and an editorial board member of international journals with a focus on biology and molecular sciences. His primary research interests are in the basic, translational, and clinical aspects of systemic sclerosis and secondarily in inflammatory myopathies, rheumatoid arthritis, and spondyloarthropathies. His academic work has resulted in 66 publications in peer-reviewed English-language, impact-factor journals with 2000+ total citations and an H-index of 26 according to the Web of Science. In addition, he has co-authored 6 books and 14 book chapters in Czech and English, and 37 publications in peer-reviewed Czech or English journals without an impact factor.

Preface to "Frontiers of Targeted Therapy and Predictors of Treatment Response in Systemic Sclerosis "

Among the group of connective tissue diseases, systemic sclerosis (scleroderma) is considered to be the most challenging one and is burdened by the highest cause-specific mortality, heterogeneity in disease course, severe impact on overall function and quality of life, and the unpredictable efficacy of the currently used pharmacological therapies. It is characterized by three main intertwined pathological mechanisms, i.e., vasculopathy, immune system activation and dysregulation, and progressive tissue fibrosis. Recent advances in basic, translational, and clinical research have shed further light on the etiology and pathophysiology of scleroderma, though it still remains incompletely understood, and have improved long-term survival.

This book is a collection of all manuscripts accepted for publication from January 2021 to June 2022 in the Special Issue of Biomedicines entitled "Frontiers of Targeted Therapy and Predictors of Treatment Response in Systemic Sclerosis" under the section "Molecular and Translational Medicine", where I served as a guest editor. It starts with my editorial, which provides a brief overview of the main outcomes demonstrated in the following six original research articles and five review articles kindly contributed by well-established scleroderma experts, addressing the most recent advancements and further elucidation of the pathophysiology, natural course of the disease, and therapeutic options for this enigmatic disease.

I would like to express my gratitude to all the contributors, colleagues, research collaborators, and support (Ministry of Health of the Czech Republic [00023728, NV18-01-00161A] and Charles University Grant Agency [GAUK-114122]), as well as my students and mentors in rheumatology and to all the patients who have taught me about systemic sclerosis and were the source of inspiration over the years. I gratefully acknowledge Yajun Li for the kind invitation to be a guest editor of this Special Issue and for her dedicated technical assistance. Most of all, I am indebted and deeply grateful to my family for their unwavering love, patience, and support.

Michal Tomcik
Editor

Editorial

Frontiers of Targeted Therapy and Predictors of Treatment Response in Systemic Sclerosis

Michal Tomcik

Institute of Rheumatology, Department of Rheumatology of the First Faculty of Medicine, Charles University, Na Slupi 4, 128 00 Prague, Czech Republic; tomcik@revma.cz

Citation: Tomcik, M. Frontiers of Targeted Therapy and Predictors of Treatment Response in Systemic Sclerosis. *Biomedicines* **2022**, *10*, 3053. https://doi.org/10.3390/biomedicines10123053

Received: 14 November 2022
Accepted: 22 November 2022
Published: 28 November 2022

Systemic sclerosis (scleroderma, SSc) is one of the most challenging rheumatic diseases, characterized by vasculopathy, dysregulation of the immune response, and progressive tissue fibrosis affecting the skin, lungs, heart, digestive tract, and kidneys. To date, it is considered incurable and carries the highest cause-specific mortality of all connective tissue diseases. This is especially true in the later stages of the disease, when irreversible damage to the musculoskeletal system and internal organs prevails, which significantly reduces overall function, the ability to perform activities of daily living, and the quality of life. Despite advances in basic, translational, and clinical research in recent years, the etiology and pathophysiology of this complex condition remain to be elucidated. In spite of a large number of clinical trials and the progress made in their design over the last decade, no approved disease-modifying therapies exist for SSc to date. Currently available pharmacological therapies predominantly target inflammatory and vascular pathways, have variable and unpredictable clinical efficacy and usually undesirable safety profiles, and only have a modest effect on long-term survival [1].

Therefore, the focus of several research groups has been to unravel novel therapeutic targets driving the main pathological processes of this disease and to identify potential predictors of the treatment response of the currently available therapies, in order to optimize the stratification of patients with this heterogeneous disease while minimizing the possible adverse events and maximizing the potential benefit of therapy. The collection of six original research manuscripts and five review articles highlights the most recent progress in this field and provides further elucidation of the pathogenesis and natural disease course of this enigmatic disease.

A large body of evidence supports the concept that the activation of the immune system drives the complex interplay between vasculopathy and fibrogenesis in SSc. In their review article, Papadimitriou et al. focus on the contribution of the innate and acquired immune system to the development and progression of tissue fibrosis in SSc [2]. They provide a complex overview of crucial cellular and soluble immune mediators implicated in SSc pathogenesis and discuss their impact on skin and lung fibrosis. Furthermore, they review the evidence supporting the use of first-line anti-inflammatory therapies (e.g., corticosteroids, methotrexate, cyclophosphamide, mycophenolate mofetil, and autologous hematopoietic stem cell transplantation), T cell-, B cell-, and cytokine-specific targeted therapies (e.g., rituximab, inebilizumab, belimumab, abatacept, tocilizumab, rilonacept, basiliximab, fresolimumab, metelimumab, and pirfenidone), as well as emerging therapies with tyrosine kinase inhibitors (e.g., imatinib mesylate, nilotinib, nintedanib, and tofacitinib) to reduce fibrosis.

Among the plethora of soluble immune mediators implicated in the pathogenesis of SSc, interleukin (IL)-6 plays a vital role both in microangiopathy and the development of tissue fibrosis. Cardoneanu et al. provide an overview of the involvement of IL-6 in the activation and apoptosis of vascular endothelial cells, the potentiation and maintenance of innate and adaptive immune response, and fibrogenesis [3]. They discuss the systemic adverse effects of IL-6 in the involvement of the skin, lungs, kidneys, gastrointestinal tract, and cardiovascular system, and present a comprehensive overview of the published data on the clinical efficacy of tocilizumab in the treatment of SSc-related manifestations.

All three of the above-mentioned processes involved in the pathogenesis of SSc can also be modulated by epigenetic mechanisms. In their review article, Szabo et al. provide an extensive overview of complex interactions of micro RNAs (miRNAs, small non-coding RNA sequences) with microangiopathy, immune activation, and tissue fibrosis [4]. They present the published serum- and tissue-specific miRNA signatures in SSc and focus on miRNA transcripts with pro- and anti-fibrotic properties, and the roles of miRNAs in fibroblast apoptosis and the pathogenesis of SSc-related interstitial lung disease (ILD), vascular damage and immune dysfunction. Furthermore, they discuss the potential use of miRNAs as diagnostic and prognostic biomarkers as well as potential novel targeted therapies of SSc.

Zaaroor Levy et al. studied the circulating miRNAs that may differentiate between SSc patients with and without pulmonary arterial hypertension (PAH) and assessed their expression in plasma, white blood cells, and dermal myofibroblasts [5]. They demonstrate that plasma levels of miR-26 and miR-let-7d are decreased in SSc-PAH patients compared to SSc patients without PAH. Furthermore, their findings suggest that miR-26 and miR-let-7d might be involved in the activation of fibroblasts by regulating the genes in the transforming growth factor (TGF)-β and endothelin-1 pathways, and their reduced plasma levels could be an early predictor of the development of PAH.

In their original article, Štorkánová et al. present compelling evidence for a potential novel therapeutic target implicated in SSc-related skin fibrosis [6]. Using a modified murine model of bleomycin-induced dermal fibrosis, they demonstrate that the inhibition of Heat shock protein (Hsp)-90 by 17-dimethylaminoethylamino-17-demethoxy-geldanamycin (17-DMAG) prevents further progression and may induce regression of the established skin fibrosis with a comparable effect to that of nintedanib. The anti-fibrotic effects of 17-DMAG were mediated by the inhibition of fibroblast activation, suppression of TGF-β/Smad signaling, and the systemic and local inflammatory response induced by bleomycin. Since several Hsp90 inhibitors have been assessed in clinical trials for oncological indications, these preclinical findings may have translational implications in the treatment of SSc-related tissue fibrosis.

Bögl et al. used targeted and untargeted metabolomic approaches to elucidate the distinct biochemical mechanisms that might be relevant to the pathophysiology of SSc [7]. Their metabolomic profiling in plasma from SSc patients compared to healthy controls revealed four significantly dysregulated metabolic pathways, namely the kynurenine pathway, urea cycle, lipid metabolism, and gut microbiome, all of which are associated with autoimmune inflammation, vascular damage, fibrosis, and intestinal dysbiosis. These four altered metabolomics networks deserve further validation; however, they could become potential therapeutic targets or biomarkers of treatment response in SSc.

The potential pathomechanisms leading to the maturation of autoreactive B cells, formation of autoantibodies, and processes underlying the potential promotion and maintenance of pathologic processes involved in SSc are not sufficiently understood. These issues are extensively discussed in the review article of Graßhoff and others [8]. They also provide a comprehensive overview of autoantibodies which might be useful as biological predictors of disease course and of therapeutic approaches that have been or might be used in the future for the treatment of autoantibody-induced pathologies in SSc.

Recent advances in the basic and clinical research on ILD, as a leading cause of mortality in SSc, have led to the discovery of novel molecular therapeutic targets. The review article by Aragona et al. analyzes phase II/III randomized clinical trials focused on SSc-ILD, which were published between January 2016 and December 2021 [9]. The authors provide supporting evidence on the phase III trials involving nintedanib, tocilizumab, and lenabasum, as well as of phase II trials involving pirfenidone, pomalidomide, romilkimab, riociguat, rituximab, and abatacept. In addition, they report on the head-to-head studies of rituximab vs. cyclophosphamide and mycophenolate mofetil vs. cyclophosphamide, as well as future directions in clinical trials on SSc-ILD.

Two studies from the research group of Veronika Müller shed some light on the disease course of SSc-ILD from two different perspectives. Nagy T et al. examined SSc-ILD patients with a normal initial spirometry and followed up with them for at least one year [10]. They reported that patients who present with cough and pulmonary hypertension were at a higher risk of developing progressive decline in lung functions. Similarly, Nagy A et al. assessed the disease course and progressive functional decline in SSc-ILD patients treated in a real-world setting and demonstrated that the most prominent functional decline occurred in untreated patients, which was as expected, but paradoxically, in those with BMI < 25 kg/m^2 [11]. Thus, overweight SSc-ILD patients appear to be at a lower risk of progressive functional decline. These predictors require further validation; however, they might be useful for defining a high-risk population of patients, who should be closely monitored with regular follow-ups and started early and aggressively on immunosuppressive and/or antifibrotic therapy.

In the last original article, Rauch et al. address the unmet need for new treatment options for a therapeutically problematic clinical manifestation of SSc, cutaneous calcinosis [12]. They retrospectively assessed the effect of a second-generation bisphosphonate, intravenous pamidronate, on cutaneous calcinosis, and demonstrated a reduction in pain, improvement in general condition, and cessation of progression in 83%, regression of calcinosis in 33% of patients, stable disease in 17% of patients, and radiological improvement or stabilization in 50% of the patients. Fever was the most common side effect and one patient developed jaw osteonecrosis.

Hopefully, the novel advancements included here covering several aspects of the pathogenesis, prognosis, and management of SSc will contribute to a reduction in suffering and disability, as well as an improvement in the morbidity and mortality of our patients in the near future.

Funding: This manuscript was supported by the Ministry of Health of the Czech Republic [00023728, NV18-01-00161A] and Ministry of Education Youth and Sports [GAUK-114122].

Acknowledgments: I would like to express my gratitude to the authors who decided to contribute their valuable work and the journal editors for their efforts to make the Special Issue a success. I would like to thank Xiao Švec for language editing.

Conflicts of Interest: The author declares no conflict of interest.

References

1. Denton, C.P.; Khanna, D. Systemic sclerosis. *Lancet* **2017**, *390*, 1685–1699. [CrossRef] [PubMed]
2. Papadimitriou, T.I.; van Caam, A.; van der Kraan, P.M.; Thurlings, R.M. Therapeutic Options for Systemic Sclerosis: Current and Future Perspectives in Tackling Immune-Mediated Fibrosis. *Biomedicines* **2022**, *10*, 316. [CrossRef] [PubMed]
3. Cardoneanu, A.; Burlui, A.M.; Macovei, L.A.; Bratoiu, I.; Richter, P.; Rezus, E. Targeting Systemic Sclerosis from Pathogenic Mechanisms to Clinical Manifestations: Why IL-6? *Biomedicines* **2022**, *10*, 318. [CrossRef] [PubMed]
4. Szabo, I.; Muntean, L.; Crisan, T.; Rednic, V.; Sirbe, C.; Rednic, S. Novel Concepts in Systemic Sclerosis Pathogenesis: Role for miRNAs. *Biomedicines* **2021**, *9*, 1471. [CrossRef] [PubMed]
5. Zaaroor Levy, M.; Rabinowicz, N.; Yamila Kohon, M.; Shalom, A.; Berl, A.; Hornik-Lurie, T.; Drucker, L.; Tartakover Matalon, S.; Levy, Y. MiRNAs in Systemic Sclerosis Patients with Pulmonary Arterial Hypertension: Markers and Effectors. *Biomedicines* **2022**, *10*, 629. [CrossRef] [PubMed]
6. Storkanova, H.; Storkanova, L.; Navratilova, A.; Becvar, V.; Hulejova, H.; Oreska, S.; Hermankova, B.; Spiritovic, M.; Becvar, R.; Pavelka, K.; et al. Inhibition of Hsp90 Counteracts the Established Experimental Dermal Fibrosis Induced by Bleomycin. *Biomedicines* **2021**, *9*, 650. [CrossRef] [PubMed]
7. Bogl, T.; Mlynek, F.; Himmelsbach, M.; Sepp, N.; Buchberger, W.; Geroldinger-Simic, M. Plasma Metabolomic Profiling Reveals Four Possibly Disrupted Mechanisms in Systemic Sclerosis. *Biomedicines* **2022**, *10*, 607. [CrossRef] [PubMed]
8. Grasshoff, H.; Fourlakis, K.; Comduhr, S.; Riemekasten, G. Autoantibodies as Biomarker and Therapeutic Target in Systemic Sclerosis. *Biomedicines* **2022**, *10*, 2150. [CrossRef] [PubMed]

9. Aragona, C.O.; Versace, A.G.; Ioppolo, C.; La Rosa, D.; Lauro, R.; Tringali, M.C.; Tomeo, S.; Ferlazzo, G.; Roberts, W.N.; Bitto, A.; et al. Emerging Evidence and Treatment Perspectives from Randomized Clinical Trials in Systemic Sclerosis: Focus on Interstitial Lung Disease. *Biomedicines* **2022**, *10*, 504. [CrossRef] [PubMed]
10. Nagy, T.; Toth, N.M.; Palmer, E.; Polivka, L.; Csoma, B.; Nagy, A.; Eszes, N.; Vincze, K.; Barczi, E.; Bohacs, A.; et al. Clinical Predictors of Lung-Function Decline in Systemic-Sclerosis-Associated Interstitial Lung Disease Patients with Normal Spirometry. *Biomedicines* **2022**, *10*, 2129. [CrossRef] [PubMed]
11. Nagy, A.; Palmer, E.; Polivka, L.; Eszes, N.; Vincze, K.; Barczi, E.; Bohacs, A.; Tarnoki, A.D.; Tarnoki, D.L.; Nagy, G.; et al. Treatment and Systemic Sclerosis Interstitial Lung Disease Outcome: The Overweight Paradox. *Biomedicines* **2022**, *10*, 434. [CrossRef] [PubMed]
12. Rauch, L.; Hein, R.; Biedermann, T.; Eyerich, K.; Lauffer, F. Bisphosphonates for the Treatment of Calcinosis Cutis-A Retrospective Single-Center Study. *Biomedicines* **2021**, *9*, 1698. [CrossRef] [PubMed]

Review

Therapeutic Options for Systemic Sclerosis: Current and Future Perspectives in Tackling Immune-Mediated Fibrosis

Theodoros-Ioannis Papadimitriou *, Arjan van Caam, Peter M. van der Kraan and Rogier M. Thurlings

Department of Rheumatic Diseases, Radboud University Medical Center, 6525 GA Nijmegen, The Netherlands; Arjan.vanCaam@radboudumc.nl (A.v.C.); Peter.vanderKraan@radboudumc.nl (P.M.v.d.K.); rogier.thurlings@radboudumc.nl (R.M.T.)

* Correspondence: theogiannis.papadimitriou@radboudumc.nl

Abstract: Systemic sclerosis (SSc) is a severe auto-immune, rheumatic disease, characterized by excessive fibrosis of the skin and visceral organs. SSc is accompanied by high morbidity and mortality rates, and unfortunately, few disease-modifying therapies are currently available. Inflammation, vasculopathy, and fibrosis are the key hallmarks of SSc pathology. In this narrative review, we examine the relationship between inflammation and fibrosis and provide an overview of the efficacy of current and novel treatment options in diminishing SSc-related fibrosis based on selected clinical trials. To do this, we first discuss inflammatory pathways of both the innate and acquired immune systems that are associated with SSc pathophysiology. Secondly, we review evidence supporting the use of first-line therapies in SSc patients. In addition, T cell-, B cell-, and cytokine-specific treatments that have been utilized in SSc are explored. Finally, the potential effectiveness of tyrosine kinase inhibitors and other novel therapeutic approaches in reducing fibrosis is highlighted.

Keywords: systemic sclerosis; immune cells; anti-inflammatory; therapy; fibrosis

Citation: Papadimitriou, T.-I.; van Caam, A.; van der Kraan, P.M.; Thurlings, R.M. Therapeutic Options for Systemic Sclerosis: Current and Future Perspectives in Tackling Immune-Mediated Fibrosis. *Biomedicines* **2022**, *10*, 316. https://doi.org/10.3390/biomedicines10020316

Academic Editor: Michal Tomcik

Received: 21 December 2021
Accepted: 26 January 2022
Published: 29 January 2022

1. Introduction

Systemic sclerosis (SSc) is a severe rheumatic, auto-immune disease which affects up to 20 people per 100,000 and women up to nine times more often than men [1–3]. Inflammation, vasculopathy, and fibrosis are the key hallmarks of SSc. Patients suffer from fibrotic skin lesions, and as the disease progresses, the function of internal organs including the heart, lungs, gastrointestinal tract, and kidneys deteriorates due to fibrosis. The disease has a high morbidity and greatly negatively affects the quality of life and life expectancy of patients.

Although the pathophysiology of SSc has not been elucidated yet, the immune system has been hypothesized as an important driver of the disease (Figure 1). SSc patients exhibit disease-specific autoantibodies, and a typical cytokine profile in their blood indicating immune cell activation. This profile is characterized by increased T helper 2 (Th2) and decreased T helper 1 (Th1) cytokines [4,5]. Furthermore, in skin biopsies of early disease patients, perivascular accumulation of immune cells, such as CD4+ and CD8+ cytotoxic T cells (CTLs), can be observed [6]. This immune activation can cause (or is maybe a response to) capillary damage, which leads to capillary breakdown, adherence of platelets, and activation of pro-fibrotic pathways. Additionally, vascular injury causes damage and apoptosis of endothelial cells. The release of internal damage-associated molecular patterns (DAMPs) increases microvascular permeability which causes additional recruitment of immune cells to the endothelium and therefore increased immune cell activation and inflammation [7].

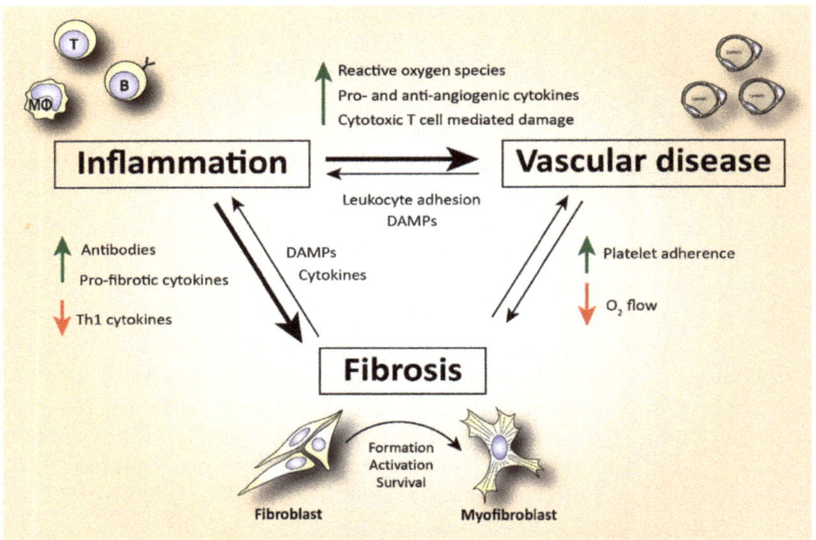

Figure 1. Simplified schematic representation of the complex interplay between inflammation, vasculopathy, and fibrosis in SSc. Abbreviations: Th1 = T helper type 1 cells; DAMPs = damage-associated molecular patterns; T = T lymphocyte; B = B lymphocyte; MΦ = macrophage.

Activation of the immune system is also linked to fibrosis. For example, Th2 cytokines, such as interleukin (IL)-4 and -13, can activate myofibroblasts [8]. Myofibroblasts are a strongly pro-fibrotic cell type which produces large amounts of extracellular matrix (ECM) molecules and matrix-strengthening enzymes. Furthermore, these cells also excrete growth factors and cytokines that worsen inflammation [9]. IL-4, IL-6, and transforming growth factor (TGF)-β are the predominant fibrogenic cytokines that cause subendothelial accumulation of fibrous tissue, leading to aberrant vascular remodeling [10]. In turn, this aberrant vascular remodeling makes the capillaries more prone to damage, fueling an immune response.

The important role of the immune system in the pathogenesis of SSc is further supported by the fact that autologous hematopoietic stem cell transplantation (ASCT) can induce long-term disease remission in patients with SSc. However, ASCT is accompanied by a high mortality rate (~10%) and is only administered in the most severe cases of SSc [11]. Currently, there is no specific and effective disease-modifying treatment available, which, regarding the severity of the disease, results in an unmet medical need. Presently, mainly broad-spectrum immunosuppressive and anti-inflammatory drugs, developed for other auto-immune diseases, are used to treat SSc. In this narrative review, we discuss the effectiveness of the currently used and/or investigated agents in diminishing skin and lung fibrosis in scleroderma patients. For this, we first provide an overview of the main pathogenic pathways that are implicated in the disease. Secondly, drugs targeting these pathways are assessed for their treatment efficacy based on results from selected clinical trials. To write this review, the databases PubMed and clinicaltrials.gov were used to search for publications up to November 2021. Combinations of Medical Subject Headings (MeSH) terms referring to SSc ("Systemic sclerosis" OR "Scleroderma, Systemic") and drugs of interest, e.g., "methotrexate", were included in the search strategy.

2. The Role of Immune Cells in SSc-Related Inflammation and Fibrosis

A large body of evidence suggests that both the innate and the adaptive immune system are involved in the fibrogenesis of SSc.

2.1. The Role of Innate Immunity in SSc

The innate immune system mediates the immediate defensive response against pathogens and chemical or mechanical damage. An increased number of macrophages, mast cells, type 2 innate-like lymphoid cells, eosinophils, and plasmacytoid dendritic cells have been identified in SSc tissues [12]. The induction of excessive numbers of myofibroblasts that is observed in SSc is partly mediated by the previously mentioned innate immune cells. Other innate immune cell types such as platelets, neutrophils, and natural killer cells have also been found to be dysregulated in SSc pathology. However, only a few studies highlighting their role in SSc have been published. Thus, in this review, we focus on the cells of the innate immune system that are well characterized in SSc.

2.1.1. Macrophages

Macrophages are effector phagocytotic cells specialized in eliminating pathogens. These cells play an essential role in homeostatic manifestations related to the disposal of internal waste products and tissue repair. Monocyte-derived macrophages can be found in the blood circulation under inflammatory conditions but are prevalent in all human tissues as tissue-resident macrophages. However, macrophages exhibit a highly heterogeneous phenotype that is regulated by their microenvironment. Under the influence of certain inflammatory mediators, macrophages can be polarized in vitro towards a classically activated (M1) inflammatory phenotype or an alternatively activated (M2) tissue repair phenotype. M2 macrophages are often characterized by the secretion of large amounts of IL-10, TGF-β, and other pro-fibrotic cytokines such as IL-4, IL-6, and IL-13 that play an essential role in wound healing and tissue repair (Figure 2). These cytokines are also known to induce fibrosis by activating myofibroblasts [13]. On the other hand, in states of chronic tissue repair, M2 macrophages can also exhibit anti-fibrotic functions that lower the progression of fibrosis by suppressing local T helper cell responses and decreasing the production of ECM by (myo)fibroblasts [14,15]. Furthermore, to a certain extent, M1 and M2 phenotypes are artificial in vitro cell states. In vivo, different types of anti-inflammatory and anti-fibrotic macrophages have been reported in the process of wound healing, repair, and fibrosis [16].

In SSc, transcriptomic and immunohistochemistry studies identified a prominent pro-fibrotic M2 macrophage signature in patients' skin and lung lesions that was correlated with an increased skin score and disease severity [17–19]. In addition, a dominant M2 monocyte signature has also been observed in SSc blood [20]. This notion suggests a role of M2 macrophages in the immunopathogenesis of SSc. The importance of macrophages in regulating fibrosis is further supported by the observation that mice lacking macrophages (with the use of liposomal chlodronate) exhibit reduced bleomycin-induced lung fibrosis [21]. In skin lesions, M2 macrophages are characterized by overexpression of the scavenger receptor CD163 or CD204 [22]. The number of CD163+ or CD204+ cells was found to be significantly expanded in SSc compared to healthy skin [23]. Strikingly, these activated macrophages were localized not only in the perivascular areas, but also between thickened collagen fibers. This indicates a potential role of tissue-infiltrating macrophages in the development of fibrosis [24]. However, it is worth mentioning that bone marrow-derived mesenchymal progenitors such as monocytes and fibrocytes may also be implicated in SSc-related fibrosis. These cells are able to migrate from the circulation to affected tissues such as the lungs, where they differentiae into activated (myo)fibroblasts [25]. Of note, it has been reported that fibrocytes were only detected in patients with connective tissue disorders (including SSc) and not in healthy donors [26]. In SSc, the frequency of circulating fibrocytes was positively correlated with an increased dermal thickness [27]. Furthermore, in a very recent single-cell RNA sequencing (sc-RNA seq) study, highly proliferating macrophages were only found in SSc and not healthy skin and were correlated with increased skin fibrosis [28]. All in all, the role of macrophages in fibrotic activation in SSc is prominent, but the signaling pathways characterizing their aberrant activation remain poorly understood. In addition, their role as regulators of the disease, in combination with deciphering a potential role of anti-fibrotic macrophages in SSc, definitely warrants further research. Unraveling the pathogenic mechanisms by which macrophages mediate fibrosis will likely contribute to targeted therapies that reduce fibrosis and ameliorate inflammation in SSc and other connective tissue disorders. To this end, future therapeutic options should opt to reduce the activation/number of activated pro-fibrotic and pro-inflammatory macrophages while boosting the activity of the anti-inflammatory and anti-fibrotic ones.

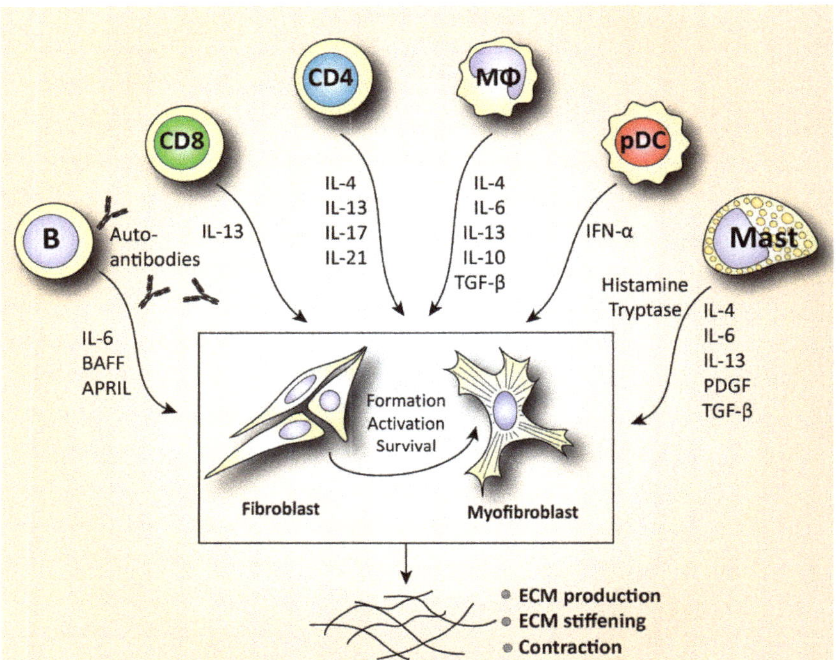

Figure 2. An overview of the main pathogenic mechanisms by which T cells, B cells, macrophages, mast cells, and plasmacytoid dendritic cells may contribute to (pro-)fibrotic manifestations in SSc immunopathology. Abbreviations: MΦ = macrophages; B = B lymphocyte; CD8 = cytotoxic T cell; CD4 = T helper cell; pDC = plasmacytoid dendritic cell; TGF-β = transforming growth factor-β; BAFF = B cell activating factor; APRIL = a proliferation-inducing ligand; PDGF = platelet-derived growth factor; ECM = extracellular matrix.

2.1.2. Eosinophils and Mast Cells

As we discussed previously, serum levels of the cytokines IL-4, IL-10, and IL-13 are elevated in SSc patients and are correlated with increased fibrosis and disease severity. Meanwhile, IL-4 and IL-13 play an essential role in eosinophil-mediated inflammation, suggesting a potential role of eosinophils in SSc pathology [29,30]. Eosinophils are granulocytes derived from the myeloid stem cells that are part of the innate immune system. Elevated levels of eosinophils have been described in SSc and other diseases of the connective tissue [31]. In the peripheral blood of early untreated SSc patients, eosinophil counts were higher compared to patients with other major collagen diseases such as dermatomyositis, Sjogren's syndrome, and systemic lupus erythematosus. Furthermore, eosinophil counts were significantly correlated with severe interstitial lung disease (ILD) and an increased skin thickness in patients with SSc, but not in patients with other collagen diseases [32]. In addition, the presence of skin ulcers in SSc has been associated with elevated counts of peripheral blood eosinophils [33]. Collectively, these results implicate eosinophils in the pathogenesis of SSc including the hallmarks of inflammation and vascular dysfunction. However, studies exploring the exact mechanism of inflammation or ulcer formation are lacking.

Another cell type of the innate immune system derived from myeloid stem cells is mast cells. Mast cells have been detected in SSc tissues and can exhibit pro-fibrotic manifestations. Elevated infiltration of mast cells has been found in various SSc tissues including the skin and salivary glands [34–36]. Mast cell infiltration has been correlated with more severe disease phenotypes. Except for their well-established role in immediate inflammatory and allergic reactions, these cells are also implicated in cardiac, renal, and pulmonary fibrosis. Mast cells have been linked to fibrosis by their ability to produce pro-fibrotic cytokines such as IL-4, IL-6, IL-13, tumor necrosis factor alpha (TNF-α), platelet-derived growth factor (PDGF), and TGF-β [37]. This way, mast cells stimulate the production and activity of myofibroblasts (Figure 2).

Serotonin is another molecule that is produced by mast cells. Serotonin can directly increase ECM deposition in primary skin fibroblasts in a TGF-β-dependent manner [38]. In addition, mast cells produce tryptase, a serine proteinase which triggers fibroblast proliferation and collagen production [39]. It was also demonstrated that fibroblast proliferation in patients' lungs can be caused by mast cell-related histamine release [40]. More specifically, human primary lung fibroblasts that were cultured in the presence of physiologically relevant concentrations of histamine exhibited increased cell proliferation that was mediated through an H2 histamine receptor on the fibroblasts. Interestingly, the observed fibroblast proliferation was inhibited by an H2 antagonist, empowering the role of histamine release in fibroblast proliferation. Of note, mast cell depletion with phototherapy ameliorates SSc fibrosis in vivo [41]. Thus, targeting mast cells in systemic sclerosis might be an effective treatment approach. We believe that exploring the immunopathogenic role of eosinophils and mast cells with the currently available novel arsenal of biomolecular techniques will unveil new innate pathogenic pathways that could pave the way for novel treatment approaches.

2.1.3. New Players: Innate Lymphoid Cells and Plasmacytoid Dendritic Cells

Among the family of innate immune cells, a highly interesting tissue-resident cell type that was recently shown to be involved in SSc is innate lymphoid cells (ILCs). These cells are derived from common lymphoid progenitors such as adaptive immune cells, but they lack rearranged antigen receptors. Therefore, ILCs do not exhibit antigen-specific responses but are characterized by a functional diversity similar to that of T lymphocytes. These cells have been categorized into distinct subtypes based on their cytokine and transcriptome profile. One of these subtypes is ILC2, which is identified by the expression of the transcription factor GATA-3. As with Th2 cells, ILC2s predominantly produce IL-4, IL-5, and IL-13 [42].

Intriguingly, locally accumulating ILC2s have emerged as a pivotal source of pro-fibrotic cytokines in inflammatory and fibrotic diseases [43]. Studies using the carbontetrachloride (CCl4)-induced liver fibrosis mouse model showed that ILC2s mediate liver fibrosis by producing the pro-fibrotic cytokine IL-13 [44]. This observation suggests that ILC2s may stimulate the activation of fibroblasts and thus increase tissue fibrosis. Indeed, ILC2 counts are significantly increased in both the skin and peripheral blood from patients with SSc compared to healthy controls, and their numbers are correlated with an increased skin thickness [45,46]. This notion is further supported by studies that have shown an elevated expression of the ILC2 cytokines IL-25, IL-33, and thymic stromal lymphopoietin (TSLP) in both the serum and skin of patients with SSc [46–48]. From a mechanistic point of view, TGF-β and IL-10 are two cytokines that have been implicated in ILC2-mediated skin fibrosis in SSc. Elevated TGF-β enhanced the in vitro pro-fibrotic function of ILC2s by increasing the activation of myofibroblasts and downregulating the levels of IL-10. Downregulation of IL-10 increased the production of collagen by dermal fibroblasts. In the same study, TGF-β inhibition combined with IL-10 administration prevented fibrotic manifestations in a mouse model recapitulating SSc [49]. Although there is a limited number of studies associating ILC2s in the pathogenesis of SSc, we believe that findings implicating

TGF-β in their potential fibrotic mechanism might be promising in utilizing alternative therapeutic strategies that are based on TGF-β blocking.

Plasmacytoid dendritic cells (pDCs) are innate immune cells that secrete large amounts of interferon (IFN) and mediate Toll-like receptor (TLR)-induced inflammation in autoimmunity. pDCs have been recently detected in the sclerotic skin of patients with SSc. These cells were found to be chronically activated and characterized by an elevated expression of chemokine (C-X-C motif) ligand 4 (CXCL4) and IFN-α in a TLR8-dependent manner [28,50]. CXCL4 has been identified as a biomarker that is associated with severe disease and lung fibrosis in SSc [51]. pDCs have also been detected in bronchoalveolar lavage fluids of patients with SSc, and their levels were correlated with increased lung fibrosis [52]. Compared to patients with idiopathic pulmonary fibrosis (IPF), SSc-ILD-infiltrating pDCs exhibited a stronger IFN and stress response gene signature, suggesting that these cells demonstrate disease-specific mechanisms in SSc [53]. Recent research has provided mounting evidence suggesting that SSc is an IFN-driven disease [54]. Indeed, an elevated expression of type 1 IFN signaling (IFN-α, IRF5, IRF7, IRF8) and its inducible genes (IL-6, STAT1, STAT3) has been illustrated in tissue biopsies, the peripheral blood, and serum of SSc patients and was correlated with disease severity [55]. In the blood circulation, on the other hand, numbers of pDCs were decreased in SSc patients compared to healthy controls, probably due to their accumulation in the fibrotic skin [56]. Additionally, pDCs are well known for their prevalent antigen-presenting role. As we discussed, SSc is characterized by auto-immunity, and the presence of multiple anti-nuclear antibodies has been reported in patients' serum. DNA topoisomerase I (topoI) is the most prevalent autoantibody in SSc and is correlated with increased disease severity and mortality [57]. Mice administered with topoI-loaded pDCs exhibited robust autoantibody production accompanied by long-term lung and skin fibrosis [58]. Furthermore, pDC depletion in the bleomycin mouse model attenuates skin and lung fibrosis and improves clinical scores. Interestingly, numbers of B and T lymphocytes were also reduced in mouse lungs, demonstrating an important role of pDCs in ongoing innate and adaptive immune abnormalities [50,52].

2.2. The Role of Adaptive Immunity in SSc

Activation of the innate immune system in SSc is essential in activating the adaptive immune response by presentation of antigens with parallel expression of danger-indicating molecules. This activation is mediated either by cell–cell interactions, or by the release of soluble mediators [59]. We will now briefly discuss the role of T and B lymphocytes in SSc pathogenesis (Table 1), in order to better explain the possible treatment options.

2.2.1. T Lymphocytes

The role of T helper lymphocytes is well characterized in SSc pathology. Th1, Th2, and Th17 subtypes have been paid the most attention. However, it should be noted that these subtype classifications have been derived from in vitro studies. In vivo, these subsets have been implicated in various types of pathogens. In chronic inflammatory diseases, they have been shown to exhibit a certain extent of plasticity and multifunctionality. In Th2-associated pathologies (including SSc), additional T cell subtypes such as Th9 and Th22 cells have also been characterized. However, the specific importance of them still remains to be determined. Studies implicating T helper cell subsets in SSc pathogenesis have mostly utilized older conventional research techniques such as immunohistochemistry in SSc skin and flow cytometry in patients' blood. Elevated production of the Th2-associated cytokines IL-4, IL-13, and IL-10 and the Th17 signature cytokine IL-17 have been demonstrated in SSc tissues and the blood circulation [4,60–63] (Figure 2). These studies suggested that SSc is driven by Th2- and Th17-type mechanisms. Elevated IL-4 levels have been associated with increased collagen production in fibroblasts and higher production of TGF-β. TGF-β directly triggers collagen production and ECM deposition but also inhibits the expression of matrix metalloproteinases [64]. An increased level of IL-17 is implicated in fibrosis and SSc-like manifestations [65]. IL-17 not only increases the proliferation of fibroblasts but also promotes the expression of TNF-α and IL-1 from macrophages which, in turn, triggers collagen, IL-6, and PDGF production from fibroblasts. In addition, IL-17 triggers the endothelial cell production of IL-1, IL-6, intracellular adhesion molecule 1 (ICAM-1), and vascular cell adhesion molecule 1 (VCAM-1) [66] (Figure 3). These adhesion molecules further interact with circulating leukocytes and facilitate their migration to and accumulation at the fibrotic sites [67]. In addition to elevated Th2 and Th17 cytokines, SSc patients exhibit a reduced expression of the anti-fibrotic Th1 cytokine IFN-γ, and this further suggests a decreased anti-fibrotic capacity [63]. A T helper cell subset which could play a protective role in SSc pathology is regulatory T cells (Tregs). Tregs maintain immunological self-tolerance, and their depletion has been associated with spontaneous auto-immunity. A plethora of animal studies and a few clinical studies have shown a potentially beneficial effect of Treg administration in various auto-immune diseases [68]. In SSc, the majority of the published literature illustrates a decreased frequency and functional ability of circulating Tregs. However, data on the frequency of tissue-resident Tregs are scarce and contradictory [69]. Overall, the mechanisms of circulating and tissue-resident T helper cell subsets that drive fibrosis are not precisely determined. It is to be proven if mechanisms such as T cell tolerance, anergy, and exhaustion are protective or deleterious in fibrogenesis.

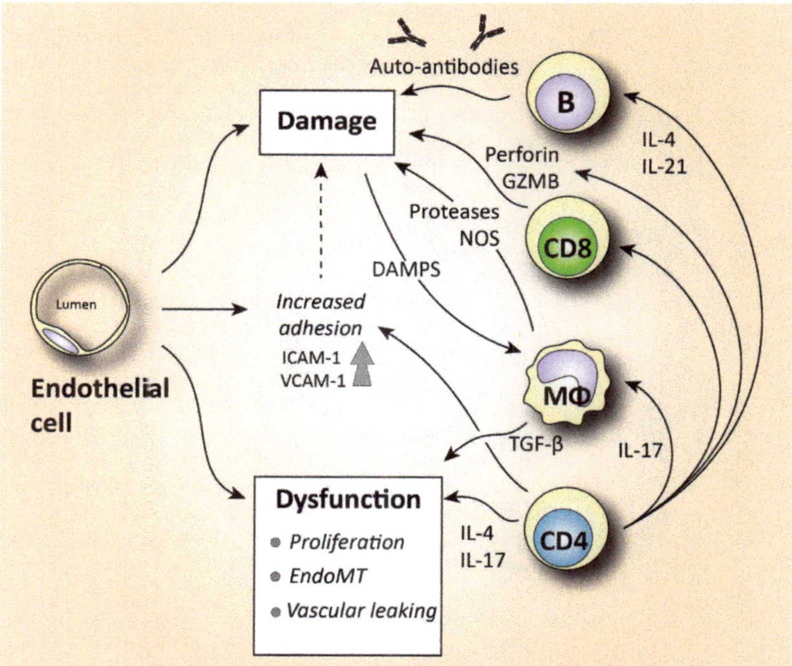

Figure 3. A simplified schematic illustration of the pathogenic mechanisms by which T cells, B cells, and macrophages may be involved in the vascular abnormalities that characterize SSc immunopathology. ICAM-1 = intercellular adhesion molecule 1; VCAM-1 = vascular cell adhesion molecule 1; EndoMT = endothelial-to-mesenchymal transition; GZMB = granzyme B; NOS = nitric oxide synthase; DAMPs = damage-associated molecular patterns; MΦ = macrophages; B = B lymphocyte; CD4 = T helper cell; CD8 = cytotoxic T cell; TGF-β = transforming growth factor-β.

Novel techniques such as sc-RNAseq and confocal immunofluorescence microscopy have provided us with essential tools to gain a deep understanding of the phenotype and function of tissue-infiltrating T cells. Utilization of these newly available techniques reveals the heterogeneity of T cell responses and opens avenues for detecting patient-specific T cell subsets. For example, quantitative analyses of skin-resident cells showed that early diffuse cutaneous systemic sclerosis (dcSSc) skin was predominantly infiltrated by granzyme A-producing CD8+ and CD4+ CTLs. Th1, Th2, and Th17 cells were also detected but in much lower amounts. This study unveiled that CD4 T cells in SSc may directly induce cell death, a function that deviates from their conventional role in promoting effector immune responses from other lymphocytes. CD4+ and CD8+ CTLs exhibited the ability to drive fibrosis and contribute to vasculopathy (Figure 3) via the secretion of pro-inflammatory cytokines such as IL-1β and/or by inducing cytotoxicity-mediated apoptosis of stromal cells, leading to exuberant tissue remodeling [6]. The importance of CTLs in SSc pathogenesis is further supported by another recent study identifying the genome-wide expression of cytotoxic genes in SSc skin such as perforin and granzymes B, K, and H. Of note, the observed cytotoxic gene signature was positively correlated with skin thickness [70,71]. Furthermore, a prominent infiltration of IL-13-producing CD8+ CTLs was observed in skin biopsies early in SSc pathogenesis (<3 years), implicating a potentially important role of CTLs in the disease onset [72] (Figure 2). This notion is further supported by earlier studies that paid attention to the role of the CTL-mediated apoptosis induced by granzyme B in the initiation of systemic auto-immunity. The unique fragments generated by granzyme B degranulation represent an exclusive source of auto-antigens, and these self-protein fragments are recognized by autoantibodies in a subset of SSc patients [73].

Another example of a recently identified CD4+ T cell subset in SSc pathology is follicular helper T (Tfh) cells. Tfh cells are specialized in providing help to B cells in lymph nodes by stimulating proliferation, class switching, and somatic hypermutation. These cells have not only been found in lymph nodes but have also recently been detected in the blood circulation where they are referred to as T peripheral helper (Tph) cells [74]. A key cytokine in Tph function is IL-21. Strikingly, in SSc patients, elevated counts of Tph cells have been shown to promote plasmablast differentiation through an IL-21-mediated pathway [75]. Furthermore, increased infiltration of inducible T cell co-stimulator (ICOS)+ Tfh-like cells has been observed in the skin of SSc patients. The presence of these cells in the skin of sclerodermous graft-versus-host disease mice was strongly linked to increased skin fibrosis. Interestingly, both the depletion of ICOS+ (including Tph depletion) cells and neutralization of IL-21 exhibited a significant reduction in skin fibrosis. Tfh cells activate fibroblasts in vitro. Co-culture of Tfh and fibroblasts resulted in increased expression of α-smooth muscle actin (α-SMA) on the activated myofibroblasts [76]. α-SMA is a key marker of activated myofibroblasts, and its expression is linked to a fibrotic phenotype. The importance of skin-resident Tfh-like cells in the pathology of SSc is also supported by an sc-RNAseq study that unraveled the heterogeneity of T cell responses in patients' skin biopsies. In this study, a distinct CXCL13+ Tfh-like subset that secretes factors promoting B cell responses and autoantibody production was only found in SSc-inflamed tissue [77]. In light of these new findings, developing drug strategies to target the newly described pathogenic T cell subsets is expected to set the milestones for potential personalized therapy in SSc.

2.2.2. B Lymphocytes

B lymphocytes are the second major component of our adaptive immune response. Mounting evidence suggests that B cell homeostasis in the blood of SSc patients is aberrant. Autoantibody production caused by loss of self-tolerance is an important hallmark in SSc pathogenesis. Autoantibodies such as anti-DNA topoisomerase I, anti-centromere, and anti-RNA polymerase antibodies have been detected in the sera of more than 95% of scleroderma patients [78].

The prominent infiltration of B cells in SSc tissues including the skin and lungs suggests their involvement in the disease pathogenesis [79]. Similar to T cells, earlier studies revealed their presence in sclerotic skin by utilizing mainly immunohistochemistry techniques. In a recent sc-RNAseq study, a prominent B cell gene signature was detected in the skin of 69% of early dcSSc patients [80]. In a different study, B cell infiltration was associated with skin progression in early diffuse disease. In this study, infiltration of plasma cells was also evident in the sclerotic skin [81]. Thus, B cell skin infiltration seems to be linked to skin progression early in the disease onset and in patients with dcSSc. However, the robustness of this correlation needs to be validated in a larger cohort of patients.

More specifically, B cells in SSc are characterized by elevated numbers of IL-6-producing effector B cells (Beffs) and decreased numbers of IL-10-producing regulatory B cells (Bregs). Because Bregs suppress and Beffs enhance the immune response through the production of cytokines, this change might impact the inflammatory process. Of note, in SSc, elevated IL-6 and reduced IL-10 levels have been detected in serum/plasma, which is possibly explained by this B cell imbalance. A possible cause for this change might be the altered presence of the cytokines B cell activating factor (BAFF) and a proliferation-inducing ligand (APRIL) in SSc blood [82]. BAFF and APRIL are potent activators of B cells. These cytokines stimulate the production of effector B cells and suppress the generation of regulatory B cells. Elevated serum levels of BAFF and APRIL have been documented in SSc and are correlated with skin thickening, disease severity, and increased IL-6 production by Beffs [83] (Figure 2). Interestingly, it has been demonstrated that BAFF inhibition decreased skin fibrosis in a murine model of SSc [84].

Importantly, not only the numbers of Beffs but also the phenotype of these cells is altered in SSc. To elaborate, SSc patients are characterized by an increased naïve (CD19+CD27-) and a reduced activated memory B (CD19+CD27+) cell phenotype that is accompanied by overexpressed pro-apoptotic and activation markers such as CD95, CD86, and human leukocyte antigen-DR isotype (HLA-DR) [4,85–88]. In addition, the cell surface phenotype of B cells is also changed in SSc. Significant overexpression of CD19 on the surface of B cells has been illustrated in SSc patients and was correlated with SSc-ILD [89,90]. However, in a recent study exploring lymphocyte subset aberrations between early dcSSc patients and healthy donors, the frequency of CD19+ B cells in the peripheral blood was similar [4]. CD19 is a positive B cell regulator that activates B cells through their B cell receptor (BCR) signaling. B cell hyperactivation through CD19-BCR signaling may be associated with the prevalent autoantibody production in SSc. Data from transgenic mouse models show that CD19-deficient mice have decreased serum autoantibodies, while overexpression of CD19 causes increased serum autoantibody production [91,92]. In addition, humans with homozygous mutations in the CD19 gene suffer from an antibody deficiency syndrome that impairs the response of mature B cells in antigen stimulation [93]. In conclusion, in addition to B cells which have been documented to stimulate fibrosis with the release of cytokines, intriguingly, autoantibodies with direct pro-fibrotic effects and autoantibodies against the PDGF receptor have also been documented in SSc [94]. These autoantibodies seem to induce fibrosis by facilitating the conversion of the fibroblast to the myofibroblast phenotype [95] (Figure 2).

Table 1. Function of the primary cell types implicated in SSc pathogenesis both in the blood circulation and at the site of fibrosis.

Cell Type	Function
Endothelial cell	Platelet adhesion activates fibrotic pathways. Increased microvascular permeability causes leukocyte adhesion to the endothelium, leading to increased inflammation [7].
Monocyte/Macrophage	A prominent M2 macrophage signature increases levels of pro-fibrotic cytokines such as IL-4, IL-6, and IL-13 and correlates with elevated tissue fibrosis [17–20].
Eosinophil	Elevated eosinophil counts in the peripheral blood are associated with severe lung disease and presence of skin ulcers [31–33].
Mast cell	Release of cytokines and growth factors such as IL-4, IL-6, IL-13, TNF-a, PDGF, and TGF-β activates myofibroblasts to produce collagen [37]. Tryptase and histamine release triggers fibroblast proliferation [38,39].
Innate lymphoid cell	Increased production of the ILC2 cytokines IL-25, IL-33, and TSLP in serum and skin mediates fibrosis in a TGF-β-dependent manner [43–49].
Plasmacytoid dendritic cell	Elevated numbers of CXCL4+- and IFN-a-producing pDCs in skin and lungs are involved in increased fibrotic manifestations mediated by TLR8 activation [28,50,52].
T lymphocyte	Skin is predominantly infiltrated by CD4+ and CD8+ cytotoxic T cells that produce pro-fibrotic cytokines and cause apoptosis to epithelial cells [6,72]. Increased IL-21-producing Tph cells promote plasmablast differentiation and increase activation of myofibroblasts [75,77]. In the peripheral blood, SSc patients are characterized by increased Th2 and Th17 numbers compared to healthy donors [4,60,63].
B lymphocyte	Elevated BAFF and APRIL are correlated with skin thickening [81,83]. Increased IL-6-producing Beffs increase inflammation, while decreased IL-10-producing Bregs exhibit a reduced capacity for immunosuppression [82]. In SSc peripheral blood, an increase in naïve and a decrease in activated memory B cells is observed compared to healthy controls [4,85,87,91].

3. SSc First-Line Anti-Inflammatory Treatment

Now that we have discussed the role of immune cells in SSc, we will address the efficacy of treatment approaches. According to current recommendations from the European League against Rheumatism (EULAR), methotrexate (MTX) is considered as a first-line treatment in early dcSSc [96]. Other therapeutic strategies include administration of synthetic corticosteroids or low-dose (2 mg/kg) administration of cyclophosphamide (CYC) for 1 year, continued by conservation therapy with mycophenolate mofetil (MMF) [97], or first-line therapy with MMF. These therapies represent the broad-spectrum anti-inflammatory drugs (Table 2) that have been extensively used in SSc, and thus we will start discussing them first.

3.1. Synthetic Corticosteroids

To begin, synthetic corticosteroids (CS) such as prednisolone, hydrocortisone, methylprednisolone, and dexamethasone are the oldest anti-inflammatory and immunosuppressive agents that have been extensively used in scleroderma and other rheumatic diseases. These compounds modulate the activation of all immune cells including macrophages and T and B lymphocytes but also affect the function of fibroblasts and endothelial cells. They still remain a cornerstone in the treatment of SSc as, due to their strong immunosuppressive and anti-inflammatory properties, they are able to halt inflammatory, vascular, and fibrotic manifestations (all key hallmarks of SSc pathogenesis) [98]. Retrospective case–control studies have shown that CS may improve muscle and joint inflammation and severe skin fibrosis [99–101], but to our knowledge, there is still no well-controlled, double-blind, placebo-controlled study to justify these indications. On the other hand, a high dose of prednisolone has been associated with an increased risk of scleroderma renal crisis [102–104]. This is a very severe and life-threatening complication and thus, in SSc, doses of prednisolone >10–15 mg are avoided. Administration of corticosteroids as monotherapy is not recommended. The addition of steroid-sparing agents such as MMF in the treatment scheme reduces doses of steroids along with their side effects and steroid requirements. In clinical practice, the use of CS in SSc is limited and restricted to the more inflammatory than pro-fibrotic manifestations such as arthritis and myositis. It is worth mentioning that CS still represent the mainstay of treatment for SSc patients with primary heart disease manifestations.

3.2. Methotrexate

MTX was developed as a cytotoxic folic acid analogue that inhibits purine and pyrimidine synthesis when administered at a high dosage. This means that it blocks cell proliferation, including that of immune cells. It was initially used to treat cancer and, at much lower dosages, rheumatoid arthritis (RA) [105]. Its extended use in RA suggests an important anti-inflammatory role. The exact mechanism of action at the dosage that is used for RA and SSc is not fully understood yet, but several mechanisms have been postulated to contribute to its anti-inflammatory function. It seems that MTX does not directly induce apoptosis of T cells and fibroblasts but rather increases their sensitivity to apoptosis by modulating cell survival signaling pathways. In contrast, MTX directly depletes monocytes in vitro by inducing their apoptosis [106]. Two small-scale randomized control trials (RCTs) in early dcSSC patients have evaluated the use of 15 mg MTX per week for 6 months [107,108]. The total skin score (TSS) was used as the primary endpoint to evaluate skin fibrosis. In both studies, a small but not statistically significant improvement in skin fibrosis was observed. Interestingly, continuation of the therapy up to 1 year showed a statistically significant improvement in the primary endpoints. Based on these observations, the EULAR [96] supported the use of MTX in skin manifestations in early dcSSC. However, a follow-up study [107] did not confirm the initial promising results in the long run. Taken together, data evaluating the effectiveness of MTX in reducing skin fibrosis are contradictory, and thus treatment guidelines supporting the use of MTX are mostly based on the positive experience of expert physicians with MTX.

3.3. Cyclophosphamide

CYC is another cytotoxic agent that targets fast-dividing cells such as tumor cells and proliferative T and B cells. Its use in auto-immune diseases is attributed to inhibition of regulatory and helper T cell proliferation, leading to their suppression. Suppression of helper and regulatory T cells leads to declined gene expression of pro-inflammatory cytokines, such as interleukin-2, and attenuated production of the fibrogenic TGF-β and the immunoregulatory IL-10, respectively [109,110]. In addition, lymphocyte depletion can decrease the production of antibodies (including autoantibodies) [111]. These observations suggest a potential role of CYC in diminishing fibrosis in patients with SSc-ILD. In the scleroderma lung study (SLC), 158 patients with SSc-ILD received 2 mg/kg of CYC or placebo orally for 12 months. The primary endpoint was the absolute difference in predicted forced vital capacity (FVC) which is a prediction of the patient's lung function. Secondary endpoints included measurement of skin thickening with the modified Rodman skin thickness score (mRSS) and monitoring of lung function with high-resolution computed tomography (CT) scans. It was observed that the CYC group exhibited reduced skin fibrosis and thoracic fibrosis after the 12-month treatment compared to baseline ($p = 0.014$) [112,113]. All in all, CYC showed a measurable but small effect on improving the lung function of SSc patients. However, the use of CYC is limited due to severe toxicity notifications including leucopenia and thrombocytopenia. Thus, according to the EULAR [96], the usage of CYC is only recommended in patients with progressive ILD.

3.4. Mycophenolate Mofetil

Another broad-spectrum agent that has been used in SSc treatment is MMF. MMF causes selective depletion of guanosine nucleotides in T and B lymphocytes. It inhibits their proliferation and therefore suppresses immune-associated responses and antibody formation [114]. Indeed, SSc patients receiving MMF exhibit decreased numbers of T helper cells (implicated in various SSc pathogenic manifestations) compared to patients with no immunosuppressive treatment [115]. Strikingly, MMF treatment inhibits the infiltration of myeloid cells including tissue-resident macrophages in the skin of patients with SSc. Furthermore, treatment with MMF is strongly associated with a reduced inflammatory gene signature in the sclerotic skin [116]. Although MMF has been traditionally regarded as a lymphocyte-targeting drug, in vitro and clinical evidence shows its anti-fibrotic capacity by decreasing the proliferation of human fibroblasts [117]. The results from the Scleroderma Lung Study I and II suggest the efficacy of MMF in SSc patients with severe lung fibrosis. More specifically, treatment with 3 gr/day of MMF for 2 years resulted in a significantly improved percentage of predicted FVC and mRSS [118]. The effect of MMF has also been evaluated on patients with mild SSc-ILD (FVC $\geq$ 70% predicted). In this double-blind, placebo-controlled, randomized clinical trial [119], MMF was well tolerated while exhibiting an observed but not statistically significant improvement in FVC and mRSS scores. In the Scleroderma Lung Study, previous results of MMF were compared with results from oral CYC administration (2.0 mg/kg per day), followed by placebo administration for 1 more year. The adjusted percentage of predicted FVC after 24 months improved by 2.19 with MMF and 2.88 with CYC. However, no statistically significant differences among the two treatment groups were observed, and thus the superiority of MMF compared to CYC is not justified. In the European Scleroderma Observational Study (ESOS), MMF was further compared with MTX, CYC, or "no immunosuppressant" [1]. This was a multicenter, prospective, observational cohort of 326 patients with early dcSSc (up to 3 years of onset of skin thickening) with a 24-month duration. After 12 months of treatment, a statistically significant reduction in mRSS was observed in all groups compared to baseline measurements. To sum up, results from clinical trials regarding the anti-fibrotic efficacy of MMF are also conflicting, and there are no signs revealing its superiority to MTX and CYC. Similar to MTX, the recommendation for first-line MMF treatment in patients with SSc-associated ILD has been based on positive clinical experience with the drug, and its beneficial safety profile [120]. The use of MMF in improving lung fibrosis of SSc-ILD has also been supported by a large number of small retrospective cohort studies [121–125].

3.5. Autologous Hematopoietic Stem Cell Transplantation

ASCT has been utilized in treating auto-immune disorders resistant to conventional immunosuppressive therapy for more than 20 years now [126]. ASCT is the only potential disease-modifying treatment in SSc, and it has been proposed for patients with early and rapidly progressive dcSSc who have a high mortality prognosis, but in whom advanced organ involvement has not started yet [127]. Autologous stem cell transplantation begins with the isolation of the patient's CD34+ cells. Then, B and T cells are depleted using a high dose of cyclophosphamide and anti-thymocyte globulin. The last step includes the transplantation of the patient's stem cells that were isolated, resulting in the repopulation of the lymphocytes [128]. According to clinical data, ASCT shows promising results in improving mortality rates, reducing skin thickness, and enhancing lung function [129,130]. Two randomized clinical trials evaluating the safety and efficacy of ASCT in SSc have been completed thus far. First, in the Cyclophosphamide or Transplantation (SCOT) trial (NCT00860548), it was observed that myeloablative CD34+-selected ASCT was more effective in diminishing skin fibrosis than CYC administration alone. However, the mortality rates of the participants during the first year after the transplantation were as high as 10%.

Treatment-related deaths as well as cancer and infection were the causes of death in these participants. Among them, infection was the leading cause, due to the suppression of the immune system. Furthermore, the possibility of relapsing was also prevalent. Secondly, in the Autologous Stem Cell Transplantation International Scleroderma (ASTIS) study [131], ASCT was compared with CYC pulse therapy in 156 rapidly progressive diffuse SSc patients. In the first year, eight patients from the ASCT group compared to none in the CYC group died. However, parameters such as long-term event-free survival, overall survival, mRSS score, and FVC were significantly improved in the ASCT-treated patients. The increased mortality rates counteract transplantation's benefit, putting experts in a difficult judging position [132]. However, to date, no disease-modifying anti-rheumatic drug has been shown to effectively reduce patients' morbidity in the long term. ASCT, on the other hand, has shown a remarkable reversal of skin fibrosis accompanied by improved lung and internal organ function. In view of the outcomes of the previously described RCTs, the new EULAR recommendations [96] suggest that experts consider ASCT for the treatment of rapidly progressive SSc patients at risk of organ failure. It is believed that treatment-related mortality can be reduced by a more careful exclusion of patients with compromised heart function and by applying ASCT in earlier stages of the disease, enabling selected patients to benefit from this treatment. There are currently four ongoing clinical trials evaluating the effectiveness and safety of ASCT in SSc (NCT01895244, NCT01413100, NCT04464434, NCT03630211). The results from these studies are expected to shed light on the safety and efficacy of autologous ASCT in a better stratified SSc patient population.

Table 2. Clinical trials conducted for the evaluation of broad-spectrum treatment in SSc.

Drug	Target	Type of Trial(s)	Duration (Months)	Patients	Results
Methotrexate (MTX)	Exact anti-inflammatory role is unknown	Multicenter, double-blind 1. RCT [108] 2. RCT [107]	1. 6 2. 12	1. 29 early dSSc 2. 71 early dSSc	1. Mean TSS 21.61 at baseline, 19.96 ($p = 0.135$) 6 months after 2. Mean TSS 18.3 at baseline, 14.5 ($p = 0.027$) 12 months post-treatment
Cyclophosphamide (CYC)	Inhibition and suppression of T helper and regulatory T cells	Double-blind, RCT (SLC) [112]	12	158 SSc-ILD	2. 53% ($p < 0.03$) improvement in predicted FVC and 3.02 ($p = 0.08$) unit improvement in mRSS in CYC's favor
Mycophenolate mofetil (MMF)	T and B cell depletion	1. Double-blind, RCT (SLC II) [118] 2. Double-blind RCT [119]	1. 24 2. 6	1. 69 SSc-ILD 2. 41 mild SSc-ILD	1. Percentage of predicted FVC improved from 67 to 75, and mRSS decreased from 14.5 at baseline to 10 24 months post-treatment 2. No statistically significant improvement in mRSS and FVC scores
Autologous hematopoietic stem cell transplantation (ASCT)	Depletion of T and B cells, followed by stem cell transplantation	1. Open-label, multicenter RCT (SCOT) (NCT00860548) 2. Open-label, multicenter RCT (ASTIS) [131]	1. 54 2. 24	1. 75 severe SSc 2. 156 early dcSSc	1. ASCT more effective in diminishing skin fibrosis compared to CYC (−19.9 vs. −8.8, $p < 0.001$) and shows greater event-free survival 2. Overall survival, mRSS, and FVC significantly improved with ASCT compared to CYC (67% of 1404 pairwise comparisons in favor of ASCT vs. 33% in CYC, $p = 0.01$)

4. Evaluation of Cell-Specific Anti-Inflammatory Treatment

On the road towards personalized therapy in SSc, specific elimination of pathogenic innate or adaptive immune cell populations could be promising in alleviating fibrosis and reducing the morbidity of the side effects of broad immunosuppression. Of note, there are no clinical trials evaluating drugs depleting innate immune cell populations. On the other hand, several trials (Table 3) have evaluated the efficacy of B and T cell-depletive therapies in SSc, and thus we begin with addressing treatments targeting these cells first [133].

4.1. B Cell-Specific Treatment

To begin with B cell-specific treatment, rituximab (RTX), belimumab, and inebilizumab are the three representative monoclonal antibodies that have been used.

4.1.1. Rituximab

Rituximab is a chimeric anti-CD20 monoclonal antibody that binds to both immature and mature B lymphocytes expressing CD20 on their surface (including the pathogenic Beffs) and eliminates them (Figure 4). Of note, RTX does not consistently succeed in depleting B cells in tissues [134,135], and antibody-producing plasma cells are not targeted by RTX, since they lack the CD20 surface antigen [136]. Indeed, in a small, randomized, double-bind, placebo-controlled trial, treatment with RTX was not associated with reduced fibrosis but caused significant depletion of naïve and memory B cells in the peripheral blood and scleroderma-associated dermal lesions [137]. As expected from RTX's mode of action, various plasma cells and autoantibody titers did not show a statistically significant change after treatment. The potential anti-fibrotic effect of RTX is supported by histopathological observations suggesting a reduced dermal hyalinized collagen content and myofibroblast count [138]. As we discussed, IL-6 is a cytokine largely produced by activated B cells that shows pro-fibrotic capacity. In a small, open-label study where RTX significantly improved patients' skin scores, it was observed that patients had elevated basal levels of IL-6 in biopsies from their skin lesions. IL-6 levels decreased from 3.7 $\pm$ 5.3 pg/mL at baseline to 0.6 $\pm$ 0.9 pg/mL (p = 0.02) six months post-treatment. Furthermore, after treatment 7/9 patients exhibited a complete depletion of B cells in skin lesions. Thus, interestingly, the RTX-associated skin thickness was correlated with B cell depletion and IL-6 reduction in patients' skin [139]. This observation further empowers the mechanism and role of RTX in skin fibrosis.

The potential efficacy of treating SSc patients with RTX has been extensively evaluated by Daoussis and colleagues. In an open-label, randomized, 1-year study, weekly administration of RTX (375 mg/m^2) for 4 weeks at baseline and after 6 months on top of each patient's standard treatment compared to standard treatment alone was assessed. Patients that received RTX showed a significantly increased median percentage of FVC accompanied by a further improvement in the diffusing capacity for carbon monoxide (DLCO). Furthermore, skin improvement was also evident when evaluated both clinically and histologically [140]. In a more recent (2017) multicenter, open-label study, the same group of researchers compared administration of 4 infusions of RTX at a dose of 375 mg/m^2 once weekly every 6 months with controls receiving traditional therapies (MTX, azathioprine, MMF). In a 4-year follow-up, the RTX group showed a reduced mRSS score of 14.72 $\pm$ 10.52 compared to 17.78 $\pm$ 9.48 in the control group (p = 0.31). Patients enrolled in this study exhibited a significant improvement in FVC and stabilization of pulmonary function tests (PFTs) [141]. In the latest open-label, randomized, head-to-head clinical trial, 60 anti-scl70-positive early dcSSc patients randomly received either monthly pulses of CYC (500 mg/m^2) or 1 g of RTX at 0 and 15 days. In the RTX group, there was a significant improvement in the percentage of FVC, while patients receiving CYC experienced deterioration of their lung function. Both groups exhibited a similar improvement in skin scores, but the safety profile of RTX was better [142].

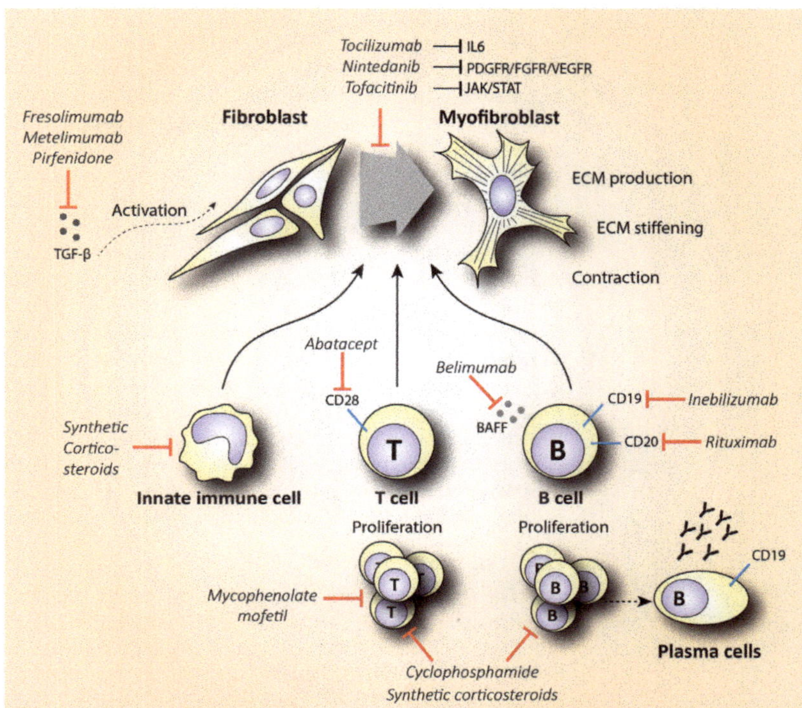

Figure 4. Schematic representation of the mechanism of action of synthetic corticosteroids (CS), mycophenolate mofetil, cyclophosphamide, inebilizumab, rituximab, belimumab, abatacept, nintedanib, tocilizumab, tofacitinib, fresolimumab, metelimumab, and pirfenidone in reducing fibrosis in SSc. Abbreviations: ECM = extracellular matrix; TGF-β = transforming growth factor-β.

Furthermore, B cell depletion may be an efficient and well-tolerated adjuvant treatment for SSc-associated pulmonary arterial hypertension (SSc-PAH). In a multicenter, double-blind, randomized, placebo-controlled, proof-of-concept trial, 57 SSc patients on standard medical therapy received two doses of either 1 gr RTX or placebo in a 2-week interval [143]. To evaluate the effect on PAH, the change in the 6 min walk distance (6MWD) was the primary outcome. Twenty-four weeks after treatment, patients in the RTX arm exhibited an estimated change of 23.6 ± 8.8 m compared to 0.4 ± 9.7 m in the control arm ($p = 0.03$). The beneficial results of RTX in diminishing skin fibrosis have also been demonstrated in a number of case or pilot studies [144–148]. Most of these studies, however, were open label, often lacked a control group, and included heterogeneous patient populations and concurrent treatments. Interestingly, the fact that RTX shows positive results in reducing skin fibrosis and alleviating PAH and ILD, together with being more potent than traditional medications and well tolerated in all studies, suggests a promising role in scleroderma medication options. However, a phase III randomized control study will be required to verify its safety and efficacy in patients.

4.1.2. Inebilizumab and Belimumab

Since depletion of CD20+ B cells does not consistently deplete B cells in tissues and does not eliminate plasma cells, developing monoclonal antibodies that target the bone marrow-resident pro- and pre-B cell co-receptor CD19 might be a more effective B cell-depletive approach (Figure 4). The efficacy of a humanized anti-CD19 monoclonal antibody, inebilizumab (MEDI-551), in reducing SSc fibrosis was assessed in a phase I randomized, double-blind, placebo-controlled study [149]. This therapy effectively depleted both B cells and plasma cells in the skin and blood of SSc patients in a dose-dependent manner. Furthermore, an improved skin score was observed in all patients, suggesting a role of inebilizumab in halting skin fibrosis. Strikingly, patients with an elevated plasma cell signature at inclusion showed greater improvement in mRSS compared to the patients that had normal or lower plasma cell counts [150].

Belimumab is another human monoclonal antibody targeting the soluble BAFF cytokine, a member of the tumor necrosis factor ligand superfamily that was a breakthrough in the treatment of systemic lupus erythematosus (Figure 4). This monoclonal antibody prevents the binding of BAFF to receptors on B cells and inhibits the survival of B cells, including the autoreactive ones. It further decreases the differentiation of B cells into immunoglobulin-producing plasma cells [151]. In a randomized, double-blind, placebo-controlled, pilot trial, 20 patients with dcSSc recently treated with MMF (1 gr twice a day) were administered 10 mg/kg of belimumab i.v., or placebo. Belimumab administration improved the mRSS score from 27 to 18 ($p = 0.039$), while the improvement in the placebo group was from 28 to 21 ($p = 0.023$). Patients with improved mRSS exhibited a significant reduction in both B cell and pro-fibrotic gene expression. However, the results of this small pilot study were not statistically significant, and larger trials should be conducted in order to make treatment recommendations [152]. A study evaluating the combination therapy of belimumab and RTX in SSc is currently pending (NCT03844061).

4.2. T Cell-Specific Treatment

The elevated infiltration of cytotoxic T cells in SSc-affected skin and lungs, along with the increased Th2 pro-fibrotic cytokine levels in the serum of SSc patients, suggests that targeting T cells could be a promising treatment option in SSc. T cell-specific treatment has shown encouraging results in other inflammatory diseases such as RA [153]. The biologicals that have been used either directly affect T cell activity by reducing their number, or they indirectly interact with cytokines expressed by T cells. Abatacept is the main representative of the former category, and it causes T cell deactivation by blocking CD28 (Figure 4), a molecule that is essential for T cell activation mediated by antigen-presenting cells (APCs) [154]. CD28 knockout mice exhibited decreased fibrosis in comparison with the wild type in bleomycin-induced fibrosis [155]. In 2015, the efficacy of abatacept in attenuating fibrosis was examined in a placebo-controlled RCT that included 10 patients with dcSSc [156]. Five out of seven patients that received abatacept showed a $\geq$30% decrease in mRSS, while this improvement was shown in one out of the three controls. Interestingly, patients with decreased mRSS also showed reduced CD28 gene expression along with decreased expression of genes associated with T cell proliferation. This empowers the proposed mechanism of abatacept's action. Recently, a phase II, multicenter, double-blind, randomized, placebo-controlled trial assessed the safety and efficacy of subcutaneous administration of abatacept in 88 dcSSc patients [157]. The results failed to show a statistically significant improvement in mRSS (primary outcome) in the abatacept arm, but secondary outcomes reflecting patients' disability and quality of life were significantly improved in favor of the abatacept group. Of note, patients were stratified into inflammatory, normal-like, and fibroproliferative subgroups based on molecular gene expression data of their skin biopsies. Patients with an inflammatory intrinsic gene expression showed a larger decrease in mRSS compared to the other groups. Interestingly, the improvement in the skin score was associated with decreased gene expression of immune pathways including cytotoxic T lymphocyte-associated protein 4 (CTLA-4) and CD28 (targets of abatacept). These results are novel regarding the potentially targeted mechanism of action of abatacept as an inflammation immunomodulator. Furthermore, abatacept exhibits a good safety profile and seems to decrease joint involvement and related disability [158,159]. Based on these studies, abatacept seems to be effective in treating scleroderma-associated joint inflammation, but studies with a larger number of participants (phase III) are required to evaluate its anti-fibrotic effect.

Table 3. Representation of the studies evaluating the role of B and T cell-specific treatments in SSc.

Drug	Target	Type of Trial(s)	Duration (Months)	Patients	Results
Rituximab (RTX)	Anti-CD20 B cell depletion	1. Double-blind, RCT [137] 2. Open-label trial [139] 3. Open-label RCT [140] 4. Multicenter, open-label trial [141] 5. Open-label RCT [142] 6. Multicenter, double-blind RCT [143]	1. 24 2. 54 3. 24 4. 48 5. 6 6. 6	1. 16 early SSc 2. 9 dcSSc 3. 14 SSc 4. 51 SSc-ILD 5. 60 early dSSc 6. 57 SSc-PAH	1. B cell depletion in blood did not improve mRSS and FVC scores 2. Median decrease in mRSS of 43.3% ($p = 0.001$) accompanied by reduction in IL-6 levels 3. FVC improved by 10.25% in RTX and reduced by 5.04% in the placebo group ($p = 0.0018$). RTX arm: mRSS improvement of 13.5 vs. 8.37 in placebo, 12 months post-treatment ($p < 0.001$) 4. No significant change in mRSS, but lung function improved significantly 5. Significant improvement in RTX vs. CYC in median percentage of FVC (67.52 vs. 58.06, $p = 0.003$), but not in mRSS scores 6. RTX arm: the improvement in median 6MWD was 25.5 m compared to 0.4 m in placebo ($p = 0.03$) after 48 weeks
Inebilizumab (MEDI-551)	Anti-CD19 B cell depletion	Multicenter, double-blind RCT [149]	3	28 SSc	Depletion of B and plasma cells was correlated with improved mRSS and reduced expression of fibrotic genes in skin biopsies
Belimumab	Inhibition of B cell survival by blocking BAFF	Double-blind RCT [152]	13	20 dcSSc	mRSS score improved from 27 to 18 ($p = 0.039$), while in placebo group from 28 to 21 ($p = 0.023$)
Abatacept	CD28 blocking T cell depletion	1. RCT [156] 2.Multicenter, double-blind RCT [157]	1. 10 2. 12	1. 6 dcSSc 2. 88 dcSSc	1. 5/7 patients and 1/3 controls showed >30% improved mRSS 2. No significant improvement in mRSS, but secondary outcomes related to quality of life improved significantly

5. Targeting Cytokine Production

An alternative option to T cell treatment may be targeting cytokines and growth factors (Table 4), such as IL-6, IL-1, IL-2, and TGF-β, that are produced by multiple pathological cell types in SSc. Tocilizumab (TCZ) is a monoclonal antibody that inhibits IL-6 signaling, by blocking the IL-6 receptor (Figure 4), and is effective in treating patients with juvenile idiopathic arthritis and RA [160,161]. Multiple studies have evaluated the use of TCZ in systemic sclerosis as well. Among these studies, there is one open-label, phase II, randomized controlled study (faSScinate) [162] that evaluated the efficacy and safety of a 24-week treatment with TCZ in reducing fibrosis. The observed difference of 2.70 units in mRSS was not statistically significant ($p = 0.0579$). However, the interesting findings of this study arise from a probable correlation between decreased IL-6 levels and TGF-β function. More specifically, researchers cultured fibroblasts from patients' skin biopsies at baseline and 24 weeks after treatment and showed that the IL-6 reduction was correlated with decreased TGF-β-related gene expression, suggesting a role of TCZ in reducing fibrosis [163]. The European Scleroderma Trials and Research (EUSTAR) observational study also evaluated the use of TCZ in SSc-associated joint and muscle inflammation, compared to abatacept. Similar to abatacept, the reduction of 3 units in mRSS between baseline and post-treatment was not significant ($p = 0.109$), but a significant improvement in joint function was observed [158]. A number of case series have shown similar results [164–168]. Post hoc analysis of the results from the faSScinate study [162] showed that considerably fewer patients treated with TCZ compared to placebo exhibited a decline in their lung function. These participants were also characterized by a reduced expression of pro-fibrotic genes, suggesting that TCZ treatment may be associated with preserved lung function. These promising results, together with the unmet medical need for effective fibrosis-reducing treatment, supported the exploration of TCZ in a phase III clinical trial (focuSSed) [169]. In this very recent randomized, double-blind, placebo-controlled, multicenter trial, 210 dcSSc patients were administered subcutaneous TCZ (162 mg) or placebo weekly for 48 weeks. The reduction in mRSS at the endpoint compared to baseline levels was −6.14 for TCZ and −4.41 for placebo, with the adjusted difference of −1.73 not being able to show statistically significant results ($p = 0.10$). Interestingly, the change in the FVC percentage predicted at week 48 was in favor of the TCZ arm, expressed as a difference in the least squares mean of 4.2 ($p = 0.0002$). These results further support the protective role of TCZ in preserving lung function. This was later investigated by stratifying the same patients according to the level of lung involvement [170]. Lung involvement was evaluated by serial spirometry, high-resolution chest CT, and quantitative interstitial lung disease (QILD) and fibrosis scores. Strikingly, TCZ was effective in stabilizing and thus preventing the progression of early SSc-ILD in all patient groups, irrespective of the severity of lung involvement. These results likely highlight the importance of targeting the immunoinflammatory, early fibrotic stage of SSc pathology. Taking everything together, TCZ does not seem to significantly improve skin fibrosis, but it seems to serve as a safe treatment option to preserve patients' lung function. Thus, TCZ has now been approved by the US Food and Drug Administration (FDA) to treat SSc-ILD.

Based on the elevated levels of IL-1 and IL-2 and their role in inflammation and fibrosis in SSc, the potential effect of the monoclonal antibodies rilonacept and basiliximab in tackling SSc-associated fibrosis has been examined lately. Rilonacept is an anti-inflammatory treatment that has been approved by the FDA as a therapy for cryopyrin-associated periodic syndromes [171]. Its potential use in SSc depends on the fact that it blocks IL-1β signaling by binding with IL-1β and preventing its reaction with cell surface receptors, therefore reducing IL-1-triggered inflammation. The results from a double-blind, placebo-controlled, randomized trial did not support this hypothesis, as no changes were observed in mRSS between treatment and placebo. In addition, the researchers measured IL-1-regulated gene expression from explants obtained from patients' skin, and similarly, no difference was found [172]. Regarding IL-2 expression, basiliximab is an anti-CD25 monoclonal antibody that inhibits the IL-2 receptor and is used for the treatment of kidney allograft rejection as it inhibits the activation and proliferation of T cells [173]. The selective inhibition of IL-2 cytokines seems promising in counteracting the Th2 predominance in SSc lesions, but is this associated with reduced fibrosis? The results from a case study in which a patient with early dcSSC was treated with a combination of CYC, prednisolone, and basiliximab showed a decrease in mRSS from 24 at baseline to 19 six months post-treatment [174]. The same research group, 4 years later, conducted a small-scale open-label study including 10 patients with dcSSC [175]. The median mRSS was reduced from 26/51 to 11/51 at week 68 ($p = 0.015$). Furthermore, the median predicted FVC between baseline and 44 weeks after treatment increased from 82.1% to 88.4%. The observed trend towards an improvement in skin disease cannot be attributed to basiliximab alone as the patients also received other immunosuppressive treatments and there was no control group to compare with. Therefore, whether basiliximab can attenuate fibrosis is still unclear.

Another important inflammatory and pro-fibrotic mediator in SSc is TGF-β. Currently, drugs targeting the TGF-β pathway are under investigation mainly for diseases such as cancer. In SSc, the possible efficacy in reducing fibrosis by blocking TGF-β has been examined with the use of the monoclonal antibodies fresolimumab and metelimumab (Figure 4). In a small-scale, open-label, single-center study, fresolimumab, a high-affinity TGF-β-inactivating monoclonal antibody, reduced TGF-β-dependent gene expression in skin biopsies. Furthermore, fresolimumab administration was accompanied by an average reduction of 8 units in mRSS at 11 weeks post-treatment [176]. On the other hand, metelimumab's administration did not show similar encouraging effects. The effect of metelimumab has been examined in a randomized, double-blind, placebo-controlled trial in 45 patients with early (<18 months) dcSSc. All patients experienced an improvement in mRSS, but this improvement was not statistically significant ($p = 0.49$). Furthermore, the drug's administration was related to an increased morbidity. A large number of severe adverse effects were reported, with skin ulceration and worsening of breathlessness being the most prominent [177]. With the provided results, TGF-β targeting exhibits conflicting results. Currently, several other compounds that target this pathway are under investigation. Those that target TGF-β signaling by inhibiting integrin expression have, thus far, shown promising results in reducing fibrosis in animal models [178].

Pirfenidone is a small molecule agent that interferes with TGF-β signaling and has potential interest in attenuating SSc-associated fibrosis. Pirfenidone is a pyridine derivative that is widely used in IPF, as it reduces fibroblast proliferation and TGF-β-induced collagen production in primary skin fibroblasts [179] (Figure 4). However, the molecular target of pirfenidone remains unknown. Of note, only one open-label, phase II study (LOTUSS) has evaluated the safety and tolerability of pirfenidone in patients with SSc-ILD. The results of this study did not confirm the IPF observed beneficial effect of pirfenidone in the treatment of ILD in scleroderma patients, as there was no difference in the predicted FVC between baseline and post-treatment [180]. This inability to improve clinical primary outcomes is also reflected molecularly, since serum levels of TNF-α and TGF-β did not significantly differ between treated and untreated patients. These outcomes are in line with the neutral effect on lung function reported by retrospective case studies of SSc patients with ILD that were treated with pirfenidone [181,182]. Given the documented efficacy of pirfenidone in IPF and the overlapping pathogenic manifestations between IPF and SSc-ILD, further investigation of the potential anti-fibrotic effect of pirfenidone in patients with SSc-ILD is needed. In this regard, two phase II and one phase III, multicenter, double-blind, randomized, placebo-controlled trials that are evaluating the efficacy of pirfenidone in diminishing skin fibrosis in SSc-ILD are currently recruiting and may reinforce the potential anti-fibrotic role of pirfenidone in the near future (NCT03068234, NCT03221257, NCT03856853).

Table 4. Clinical trials and other studies that evaluated the role of cytokine targeting treatment in SSc.

Drug	Target	Type of Trial(s)	Duration (Months)	Patients	Results
Tocilizumab (TCZ)	Inhibits IL-6 signaling	1. Open-label RCT (faSScinate) [162] 2. EUSTAR observational study [158] 3. Multicenter, double-blind RCT (focuSSed) [169]	1. 24 2. 5 3. 12	1. 87 early SSc 2. 189 SSc-polyarthritis and myopathy 3. 210 dcSSc	1. Insignificant change in mRSS, but IL-6 reduction was correlated with decreased TGF-β expression 2. No statistically significant change in mRSS, but remarkable improvement in joint function 3. Change of −1.73 in mRSS between treated and placebo groups was not statistically significant ($p = 0.10$), but the 4.2% improvement in predicted FVC was ($p = 0.0002$)
Rilonacept	Blocks IL-1b signaling	Double-blind RCT [172]	5 weeks	19 dcSSc	No changes in mRSS between treatment and placebo
Basiliximab	Anti-CD25-mediated inhibition of IL-2 inhibits T cell activation and proliferation	Small-scale, open-label study [175]	17	10 dcSSc	Median mRSS reduced from 26/51 to 11/51 at week 68 ($p = 0.015$) Median predicted FVC between baseline and 44 weeks after treatment increased from 82.1% to 88.4%
Fresolimumab	Blocks TGF-β signaling	Small-scale, open-label study [176]	6	15 dcSSc	Reduction of 8 units in mRSS score ($p < 0.001$) Reduced expression of TGF-β-regulated genes in skin biopsies ($p < 0.049$)
Metelimumab	Blocks TGF-β signaling	Double-blind RCT [177]	6	45 early SSc	No statistically significant change in mRSS scores
Pirfenidone	Reduces fibroblast proliferation and TGF-β-induced collagen production in primary skin fibroblasts	Open-label, phase II study (LOTUSS) [180]	4	63 SSc-ILD	No difference in the predicted FVC between baseline and post-treatment

6. Emerging Therapies with Tyrosine Kinase Inhibitors

Tyrosine kinases are enzymes that use adenosine triphosphate (ATP) to transfer a phosphate group to intracellular proteins. Phosphorylation of proteins by kinases is vital for signal transduction and regulation of cell proliferation, differentiation, migration, and development. The pathological activation of tyrosine kinases (TKs) is crucial in multiple disease processes such as carcinogenesis, vascular remodeling, and fibrogenesis [183,184]. Furthermore, in inflammatory lesions, tyrosine kinases are also activated through overproduction of growth factors and/or cytokines from the tissues [185]. Several current studies suggest that inhibiting tyrosine kinases may lead to an effective anti-inflammatory therapy. Of note, targeting TK activity is feasible with monoclonal antibodies designed to target the extracellular domains of the receptors, or with small molecules that enter the cytoplasm and bind to intracellular catalytic domains of both receptor and non-receptor TKs [186] (Table 5).

6.1. Indolinone-Derived Small Molecule Tyrosine Kinase Inhibitors

Imatinib mesylate is the principal compound among the small molecule TK inhibitors and was approved by the FDA in 2001 for the treatment of chronic myelogenous leukemia, in order to block the Ab1 kinase activity [187]. Its use in SSc has been examined due to its ability to inhibit the PDGF receptor and TGF-β signaling pathways [188]. Several case studies [188–192] have examined the efficacy and tolerability of imatinib in scleroderma patients. Moderate improvement in skin scores and predicted FVC has been reported, along with a large number of adverse effects. From a mechanistic point of view, imatinib successfully depleted patients' pDCs, immune cells that play an important role in the disease's pathogenesis [192,193]. Of note, treatment with imatinib exhibited a reduced amount of pathogenic IL-4-producing T cells in bronchoalveolar lavage of SSc patients, which was accompanied by elevated numbers of total T helper cells. Based on this observation, it could be suggested that imatinib's anti-fibrotic capacity could be mediated via shifting the Th2 response to a non-type 2 T helper phenotype [190]. In an open-label, multicenter study, 16 out of 27 patients with early dcSSc completed a six-month imatinib administration. The mean decrease in mRSS was 21% compared to baseline; however, five patients exhibited severe adverse effects including generalized edema, erosive gastritis, anemia, upper respiratory tract infection, neutropenia, and neutropenic infection [193]. Therefore, based on the moderate improvement and the severe adverse effects that were observed, the use of the imatinib's analog nilotinib was proposed.

Nilotinib has the same mechanism of action as imatinib, but it is 20–30-fold more potent and has a more favorable toxicity profile [194]. The use of nilotinib in SSc has been tested in an open-label, single-arm, phase II trial that included 10 scleroderma patients. In this trial, nilotinib was well tolerated by the majority of the participants (7/10), with only a few cases of increased edema and mild QTc prolongation being reported. Furthermore, it was observed to be promising in reducing mRSS scores in patients with early and active disease (reduction of 16%, $p = 0.02$, and 23%, $p = 0.01$, at 6 and 12 months post-treatment, respectively) [195]. Furthermore, data from skin gene expression profiling showed that patients responding to treatment were characterized by elevated TGF-β signaling. Interestingly, the expression of TGF-β signaling genes was significantly reduced after these patients were treated with nilotinib. This observation empowers the mechanism of action of nilotinib and supports its potential anti-fibrotic effect. However, the progression into phase III trials will help to validate its efficacy in reducing fibrosis and also determine its limitations concerning potential side effects.

The last indolinone-derived small molecule inhibitor that will be discussed is nintedanib Nintedanib inhibits a plethora of molecules implicated in fibroblast activation such as PDGF receptor, fibroblast growth factor receptor (FGFR), and vascular endothelial growth factor receptor (VEGFR) (Figure 4). In vitro and animal data support the anti-fibrotic effect of nintedanib in inhibiting crucial pathways in the initiation and progression of lung fibrosis. For instance, nintedanib inhibits the secretion of the pro-fibrotic cytokines IL-4, IL-5, and

IL-13 in the peripheral blood of SSc patients. In addition, its anti-fibrotic effect is partly mediated by preventing the polarization towards the pro-fibrotic M2 macrophages [196]. Nintedanib may also exert its anti-fibrotic efficacy by restraining the migration and differentiation of fibroblasts and fibrocytes. Interestingly, in vitro data have shown that nintedanib downregulates the transition of fibrocytes towards myofibroblasts, and thus the pro-fibrotic effect of the latter is reduced [197,198]. Nintedanib has received FDA approval for the treatment of IPF. Since IPF shares pathogenic commonalities with SSc-ILD, the use of nintedanib to reduce lung fibrosis in patients with SSc-ILD has been recently evaluated in the SENSCIS trial [199]. This trial included 576 patients with SSc-ILD that were randomly assigned (1:1 ratio) to receive 150 mg of nintedanib two times daily or placebo for 52 weeks. Deterioration of lung function in the nintedanib arm was significantly lower compared to the control group. More specifically, the adjusted annual rate of decline in FVC was −52.4 mL in patients treated with nintedanib compared to −93.3 mL in the placebo group ($p = 0.04$). Based on the results of this study, in 2019, nintedanib was the first FDA-approved treatment for patients with SSc-ILD [200]. It is worth mentioning that no significant clinical benefit was observed in skin fibrosis measured by mRSS over a 1-year period. To our knowledge, no clinical data verifying the in vitro and in vivo anti-fibrotic mechanisms have been published to date. An open-label, extension trial, assessing the long-term safety and efficacy of nintedanib in 444 SSc-ILD patients, is ongoing and expected to provide data on prolonged nintedanib therapy and its exact mechanism of action in SSc patients (NCT03313180)

6.2. Tofacitinib

Except for the Ab1 and PDGF tyrosine kinases, inhibition of Janus kinases (JAK) is another potential treatment with anti-inflammatory properties. Recent data provide evidence that the JAK/STAT signaling pathway is highly activated in SSc skin biopsies [201]. Tofacitinib is a small molecule JAK inhibitor, with a structure similar to ATP, and has been used in the treatment of RA. Regarding its mechanism of action, it enters the cell with passive diffusion and binds to the ATP site of the JAK1 and JAK3 receptors. As a result, it inhibits the phosphorylation and activation of signal transducer and activator of transcription proteins (STATS), which leads to the downregulation of IL-6 expression [202] (Figure 4). In view of this mechanism, the use of tofacitinib in SSc might be of interest. The results from an observational study suggest tofacitinib as a potential treatment in reducing skin thickening of patients with refractory dcSSc [203]. More specifically, tofacitinib-treated patients showed a significantly shorter response time compared to the conventional immunosuppressive group, with a mean change of −3.7 ($p = 0.001$) in mRSS already 1 month post-treatment. After 6 months of treatment, the mean change in mRSS was even greater between the two groups (−10.0 tofacitinib vs. −4.1 conventional immunosuppressants, $p = 0.026$). Similarly, in another pilot study, patients treated with tofacitinib showed a significantly higher improvement in skin fibrosis compared to patients treated with MTX [204]. These results indicate that tofacitinib might be even more effective in reducing fibrosis than conventional immunosuppressants, further showing a quicker and higher response rate. The tolerability and efficacy of tofacitinib have been recently evaluated in a small clinical phase I/II trial in 15 SSc patients. According to the preliminary results, tofacitinib is well tolerated and shows a trend towards improving fibrosis (NCT03274076). Further evaluation of tofacitinib in dcSSc seems warranted.

Table 5. Clinical trials evaluating the efficacy of tyrosine kinase inhibitors in SSc treatment.

Drug	Target	Type of Trial(s)	Duration (Months)	Patients	Results
Imatinib mesylate	Inhibits PDGFR and TGF-β signaling	Multicenter, open-label RCT [193]	6	27 dcSSc	Mean decrease in mRSS was 21% compared to baseline ($p < 0.001$)
Nilotinib	Same as imatinib, but 20–30-fold more potent	Open-label, single-arm trial [195]	8	10 early dcSSc	Promising reduction of 23% in mRSS 12 months post-treatment ($p = 0.01$)
Nintedanib	Inhibits PDGFR, FGFR, and VEGFR signaling	Double-blind RCT (SENSCIS) [199]	13	576 SSc-ILD	Adjusted annual rate of decline in FVC −52.4 mL compared to −93.3 mL in the placebo group ($p = 0.04$)
Tofacitinib	Inhibits JAK/STAT signaling	Double-blind RCT (NCT03274076)	6	15 dcSSc	Preliminary data show a trend towards improved fibrosis

7. Conclusions and Future Perspectives

SSc is a disabling, chronic, auto-immune disease accompanied by high mortality and morbidity rates. In this review, we examined the potential effects of anti-inflammatory and immunosuppressive treatments in attenuating fibrosis. To do this, we evaluated the first-line, generalized as well as cell- and cytokine-specific anti-inflammatory and anti-fibrotic treatments, including the main therapies that have been used or are currently being tested in clinical trials.

To begin, among the broad-spectrum treatments, MTX has been well tolerated, but its efficacy in reducing fibrosis is still controversial. Although CYC's use seems efficient in reducing fibrosis, its high toxicity limits its use, and it has now been replaced with MMF. MMF is well tolerated, but its use has not been fully examined. On the other hand, cellular targeting of inflammation with molecules that reduce the number of B or T cells, such as rituximab, belimumab, and abatacept, has exhibited a potential effect in diminishing fibrosis compared to broad-spectrum immunosuppressants. Furthermore, the cytokine signaling-specific antibodies rilonacept, basiliximab, fresolimumab, and metelimumab show a trend towards reducing fibrosis, but the lack of large phase III trials limits their potential addition to SSc treatment guidelines. Among the TK inhibitors, nintedanib and tofacitinib are two promising therapies in halting SSc-related fibrosis. Based on encouraging outcomes from large phase III RCTs, tocilizumab and nintedanib have been FDA approved for the treatment of SSc-ILD. On the positive side, ASCT shows increased event-free and improved overall survival rates, but its use is limited to younger patients with early dcSSc that fulfill a list of very strict inclusion criteria. Careful patient selection is vital in reducing the relatively high mortality rates that accompany ASCT.

In conclusion, we are in a new era of a multitude of clinical trials with drugs targeting specific pathogenic cells and biological pathways related to SSc. Several anti-inflammatory treatments have been used or tested in SSc patients, but only a mild to moderate improvement in reducing fibrosis has been demonstrated. However, the level of published evidence on the effectiveness of each tested drug varies greatly among case series and observational studies, and only a few RCTs have been conducted [205]. Thus, conclusions and comparisons between the efficacy of different drugs should be handled with caution. The anti-inflammatory approaches described show a trend towards reducing fibrosis, but the effect is moderate and, in many cases, controversial. The lack of complete understanding of the pathophysiology and the rare frequency of the disease are two obvious arguments that support this conclusion. Furthermore, SSc is a very complex and heterogeneous disease, and the traditional classification of the patients into the limited or diffuse form based on the severity of skin involvement is an oversimplification. Lately, multiple studies have utilized intrinsic gene expression analyses to classify SSc patients into four categories: the inflammatory, fibroproliferative, limited, and normal-like subsets [206]. It is currently not certain if these categories are reflective of stable disease states that differ between patients. Other studies suggest that they concern different stages of the SSc disease process [80]. Various disease manifestations of SSc, such as ILD, pulmonary hypertension, and gastrointestinal disease, develop with a time course that differs from that of the skin manifestations. The underlying maladaptive innate and adaptive immune responses likely differ between these pathologies. Taken together, we believe that in-depth insight and measurement tools of the pathological processes may yield markers to determine the biological processes driving specific disease manifestations. Such markers will help to determine the best treatment approach in individual cases.

Furthermore, targeted drug selection is expected to show more remarkable results in the anti-inflammatory therapies that have been mentioned. Additionally, to understand if new drugs are effective, there is also a need for a better understanding of the pathological processes driving disease features. The use of novel and advanced molecular tools such as single-cell RNA sequencing is constantly advancing our knowledge about novel pathogenic cytokines, antibodies, and genes implicated in the pathogenesis of SSc. This knowledge will facilitate more personalized treatments. For example, scleroderma patients that exhibit high amounts of the Th2 cytokines IL-4 and IL-13 early after diagnosis, or those with prominent ILD could be treated with immunotoxins or monoclonal antibodies that have been designed to block the IL-4 or IL-13 pathway and have shown promising results in anti-tumor, IPF, and asthma treatment. Another suggestion for future studies would be to evaluate the outcome of combined biological therapies.

Author Contributions: Conceptualization, T.-I.P., A.v.C. and R.M.T.; writing of the original draft preparation, T.-I.P.; writing of review and editing T.-I.P., A.v.C., P.M.v.d.K. and R.M.T. All authors have read and agreed to the published version of the manuscript.

Funding: This research received no external funding.

Institutional Review Board Statement: Not applicable.

Informed Consent Statement: Not applicable.

Data Availability Statement: Not applicable. No new data were generated or analyzed in this study.

Conflicts of Interest: The authors declare no conflict of interest.

References

1. Herrick, A.L.; Pan, X.; Peytrignet, S.; Lunt, M.; Hesselstrand, R.; Mouthon, L.; Silman, A.; Brown, E.; Czirják, L.; Distler, J.H.W.; et al. Treatment outcome in early diffuse cutaneous systemic sclerosis: The European Scleroderma Observational Study (ESOS). *Ann. Rheum. Dis.* **2017**, *76*, 1207–1218. [CrossRef] [PubMed]
2. Pellar, R.E.; Pope, J.E. Evidence-based management of systemic sclerosis: Navigating recommendations and guidelines. *Semin. Arthritis Rheum.* **2017**, *46*, 767–774. [CrossRef] [PubMed]
3. Lee, J.J.; Pope, J.E. Emerging drugs and therapeutics for systemic sclerosis. *Expert Opin. Emerg. Drugs* **2016**, *21*, 421–430. [CrossRef] [PubMed]
4. Fox, D.A.; Lundy, S.K.; Whitfield, M.L.; Berrocal, V.; Campbell, P.; Rasmussen, S.; Ohara, R.; Stinson, A.; Gurrea-Rubio, M.; Wiewiora, E.; et al. Lymphocyte subset abnormalities in early diffuse cutaneous systemic sclerosis. *Arthritis Res. Ther.* **2021**, *23*, 10. [CrossRef]
5. Liu, M.; Wu, W.; Sun, X.; Yang, J.; Xu, J.; Fu, W.; Li, M. New insights into CD4+ T cell abnormalities in systemic sclerosis. *Cytokine Growth Factor Rev.* **2016**, *28*, 31–36. [CrossRef]
6. Maehara, T.; Kaneko, N.; Perugino, C.A.; Mattoo, H.; Kers, J.; Allard-Chamard, H.; Mahajan, V.S.; Liu, H.; Murphy, S.J.; Ghebremichael, M.; et al. Cytotoxic CD4+ T lymphocytes may induce endothelial cell apoptosis in systemic sclerosis. *J. Clin. Investig.* **2020**, *130*, 2451–2464. [CrossRef]
7. Henderson, J.; Bhattacharyya, S.; Varga, J.; O'Reilly, S. Targeting TLRs and the inflammasome in systemic sclerosis. *Pharmacol. Ther.* **2018**, *192*, 163–169. [CrossRef]
8. Fuschiotti, P. T cells and cytokines in systemic sclerosis. *Curr. Opin. Rheumatol.* **2018**, *30*, 594–599. [CrossRef]
9. Romano, E.; Rosa, I.; Fioretto, B.S.; Matucci-Cerinic, M.; Manetti, M. New Insights into Profibrotic Myofibroblast Formation in Systemic Sclerosis: When the Vascular Wall Becomes the Enemy. *Life* **2021**, *11*, 610. [CrossRef]

10. Bruni, C.; Frech, T.; Manetti, M.; Rossi, F.W.; Furst, D.E.; De Paulis, A.; Rivellese, F.; Guiducci, S.; Matucci-Cerinic, M.; Bellando-Randone, S. Vascular Leaking, a Pivotal and Early Pathogenetic Event in Systemic Sclerosis: Should the Door Be Closed? *Front. Immunol.* **2018**, *9*, 2045. [CrossRef]
11. Host, L.; Nikpour, M.; Calderone, A.; Cannell, P.; Roddy, J. Autologous stem cell transplantation in systemic sclerosis: A systematic review. *Clin. Exp. Rheumatol.* **2017**, *35*, S198–S207.
12. Laurent, P.; Sisirak, V.; Lazaro, E.; Richez, C.; Duffau, P.; Blanco, P.; Truchetet, M.-E.; Contin-Bordes, C. Innate Immunity in Systemic Sclerosis Fibrosis: Recent Advances. *Front. Immunol.* **2018**, *9*, 1702. [CrossRef]
13. van Caam, A.; Vonk, M.; van den Hoogen, F.; van Lent, P.; van der Kraan, P. Unraveling SSc Pathophysiology; The Myofibroblast. *Front. Immunol.* **2018**, *9*, 2452. [CrossRef] [PubMed]
14. Pesce, J.T.; Ramalingam, T.R.; Mentink-Kane, M.M.; Wilson, M.S.; El Kasmi, K.C.; Smith, A.M.; Thompson, R.W.; Cheever, A.W.; Murray, P.J.; Wynn, T.A. Arginase-1-expressing macrophages suppress Th2 cytokine-driven inflammation and fibrosis. *PLoS Pathog.* **2009**, *5*, e1000371. [CrossRef] [PubMed]
15. Ostuni, R.; Kratochvill, F.; Murray, P.J.; Natoli, G. Macrophages and cancer: From mechanisms to therapeutic implications. *Trends Immunol.* **2015**, *36*, 229–239. [CrossRef] [PubMed]
16. Wynn, T.A.; Vannella, K.M. Macrophages in Tissue Repair, Regeneration, and Fibrosis. *Immunity* **2016**, *44*, 450–462. [CrossRef]
17. Christmann, R.B.; Sampaio-Barros, P.; Stifano, G.; Borges, C.L.; de Carvalho, C.R.; Kairalla, R.; Parra, E.R.; Spira, A.; Simms, R.; Capellozzi, V.L.; et al. Association of Interferon- and transforming growth factor β-regulated genes and macrophage activation with systemic sclerosis-related progressive lung fibrosis. *Arthritis Rheumatol.* **2014**, *66*, 714–725. [CrossRef]
18. Assassi, S.; Swindell, W.R.; Wu, M.; Tan, F.D.; Khanna, D.; Furst, D.E.; Tashkin, D.P.; Jahan-Tigh, R.R.; Mayes, M.D.; Gudjonsson, J.E.; et al. Dissecting the heterogeneity of skin gene expression patterns in systemic sclerosis. *Arthritis Rheumatol.* **2015**, *67*, 3016–3026. [CrossRef]
19. Rudnik, M.; Hukara, A.; Kocherova, I.; Jordan, S.; Schniering, J.; Milleret, V.; Ehrbar, M.; Klingel, K.; Feghali-Bostwick, C.; Distler, O.; et al. Elevated Fibronectin Levels in Profibrotic CD14(+) Monocytes and CD14(+) Macrophages in Systemic Sclerosis. *Front. Immunol.* **2021**, *12*, 642891. [CrossRef]
20. Soldano, S.; Trombetta, A.C.; Contini, P.; Tomatis, V.; Ruaro, B.; Brizzolara, R.; Montagna, P.; Sulli, A.; Paolino, S.; Pizzorni, C.; et al. Increase in circulating cells coexpressing M1 and M2 macrophage surface markers in patients with systemic sclerosis. *Ann. Rheum. Dis.* **2018**, *77*, 1842–1845. [CrossRef]
21. Gibbons, M.A.; MacKinnon, A.C.; Ramachandran, P.; Dhaliwal, K.; Duffin, R.; Phythian-Adams, A.T.; Van Rooijen, N.; Haslett, C.; Howie, S.E.; Simpson, A.J.; et al. Ly6Chimonocytes direct alternatively activated profibrotic macrophage regulation of lung fibrosis. *Am. J. Respir. Crit. Care Med.* **2011**, *184*, 569–581. [CrossRef] [PubMed]
22. Cutolo, M.; Soldano, S.; Smith, V. Pathophysiology of systemic sclerosis: Current understanding and new insights. *Expert Rev. Clin. Immunol.* **2019**, *15*, 753–764. [CrossRef] [PubMed]
23. Higashi-Kuwata, N.; Jinnin, M.; Makino, T.; Fukushima, S.; Inoue, Y.; Muchemwa, F.C.; Yonemura, Y.; Komohara, Y.; Takeya, M.; Mitsuya, H.; et al. Characterization of monocyte/macrophage subsets in the skin and peripheral blood derived from patients with systemic sclerosis. *Arthritis Res. Ther.* **2010**, *12*, R128. [CrossRef] [PubMed]
24. Manetti, M. Deciphering the alternatively activated (M2) phenotype of macrophages in scleroderma. *Exp. Dermatol.* **2015**, *24*, 576–578. [CrossRef]
25. Bhattacharyya, S.; Wei, J.; Tourtellotte, W.G.; Hinchcliff, M.; Gottardi, C.G.; Varga, J. Fibrosis in systemic sclerosis: Common and unique pathobiology. *Fibrogenesis Tissue Repair* **2012**, *5* (Suppl. S1), S18. [CrossRef]
26. Borie, R.; Quesnel, C.; Phin, S.; Debray, M.-P.; Marchal-Somme, J.; Tiev, K.; Bonay, M.; Fabre, A.; Soler, P.; Dehoux, M.; et al. Detection of alveolar fibrocytes in idiopathic pulmonary fibrosis and systemic sclerosis. *PLoS ONE* **2013**, *8*, e53736. [CrossRef]
27. Ruaro, B.; Soldano, S.; Smith, V.; Paolino, S.; Contini, P.; Montagna, P.; Pizzorni, C.; Casabella, A.; Tardito, S.; Sulli, A.; et al. Correlation between circulating fibrocytes and dermal thickness in limited cutaneous systemic sclerosis patients: A pilot study. *Rheumatol. Int.* **2019**, *39*, 1369–1376. [CrossRef]
28. Xue, D.; Tabib, T.; Morse, C.; Yang, Y.; Domsic, R.; Khanna, D.; Lafyatis, R. Expansion of FCGR3A(+) macrophages, FCN1(+) mo-DC, and plasmacytoid dendritic cells associated with severe skin disease in systemic sclerosis. *Arthritis Rheumatol.* **2021**, *74*, 329–341. [CrossRef]
29. Pope, S.M.; Brandt, E.B.; Mishra, A.; Hogan, S.P.; Zimmermann, N.; Matthaei, K.I.; Foster, P.S.; Rothenberg, M.E. IL-13 induces eosinophil recruitment into the lung by an IL-5– and eotaxin-dependent mechanism. *J. Allergy Clin. Immunol.* **2001**, *108*, 594–601. [CrossRef]
30. Chen, L.; Grabowski, K.A.; Xin, J.-P.; Coleman, J.; Huang, Z.; Espiritu, B.; Alkan, S.; Xie, H.B.; Zhu, Y.; White, F.A.; et al. IL-4 induces differentiation and expansion of Th2 cytokine-producing eosinophils. *J. Immunol.* **2004**, *172*, 2059–2066. [CrossRef]
31. Falanga, V.; Medsger, T.A.J. Frequency, levels, and significance of blood eosinophilia in systemic sclerosis, localized scleroderma, and eosinophilic fasciitis. *J. Am. Acad. Dermatol.* **1987**, *17*, 648–656. [CrossRef]
32. Ando, K.; Nakashita, T.; Kaneko, N.; Takahashi, K.; Motojima, S. Associations between peripheral blood eosinophil counts in patients with systemic sclerosis and disease severity. *Springerplus* **2016**, *5*, 1401. [CrossRef] [PubMed]
33. Hara, T.; Ikeda, T.; Inaba, Y.; Kunimoto, K.; Mikita, N.; Kaminaka, C.; Kanazawa, N.; Yamamoto, Y.; Tabata, K.; Fujii, T.; et al. Peripheral blood eosinophilia is associated with the presence of skin ulcers in patients with systemic sclerosis. *J. Dermatol.* **2019**, *46*, 334–337. [CrossRef] [PubMed]

34. Saigusa, R.; Asano, Y.; Yamashita, T.; Takahashi, T.; Nakamura, K.; Miura, S.; Ichimura, Y.; Toyama, T.; Taniguchi, T.; Sumida, H.; et al. Systemic sclerosis complicated with localized scleroderma-like lesions induced by Köbner phenomenon. *J. Dermatol. Sci.* **2018**, *89*, 282–289. [CrossRef] [PubMed]
35. Hebbar, M.; Gillot, J.M.; Hachulla, E.; Lassalle, P.; Hatron, P.Y.; Devulder, B.; Janin, A. Early expression of E-selectin, tumor necrosis factor alpha, and mast cell infiltration in the salivary glands of patients with systemic sclerosis. *Arthritis Rheum.* **1996**, *39*, 1161–1165. [CrossRef] [PubMed]
36. Bagnato, G.; Roberts, W.N.; Sciortino, D.; Sangari, D.; Cirmi, S.; Ravenell, R.L.; Navarra, M.; Bagnato, G.; Gangemi, S. Mastocytosis and systemic sclerosis: A clinical association. *Clin. Mol. Allergy* **2016**, *14*, 13. [CrossRef] [PubMed]
37. Yukawa, S.; Yamaoka, K.; Sawamukai, N.; Shimajiri, S.; Kubo, S.; Miyagawa, I.; Sonomoto, K.; Saito, K.; Tanaka, Y. Dermal mast cell density in fingers reflects severity of skin sclerosis in systemic sclerosis. *Mod. Rheumatol.* **2013**, *23*, 1151–1157. [CrossRef]
38. Dees, C.; Akhmetshina, A.; Zerr, P.; Reich, N.; Palumbo, K.; Horn, A.; Jüngel, A.; Beyer, C.; Krönke, G.; Zwerina, J.; et al. Platelet-derived serotonin links vascular disease and tissue fibrosis. *J. Exp. Med.* **2011**, *208*, 961–972. [CrossRef]
39. Prieto-García, A.; Zheng, D.; Adachi, R.; Xing, W.; Lane, W.S.; Chung, K.; Anderson, P.; Hansbro, P.M.; Castells, M.; Stevens, R.L. Mast cell restricted mouse and human tryptase·heparin complexes hinder thrombin-induced coagulation of plasma and the generation of fibrin by proteolytically destroying fibrinogen. *J. Biol. Chem.* **2012**, *287*, 7834–7844. [CrossRef]
40. Jordana, M.; Befus, A.D.; Newhouse, M.T.; Bienenstock, J.; Gauldie, J. Effect of histamine on proliferation of normal human adult lung fibroblasts. *Thorax* **1988**, *43*, 552–558. [CrossRef]
41. Karpec, D.; Rudys, R.; Leonaviciene, L.; Mackiewicz, Z.; Bradunaite, R.; Kirdaite, G.; Venalis, A. The safety and efficacy of light emitting diodes-based ultraviolet A1 phototherapy in bleomycin-induced scleroderma in mice. *Adv. Med. Sci.* **2018**, *63*, 152–159. [CrossRef] [PubMed]
42. Kabata, H.; Moro, K.; Koyasu, S. The group 2 innate lymphoid cell (ILC2) regulatory network and its underlying mechanisms. *Immunol. Rev.* **2018**, *286*, 37–52. [CrossRef] [PubMed]
43. Klose, C.S.N.; Artis, D. Innate lymphoid cells as regulators of immunity, inflammation and tissue homeostasis. *Nat. Immunol.* **2016**, *17*, 765–774. [CrossRef] [PubMed]
44. McHedlidze, T.; Waldner, M.; Zopf, S.; Walker, J.; Rankin, A.L.; Schuchmann, M.; Voehringer, D.; McKenzie, A.N.J.; Neurath, M.F.; Pflanz, S.; et al. Interleukin-33-dependent innate lymphoid cells mediate hepatic fibrosis. *Immunity* **2013**, *39*, 357–371. [CrossRef] [PubMed]
45. Wohlfahrt, T.; Usherenko, S.; Englbrecht, M.; Dees, C.; Weber, S.; Beyer, C.; Gelse, K.; Distler, O.; Schett, G.; Distler, J.H.W.; et al. Type 2 innate lymphoid cell counts are increased in patients with systemic sclerosis and correlate with the extent of fibrosis. *Ann. Rheum. Dis.* **2016**, *75*, 623–626. [CrossRef]
46. Truchetet, M.-E.; Demoures, B.; Eduardo Guimaraes, J.; Bertrand, A.; Laurent, P.; Jolivel, V.; Douchet, I.; Jacquemin, C.; Khoryati, L.; Duffau, P.; et al. Platelets Induce Thymic Stromal Lymphopoietin Production by Endothelial Cells: Contribution to Fibrosis in Human Systemic Sclerosis. *Arthritis Rheumatol.* **2016**, *68*, 2784–2794. [CrossRef]
47. Manetti, M.; Guiducci, S.; Ceccarelli, C.; Romano, E.; Bellando-Randone, S.; Conforti, M.L.; Ibba-Manneschi, L.; Matucci-Cerinic, M. Increased circulating levels of interleukin 33 in systemic sclerosis correlate with early disease stage and microvascular involvement. *Ann. Rheum. Dis.* **2011**, *70*, 1876–1878. [CrossRef]
48. La Barbera, L.; Lo Pizzo, M.; DI Liberto, D.; Schinocca, C.; Ruscitti, P.; Giacomelli, R.; Dieli, F.; Ciccia, F.; Guggino, G. POS0326 role of the IL-25 / IL-17RB Axis in TH9 Polarization in Patients with Progressive Systemic Sclerosis. *Ann. Rheum. Dis.* **2021**, *80*, 390–391. [CrossRef]
49. Laurent, P.; Allard, B.; Manicki, P.; Jolivel, V.; Levionnois, E.; Jeljeli, M.; Henrot, P.; Izotte, J.; Leleu, D.; Groppi, A.; et al. TGFβ promotes low IL10-producing ILC2 with profibrotic ability involved in skin fibrosis in systemic sclerosis. *Ann. Rheum. Dis.* **2021**, *80*, 1594–1603. [CrossRef]
50. Ah Kioon, M.D.; Tripodo, C.; Fernandez, D.; Kirou, K.A.; Spiera, R.F.; Crow, M.K.; Gordon, J.K.; Barrat, F.J. Plasmacytoid dendritic cells promote systemic sclerosis with a key role for TLR8. *Sci. Transl. Med.* **2018**, *10*, 8458. [CrossRef]
51. van Bon, L.; Affandi, A.J.; Broen, J.; Christmann, R.B.; Marijnissen, R.J.; Stawski, L.; Farina, G.A.; Stifano, G.; Mathes, A.L.; Cossu, M.; et al. Proteome-wide analysis and CXCL4 as a biomarker in systemic sclerosis. *N. Engl. J. Med.* **2014**, *370*, 433–443. [CrossRef] [PubMed]
52. Kafaja, S.; Valera, I.; Divekar, A.A.; Saggar, R.; Abtin, F.; Furst, D.E.; Khanna, D.; Singh, R.R. pDCs in lung and skin fibrosis in a bleomycin-induced model and patients with systemic sclerosis. *JCI Insight* **2018**, *3*, 98380. [CrossRef] [PubMed]
53. Valenzi, E.; Tabib, T.; Papazoglou, A.; Sembrat, J.; Trejo Bittar, H.E.; Rojas, M.; Lafyatis, R. Disparate Interferon Signaling and Shared Aberrant Basaloid Cells in Single-Cell Profiling of Idiopathic Pulmonary Fibrosis and Systemic Sclerosis-Associated Interstitial Lung Disease. *Front. Immunol.* **2021**, *12*, 595811. [CrossRef] [PubMed]
54. Skaug, B.; Assassi, S. Type I interferon dysregulation in Systemic Sclerosis. *Cytokine* **2020**, *132*, 154635. [CrossRef] [PubMed]
55. Wu, M.; Assassi, S. The Role of Type 1 Interferon in Systemic Sclerosis. *Front. Immunol.* **2013**, *4*, 266. [CrossRef] [PubMed]
56. van der Kroef, M.; van den Hoogen, L.L.; Mertens, J.S.; Blokland, S.L.M.; Haskett, S.; Devaprasad, A.; Carvalheiro, T.; Chouri, E.; Vazirpanah, N.; Cossu, M.; et al. Cytometry by time of flight identifies distinct signatures in patients with systemic sclerosis, systemic lupus erythematosus and Sjögrens syndrome. *Eur. J. Immunol.* **2020**, *50*, 119–129. [CrossRef]

57. Alkema, W.; Koenen, H.; Kersten, B.E.; Kaffa, C.; Dinnissen, J.W.B.; Damoiseaux, J.G.M.C.; Joosten, I.; Driessen-Diks, S.; van der Molen, R.G.; Vonk, M.C.; et al. Autoantibody profiles in systemic sclerosis; a comparison of diagnostic tests. *Autoimmunity* **2021**, *54*, 148–155. [CrossRef] [PubMed]
58. Mehta, H.; Goulet, P.-O.; Nguyen, V.; Pérez, G.; Koenig, M.; Senécal, J.-L.; Sarfati, M. Topoisomerase I peptide-loaded dendritic cells induce autoantibody response as well as skin and lung fibrosis. *Autoimmunity* **2016**, *49*, 503–513. [CrossRef]
59. Asano, Y. Systemic sclerosis. *J. Dermatol.* **2018**, *45*, 128–138. [CrossRef]
60. O'Reilly, S.; Hugle, T.; van Laar, J.M. T cells in systemic sclerosis: A reappraisal. *Rheumatology* **2012**, *51*, 1540–1549. [CrossRef]
61. Boin, F.; De Fanis, U.; Bartlett, S.J.; Wigley, F.M.; Rosen, A. T Cell polarization indentifies distinct clinical phenotypes in scleroderma lung disease. *Arththritis Rheum.* **2008**, *58*, 1165–1174. [CrossRef] [PubMed]
62. Lonati, P.A.; Brembilla, N.C.; Montanari, E.; Fontao, L.; Gabrielli, A.; Vettori, S.; Valentini, G.; Laffitte, E.; Kaya, G.; Meroni, P.L.; et al. High IL-17E and Low IL-17C dermal expression identifies a fibrosis-specific motif common to morphea and systemic sclerosis. *PLoS ONE* **2014**, *9*, e0105008. [CrossRef] [PubMed]
63. Budzyńska-Włodarczyk, J.; Michalska-Jakubus, M.M.; Kowal, M.; Krasowska, D. Evaluation of serum concentrations of the selected cytokines in patients with localized scleroderma. *Postep. Dermatol. I Alergol.* **2016**, *33*, 47–51. [CrossRef] [PubMed]
64. Ayers, N.B.; Sun, C.-M.; Chen, S.-Y. Transforming growth factor-β signaling in systemic sclerosis. *J. Biomed. Res.* **2018**, *32*, 3–12. [CrossRef]
65. Gonçalves, R.S.G.; Pereira, M.C.; Dantas, A.T.; de Almeida, A.R.; Marques, C.D.L.; Rego, M.J.B.M.; Pitta, I.R.; Duarte, A.L.B.P.; Pitta, M.G.R. IL-17 and related cytokines involved in systemic sclerosis: Perspectives. *Autoimmunity* **2018**, *51*, 1–9. [CrossRef]
66. Rafael-Vidal, C.; Pérez, N.; Altabás, I.; Garcia, S.; Pego-Reigosa, J.M. Blocking IL-17: A Promising Strategy in the Treatment of Systemic Rheumatic Diseases. *Int. J. Mol. Sci.* **2020**, *21*, 7100. [CrossRef]
67. Kalfin, R.; Righi, A.; Del Rosso, A.; Bagchi, D.; Generini, S.; Guiducci, S.; Cerinic, M.M.; Das, D.K. Activin, a grape seed-derived proanthocyanidin extract, reduces plasma levels of oxidative stress and adhesion molecules (ICAM-1, VCAM-1 and E-selectin) in systemic sclerosis. *Free Radic. Res.* **2002**, *36*, 819–825. [CrossRef]
68. Romano, M.; Fanelli, G.; Albany, C.J.; Giganti, G.; Lombardi, G. Past, Present, and Future of Regulatory T Cell Therapy in Transplantation and Autoimmunity. *Front. Immunol.* **2019**, *10*, 43. [CrossRef]
69. Frantz, C.; Auffray, C.; Avouac, J.; Allanore, Y. Regulatory T Cells in Systemic Sclerosis. *Front. Immunol.* **2018**, *9*, 2358. [CrossRef]
70. Milano, A.; Pendergrass, S.A.; Sargent, J.L.; George, L.K.; McCalmont, T.H.; Connolly, M.K.; Whitfield, M.L. Molecular subsets in the gene expression signatures of scleroderma skin. *PLoS ONE* **2008**, *3*, e2696. [CrossRef]
71. Rice, L.M.; Stifano, G.; Ziemek, J.; Lafyatis, R. Local skin gene expression reflects both local and systemic skin disease in patients with systemic sclerosis. *Rheumatology* **2016**, *55*, 377–379. [CrossRef] [PubMed]
72. Fuschiotti, P. Current perspectives on the role of CD8+ T cells in systemic sclerosis. *Immunol. Lett.* **2018**, *195*, 55–60. [CrossRef] [PubMed]
73. Casciola-Rosen, L.; Andrade, F.; Ulanet, D.; Wong, W.B.; Rosen, A. Cleavage by granzyme B is strongly predictive of autoantigen status: Implications for initiation of autoimmunity. *J. Exp. Med.* **1999**, *190*, 815–826. [CrossRef] [PubMed]
74. Morita, R.; Schmitt, N.; Bentebibel, S.-E.; Ranganathan, R.; Bourdery, L.; Zurawski, G.; Foucat, E.; Dullaers, M.; Oh, S.; Sabzghabaei, N.; et al. Human blood CXCR5(+)CD4(+) T cells are counterparts of T follicular cells and contain specific subsets that differentially support antibody secretion. *Immunity* **2011**, *34*, 108–121. [CrossRef] [PubMed]
75. Ricard, L.; Jachiet, V.; Malard, F.; Ye, Y.; Stocker, N.; Rivière, S.; Senet, P.; Monfort, J.-B.; Fain, O.; Mohty, M.; et al. Circulating follicular helper T cells are increased in systemic sclerosis and promote plasmablast differentiation through the IL-21 pathway which can be inhibited by ruxolitinib. *Ann. Rheum. Dis.* **2019**, *78*, 539–550. [CrossRef]
76. Taylor, D.K.; Mittereder, N.; Kuta, E.; Delaney, T.; Burwell, T.; Dacosta, K.; Zhao, W.; Cheng, L.I.; Brown, C.; Boutrin, A.; et al. T follicular helper-like cells contribute to skin fibrosis. *Sci. Transl. Med.* **2018**, *10*, 5307. [CrossRef]
77. Gaydosik, A.M.; Tabib, T.; Domsic, R.; Khanna, D.; Lafyatis, R.; Fuschiotti, P. Single-cell transcriptome analysis identifies skin-specific T-cell responses in systemic sclerosis. *Ann. Rheum. Dis.* **2021**, *80*, 1453–1460. [CrossRef]
78. Mahler, M.; Hudson, M.; Bentow, C.; Roup, F.; Beretta, L.; Pilar Simeón, C.; Guillén-Del-Castillo, A.; Casas, S.; Fritzler, M.J. Autoantibodies to stratify systemic sclerosis patients into clinically actionable subsets. *Autoimmun. Rev.* **2020**, *19*, 102583. [CrossRef]
79. Melissaropoulos, K.; Daoussis, D. B cells in systemic sclerosis: From pathophysiology to treatment. *Clin. Rheumatol.* **2021**, *40*, 2621–2631. [CrossRef]
80. Skaug, B.; Khanna, D.; Swindell, W.R.; Hinchcliff, M.E.; Frech, T.M.; Steen, V.D.; Hant, F.N.; Gordon, J.K.; Shah, A.A.; Zhu, L.; et al. Global skin gene expression analysis of early diffuse cutaneous systemic sclerosis shows a prominent innate and adaptive inflammatory profile. *Ann. Rheum. Dis.* **2020**, *79*, 379–386. [CrossRef]
81. Bosello, S.; Angelucci, C.; Lama, G.; Alivernini, S.; Proietti, G.; Tolusso, B.; Sica, G.; Gremese, E.; Ferraccioli, G. Characterization of inflammatory cell infiltrate of scleroderma skin: B cells and skin score progression. *Arthritis Res. Ther.* **2018**, *20*, 75. [CrossRef] [PubMed]
82. Matsushita, T.; Kobayashi, T.; Mizumaki, K.; Kano, M.; Sawada, T.; Tennichi, M.; Okamura, A.; Hamaguchi, Y.; Iwakura, Y.; Hasegawa, M.; et al. BAFF inhibition attenuates fibrosis in scleroderma by modulating the regulatory and effector B cell balance. *Sci. Adv.* **2018**, *4*, eaas9944. [CrossRef] [PubMed]

83. Matsushita, T.; Hasegawa, M.; Yanaba, K.; Kodera, M.; Takehara, K.; Sato, S. Elevated serum BAFF levels in patients with systemic sclerosis: Enhanced BAFF signaling in systemic sclerosis B lymphocytes. *Arthritis Rheum.* **2006**, *54*, 192–201. [CrossRef] [PubMed]
84. Matsushita, T.; Fujimoto, M.; Hasegawa, M.; Matsushita, Y.; Komura, K.; Ogawa, F.; Watanabe, R.; Takehara, K.; Sato, S. BAFF antagonist attenuates the development of skin fibrosis in tight-skin mice. *J. Investig. Dermatol.* **2007**, *127*, 2772–2780. [CrossRef] [PubMed]
85. Soto, L.; Ferrier, A.; Aravena, O.; Fonseca, E.; Berendsen, J.; Biere, A.; Bueno, D.; Ramos, V.; Aguillón, J.C.; Catalán, D. Systemic Sclerosis Patients Present Alterations in the Expression of Molecules Involved in B-Cell Regulation. *Front. Immunol.* **2015**, *6*, 496. [CrossRef] [PubMed]
86. Dumoitier, N.; Chaigne, B.; Régent, A.; Lofek, S.; Mhibik, M.; Dorfmüller, P.; Terrier, B.; London, J.; Bérezné, A.; Tamas, N.; et al. Scleroderma Peripheral B Lymphocytes Secrete Interleukin-6 and Transforming Growth Factor β and Activate Fibroblasts. *Arthritis Rheumatol.* **2017**, *69*, 1078–1089. [CrossRef] [PubMed]
87. Forestier, A.; Guerrier, T.; Jouvray, M.; Giovannelli, J.; Lefèvre, G.; Sobanski, V.; Hauspie, C.; Hachulla, E.; Hatron, P.-Y.; Zéphir, H.; et al. Altered B lymphocyte homeostasis and functions in systemic sclerosis. *Autoimmun. Rev.* **2018**, *17*, 244–255. [CrossRef] [PubMed]
88. Sato, S.; Fujimoto, M.; Hasegawa, M.; Takehara, K. Altered blood B lymphocyte homeostasis in systemic sclerosis: Expanded naive B cells and diminished but activated memory B cells. *Arthritis Rheum.* **2004**, *50*, 1918–1927. [CrossRef]
89. Sato, S.; Hasegawa, M.; Fujimoto, M.; Tedder, T.F.; Takehara, K. Quantitative genetic variation in CD19 expression correlates with autoimmunity. *J. Immunol.* **2000**, *165*, 6635–6643. [CrossRef]
90. Wilfong, E.M.; Vowell, K.N.; Bunn, K.E.; Rizzi, E.; Annapureddy, N.; Dudenhofer, R.B.; Barnado, A.; Bonami, R.H.; Johnson, J.E.; Crofford, L.J.; et al. CD19 + CD21lo/neg cells are increased in systemic sclerosis-associated interstitial lung disease. *Clin. Exp. Med.* **2021**. [CrossRef]
91. Asano, N.; Fujimoto, M.; Yazawa, N.; Shirasawa, S.; Hasegawa, M.; Okochi, H.; Tamaki, K.; Tedder, T.F.; Sato, S. B Lymphocyte signaling established by the CD19/CD22 loop regulates autoimmunity in the tight-skin mouse. *Am. J. Pathol.* **2004**, *165*, 641–650. [CrossRef]
92. Sato, S.; Ono, N.; Steeber, D.A.; Pisetsky, D.S.; Tedder, T.F. CD19 regulates B lymphocyte signaling thresholds critical for the development of B-1 lineage cells and autoimmunity. *J. Immunol.* **1996**, *157*, 4371–4378. [PubMed]
93. van Zelm, M.C.; Reisli, I.; van der Burg, M.; Castaño, D.; van Noesel, C.J.M.; van Tol, M.J.D.; Woellner, C.; Grimbacher, B.; Patiño, P.J.; van Dongen, J.J.M.; et al. An antibody-deficiency syndrome due to mutations in the CD19 gene. *N. Engl. J. Med.* **2006**, *354*, 1901–1912. [CrossRef] [PubMed]
94. Terrier, B.; Tamby, M.C.; Camoin, L.; Guilpain, P.; Bérezné, A.; Tamas, N.; Broussard, C.; Hotellier, F.; Humbert, M.; Simonneau, G.; et al. Antifibroblast antibodies from systemic sclerosis patients bind to {alpha}-enolase and are associated with interstitial lung disease. *Ann. Rheum. Dis.* **2010**, *69*, 428–433. [CrossRef]
95. Baroni, S.S.; Santillo, M.; Bevilacqua, F.; Luchetti, M.; Spadoni, T.; Mancini, M.; Fraticelli, P.; Sambo, P.; Funaro, A.; Kazlauskas, A.; et al. Stimulatory autoantibodies to the PDGF receptor in systemic sclerosis. *N. Engl. J. Med.* **2006**, *354*, 2667–2676. [CrossRef]
96. Kowal-Bielecka, O.; Fransen, J.; Avouac, J.; Becker, M.; Kulak, A.; Allanore, Y.; Distler, O.; Clements, P.; Cutolo, M.; Czirjak, L.; et al. Update of EULAR recommendations for the treatment of systemic sclerosis. *Ann. Rheum. Dis.* **2017**, *76*, 1327–1339. [CrossRef]
97. Denton, C.P.; Hughes, M.; Gak, N.; Vila, J.; Buch, M.H.; Chakravarty, K.; Fligelstone, K.; Gompels, L.L.; Griffiths, B.; Herrick, A.L.; et al. BSR and BHPR guideline for the treatment of systemic sclerosis. *Rheumatology* **2016**, *55*, 1906–1910. [CrossRef]
98. Iudici, M. What should clinicians know about the use of glucocorticoids in systemic sclerosis? *Mod. Rheumatol.* **2017**, *27*, 919–923. [CrossRef]
99. Shah, A.A.; Wigley, F.M. My approach to the treatment of scleroderma. *Mayo Clin. Proc.* **2013**, *88*, 377–393. [CrossRef]
100. Uy, G.L.; Duncavage, E.J.; Chang, G.S.; Jacoby, M.A.; Miller, C.A.; Shao, J.; Heath, S.; Elliott, K.; Fulton, R.S.; Fronick, C.C.; et al. Dynamic changes in the clonal structure of MDS and AML in response to epigenetic therapy. *HHS Public Access* **2017**, *31*, 872–881. [CrossRef]
101. Lundberg, I.; Kratz, A.K.; Alexanderson, H.; Patarroyo, M. Decreased expression of interleukin-1α, interleukin-1β, and cell adhesion molecules in muscle tissue following corticosteroid treatment in patients with polymyositis and dermatomyositis. *Arthritis Rheum.* **2000**, *43*, 336–348. [CrossRef]
102. Steen, V.D.; Medsger, T.A. Case-control study of corticosteroids and other drugs that either precipitate or protect from the development of scleroderma renal crisis. *Arthritis Rheum.* **1998**, *41*, 1613–1619. [CrossRef]
103. Cozzi, F.; Marson, P.; Cardarelli, S.; Favaro, M.; Tison, T.; Tonello, M.; Pigatto, E.; De Silvestro, G.; Punzi, L.; Doria, A. Prognosis of scleroderma renal crisis: A long-term observational study. *Nephrol. Dial. Transplant.* **2012**, *27*, 4398–4403. [CrossRef] [PubMed]
104. Zanatta, E.; Polito, P.; Favaro, M.; Larosa, M.; Marson, P.; Cozzi, F.; Doria, A. Therapy of scleroderma renal crisis: State of the art. *Autoimmun. Rev.* **2018**, *17*, 882–889. [CrossRef] [PubMed]
105. Maksimovic, V.; Pavlovic-Popovic, Z.; Vukmirovic, S.; Cvejic, J.; Mooranian, A.; Al-Salami, H.; Mikov, M.; Golocorbin-Kon, S. Molecular mechanism of action and pharmacokinetic properties of methotrexate. *Mol. Biol. Rep.* **2020**, *47*, 4699–4708. [CrossRef] [PubMed]
106. Cronstein, B.N.; Aune, T.M. Methotrexate and its mechanisms of action in inflammatory arthritis. *Nat. Rev. Rheumatol.* **2020**, *16*, 145–154. [CrossRef]

107. Pope, J.E.; Bellamy, N.; Seibold, J.R.; Baron, M.; Ellman, M.; Carette, S.; Smith, C.D.; Chalmers, I.M.; Hong, P.; O'Hanlon, D.; et al. A randomized, controlled trial of methotrexate versus placebo in early diffuse scleroderma. *Arthritis Rheum.* **2001**, *44*, 1351–1358. [CrossRef]
108. van den Hoogen, F.H.; Boerbooms, A.M.; Swaak, A.J.; Rasker, J.J.; van Lier, H.J.; van de Putte, L.B. Comparison of methotrexate with placebo in the treatment of systemic sclerosis: A 24 week randomized double-blind trial, followed by a 24 week observational trial. *Br. J. Rheumatol.* **1996**, *35*, 364–372. [CrossRef]
109. Ghiringhelli, F.; Larmonier, N.; Schmitt, E.; Parcellier, A.; Cathelin, D.; Garrido, C.; Chauffert, B.; Solary, E.; Bonnotte, B.; Martin, F. CD4+CD25+ regulatory T cells suppress tumor immunity but are sensitive to cyclophosphamide which allows immunotherapy of established tumors to be curative. *Eur. J. Immunol.* **2004**, *34*, 336–344. [CrossRef]
110. Wachsmuth, L.P.; Patterson, M.T.; Eckhaus, M.A.; Venzon, D.J.; Gress, R.E.; Kanakry, C.G. Posttransplantation cyclophosphamide prevents graft-versus-host disease by inducing alloreactive T cell dysfunction and suppression. *J. Clin. Investig.* **2019**, *129*, 2357–2373. [CrossRef]
111. Bruni, C.; Shirai, Y.; Kuwana, M.; Matucci-Cerinic, M. Cyclophosphamide: Similarities and differences in the treatment of SSc and SLE. *Lupus* **2019**, *28*, 571–574. [CrossRef]
112. Tashkin, D.P.; Elashoff, R.; Clements, P.J.; Goldin, J.; Roth, M.D.; Furst, D.E.; Arriola, E.; Silver, R.; Strange, C.; Bolster, M.; et al. Cyclophosphamide versus Placebo in Scleroderma Lung Disease. *N. Engl. J. Med.* **2006**, *354*, 2655–2666. [CrossRef]
113. Goldin, J.; Elashoff, R.; Kim, H.J.; Yan, X.; Lynch, D.; Strollo, D.; Roth, M.D.; Clements, P.; Furst, D.E.; Khanna, D.; et al. Treatment of scleroderma-interstitial lung disease with cyclophosphamide is associated with less progressive fibrosis on serial thoracic high-resolution CT scan than placebo: Findings from the scleroderma lung study. *Chest* **2009**, *136*, 1333–1340. [CrossRef]
114. Broen, J.C.A.; van Laar, J.M. Mycophenolate mofetil, azathioprine and tacrolimus: Mechanisms in rheumatology. *Nat. Rev. Rheumatol.* **2020**, *16*, 167–178. [CrossRef]
115. Gernert, M.; Tony, H.-P.; Schwaneck, E.C.; Gadeholt, O.; Fröhlich, M.; Portegys, J.; Strunz, P.-P.; Schmalzing, M. Lymphocyte subsets in the peripheral blood are disturbed in systemic sclerosis patients and can be changed by immunosuppressive medication. *Rheumatol. Int.* **2021**, 1–9. [CrossRef]
116. Hinchcliff, M.; Toledo, D.M.; Taroni, J.N.; Wood, T.A.; Franks, J.M.; Ball, M.S.; Hoffmann, A.; Amin, S.M.; Tan, A.U.; Tom, K.; et al. Mycophenolate Mofetil Treatment of Systemic Sclerosis Reduces Myeloid Cell Numbers and Attenuates the Inflammatory Gene Signature in Skin. *J. Investig. Dermatol.* **2018**, *138*, 1301–1310. [CrossRef]
117. Morath, C.; Reuter, H.; Simon, V.; Krautkramer, E.; Muranyi, W.; Schwenger, V.; Goulimari, P.; Grosse, R.; Hahn, M.; Lichter, P.; et al. Effects of mycophenolic acid on human fibroblast proliferation, migration and adhesion in vitro and in vivo. *Am. J. Transplant. Off. J. Am. Soc. Transplant. Am. Soc. Transpl. Surg.* **2008**, *8*, 1786–1797. [CrossRef]
118. Volkmann, E.R.; Tashkin, D.P.; Li, N.; Roth, M.D.; Khanna, D.; Hoffmann-Vold, A.-M.; Kim, G.; Goldin, J.; Clements, P.J.; Furst, D.E.; et al. Mycophenolate Mofetil Versus Placebo for Systemic Sclerosis-Related Interstitial Lung Disease: An Analysis of Scleroderma Lung Studies I and II. *Arthritis Rheumatol.* **2017**, *69*, 1451–1460. [CrossRef]
119. Naidu, G.S.R.S.N.K.; Sharma, S.K.; Adarsh, M.B.; Dhir, V.; Sinha, A.; Dhooria, S.; Jain, S. Effect of mycophenolate mofetil (MMF) on systemic sclerosis-related interstitial lung disease with mildly impaired lung function: A double-blind, placebo-controlled, randomized trial. *Rheumatol. Int.* **2020**, *40*, 207–216. [CrossRef]
120. Tashkin, D.P.; Roth, M.D.; Clements, P.J.; Furst, D.E.; Khanna, D.; Kleerup, E.C.; Goldin, J.; Arriola, E.; Volkmann, E.R.; Kafaja, S.; et al. Mycophenolate mofetil versus oral cyclophosphamide in scleroderma-related interstitial lung disease (SLS II): A randomised controlled, double-blind, parallel group trial. *Lancet. Respir. Med.* **2016**, *4*, 708–719. [CrossRef]
121. Gerbino, A.J.; Goss, C.H.; Molitor, J.A. Effect of mycophenolate mofetil on pulmonary function in scleroderma-associated interstitial lung disease. *Chest* **2008**, *133*, 455–460. [CrossRef]
122. Morganroth, P.A.; Kreider, M.E.; Werth, V.P. Mycophenolate mofetil for interstitial lung disease in dermatomyositis. *Arthritis Care Res.* **2010**, *62*, 1496–1501. [CrossRef]
123. Stratton, R.J.; Wilson, H.; Black, C.M. Pilot study of anti-thymocyte globulin plus mycophenolate mofetil in recent-onset diffuse scleroderma. *Rheumatology* **2001**, *40*, 84–88. [CrossRef]
124. Simeón-Aznar, C.P.; Fonollosa-Plá, V.; Tolosa-Vilella, C.; Selva-O'Callaghan, A.; Solans-Laqué, R.; Vilardell-Tarrés, M. Effect of mycophenolate sodium in scleroderma-related interstitial lung disease. *Clin. Rheumatol.* **2011**, *30*, 1393–1398. [CrossRef]
125. Yilmaz, N.; Can, M.; Kocakaya, D.; Karakurt, S.; Yavuz, S. Two-year experience with mycophenolate mofetil in patients with scleroderma lung disease: A case series. *Int. J. Rheum. Dis.* **2014**, *17*, 923–928. [CrossRef]
126. Binks, M.; Passweg, J.R.; Furst, D.; McSweeney, P.; Sullivan, K.; Besenthal, C.; Finke, J.; Peter, H.H.; van Laar, J.; Breedveld, F.C.; et al. Phase I/II trial of autologous stem cell transplantation in systemic sclerosis: Procedure related mortality and impact on skin disease. *Ann. Rheum. Dis.* **2001**, *60*, 577–584. [CrossRef]
127. Walker, U.A.; Saketkoo, L.A.; Distler, O. Haematopoietic stem cell transplantation in systemic sclerosis. *RMD Open* **2018**, *4*, e000533. [CrossRef]
128. Zeher, M.; Papp, G.; Nakken, B.; Szodoray, P. Hematopoietic stem cell transplantation in autoimmune disorders: From immune-regulatory processes to clinical implications. *Autoimmun. Rev.* **2017**, *16*, 817–825. [CrossRef]
129. Farge, D.; Passweg, J.; van Laar, J.M.; Marjanovic, Z.; Besenthal, C.; Finke, J.; Peter, H.H.; Breedveld, F.C.; Fibbe, W.E.; Black, C.; et al. Autologous stem cell transplantation in the treatment of systemic sclerosis: Report from the EBMT/EULAR Registry. *Ann. Rheum. Dis.* **2004**, *63*, 974–981. [CrossRef]

130. Burt, R.K.; Shah, S.J.; Dill, K.; Grant, T.; Gheorghiade, M.; Schroeder, J.; Craig, R.; Hirano, I.; Marshall, K.; Ruderman, E.; et al. Autologous non-myeloablative haemopoietic stem-cell transplantation compared with pulse cyclophosphamide once per month for systemic sclerosis (ASSIST): An open-label, randomised phase 2 trial. *Lancet* **2011**, *378*, 498–506. [CrossRef]
131. van Laar, J.M.; Farge, D.; Sont, J.K.; Naraghi, K.; Marjanovic, Z.; Larghero, J.; Schuerwegh, A.J.; Marijt, E.W.A.; Vonk, M.C.; Schattenberg, A.V.; et al. Autologous hematopoietic stem cell transplantation vs intravenous pulse cyclophosphamide in diffuse cutaneous systemic sclerosis: A randomized clinical trial. *JAMA* **2014**, *311*, 2490–2498. [CrossRef]
132. Sullivan, K.M.; Goldmuntz, E.A.; Keyes-Elstein, L.; McSweeney, P.A.; Pinckney, A.; Welch, B.; Mayes, M.D.; Nash, R.A.; Crofford, L.J.; Eggleston, B.; et al. Myeloablative Autologous Stem-Cell Transplantation for Severe Scleroderma. *N. Engl. J. Med.* **2018**, *378*, 35–47. [CrossRef]
133. Katsiari, C.G.; Simopoulou, T.; Alexiou, I.; Sakkas, L.I. Immunotherapy of systemic sclerosis. *Hum. Vaccin. Immunother.* **2018**, *14*, 2559–2567. [CrossRef]
134. Ramwadhdoebe, T.H.; van Baarsen, L.G.M.; Boumans, M.J.H.; Bruijnen, S.T.G.; Safy, M.; Berger, F.H.; Semmelink, J.F.; van der Laken, C.J.; Gerlag, D.M.; Thurlings, R.M.; et al. Effect of rituximab treatment on T and B cell subsets in lymph node biopsies of patients with rheumatoid arthritis. *Rheumatology* **2019**, *58*, 1075–1085. [CrossRef]
135. Thurlings, R.M.; Vos, K.; Wijbrandts, C.A.; Zwinderman, A.H.; Gerlag, D.M.; Tak, P.P. Synovial tissue response to rituximab: Mechanism of action and identification of biomarkers of response. *Ann. Rheum. Dis.* **2008**, *67*, 917–925. [CrossRef]
136. Bergantini, L.; d'Alessandro, M.; Cameli, P.; Vietri, L.; Vagaggini, C.; Perrone, A.; Sestini, P.; Frediani, B.; Bargagli, E. Effects of rituximab therapy on B cell differentiation and depletion. *Clin. Rheumatol.* **2020**, *39*, 1415–1421. [CrossRef]
137. Boonstra, M.; Meijs, J.; Dorjée, A.L.; Marsan, N.A.; Schouffoer, A.; Ninaber, M.K.; Quint, K.D.; Bonte-Mineur, F.; Huizinga, T.W.J.; Scherer, H.U.; et al. Rituximab in early systemic sclerosis. *RMD Open* **2017**, *3*, e000384. [CrossRef]
138. Smith, V.; Van Praet, J.T.; Vandooren, B.; Van Der Cruyssen, B.; Naeyaert, J.M.; Decuman, S.; Elewaut, D.; De Keyser, F. Rituximab in diffuse cutaneous systemic sclerosis: An open-label clinical and histopathological study. *Ann. Rheum. Dis.* **2010**, *69*, 193–197. [CrossRef]
139. Bosello, S.; De Santis, M.; Lama, G.; Spanò, C.; Angelucci, C.; Tolusso, B.; Sica, G.; Ferraccioli, G. B cell depletion in diffuse progressive systemic sclerosis: Safety, skin score modification and IL-6 modulation in an up to thirty-six months follow-up open-label trial. *Arthritis Res. Ther.* **2010**, *12*, R54. [CrossRef]
140. Daoussis, D.; Liossis, S.-N.C.; Tsamandas, A.C.; Kalogeropoulou, C.; Kazantzi, A.; Sirinian, C.; Karampetsou, M.; Yiannopoulos, G.; Andonopoulos, A.P. Experience with rituximab in scleroderma: Results from a 1-year, proof-of-principle study. *Rheumatology* **2010**, *49*, 271–280. [CrossRef]
141. Daoussis, D.; Melissaropoulos, K.; Sakellaropoulos, G.; Antonopoulos, I.; Markatseli, T.E.; Simopoulou, T.; Georgiou, P.; Andonopoulos, A.P.; Drosos, A.A.; Sakkas, L.; et al. A multicenter, open-label, comparative study of B-cell depletion therapy with Rituximab for systemic sclerosis-associated interstitial lung disease. *Semin. Arthritis Rheum.* **2017**, *46*, 625–631. [CrossRef]
142. Sircar, G.; Goswami, R.P.; Sircar, D.; Ghosh, A.; Ghosh, P. Intravenous cyclophosphamide vs rituximab for the treatment of early diffuse scleroderma lung disease: Open label, randomized, controlled trial. *Rheumatology* **2018**, *57*, 2106–2113. [CrossRef]
143. Zamanian, R.T.; Badesch, D.; Chung, L.; Domsic, R.T.; Medsger, T.; Pinckney, A.; Keyes-Elstein, L.; D'Aveta, C.; Spychala, M.; White, R.J.; et al. Safety and Efficacy of B-Cell Depletion with Rituximab for the Treatment of Systemic Sclerosis-associated Pulmonary Arterial Hypertension: A Multicenter, Double-Blind, Randomized, Placebo-controlled Trial. *Am. J. Respir. Crit. Care Med.* **2021**, *204*, 209–221. [CrossRef]
144. Lafyatis, R.; Kissin, E.; York, M.; Farina, G.; Viger, K.; Fritzler, M.J.; Merkel, P.A.; Simms, R.W. B cell depletion with rituximab in patients with diffuse cutaneous systemic sclerosis. *Arthritis Rheum.* **2009**, *60*, 578–583. [CrossRef]
145. Jordan, S.; Distler, J.H.W.; Maurer, B.; Huscher, D.; Van Laar, J.M.; Allanore, Y.; Distler, O.; Kvien, T.K.; Airo, P.; Sancho, J.J.A.; et al. Effects and safety of rituximab in systemic sclerosis: An analysis from the European Scleroderma Trial and Research (EUSTAR) group. *Ann. Rheum. Dis.* **2015**, *74*, 1188–1194. [CrossRef]
146. Moazedi-Fuerst, F.C.; Kielhauser, S.M.; Brickmann, K.; Hermann, J.; Lutfi, A.; Meilinger, M.; Brezinschek, H.P.; Graninger, W.B. Rituximab for systemic sclerosis: Arrest of pulmonary disease progression in five cases. Results of a lower dosage and shorter interval regimen. *Scand. J. Rheumatol.* **2014**, *43*, 257–258. [CrossRef]
147. Yılmaz, D.D.; Borekci, S.; Musellim, B. Comparison of the effectiveness of cyclophosphamide and rituximab treatment in patients with systemic sclerosis-related interstitial lung diseases: A retrospective, observational cohort study. *Clin. Rheumatol.* **2021**, *40*, 4071–4079. [CrossRef]
148. Giuggioli, D.; Lumetti, F.; Colaci, M.; Fallahi, P.; Antonelli, A.; Ferri, C. Rituximab in the treatment of patients with systemic sclerosis. Our experience and review of the literature. *Autoimmun. Rev.* **2015**, *14*, 1072–1078. [CrossRef]
149. Schiopu, E.; Chatterjee, S.; Hsu, V.; Flor, A.; Cimbora, D.; Patra, K.; Yao, W.; Li, J.; Streicher, K.; McKeever, K.; et al. Safety and tolerability of an anti-CD19 monoclonal antibody, MEDI-551, in subjects with systemic sclerosis: A phase I, randomized, placebo-controlled, escalating single-dose study. *Arthritis Res. Ther.* **2016**, *18*, 131. [CrossRef]
150. Streicher, K.; Sridhar, S.; Kuziora, M.; Morehouse, C.A.; Higgs, B.W.; Sebastian, Y.; Groves, C.J.; Pilataxi, F.; Brohawn, P.Z.; Herbst, R.; et al. Baseline Plasma Cell Gene Signature Predicts Improvement in Systemic Sclerosis Skin Scores Following Treatment With Inebilizumab (MEDI-551) and Correlates with Disease Activity in Systemic Lupus Erythematosus and Chronic Obstructive Pulmonary Disease. *Arthritis Rheumatol.* **2018**, *70*, 2087–2095. [CrossRef]

151. Guerreiro Castro, S.; Isenberg, D.A. Belimumab in systemic lupus erythematosus (SLE): Evidence-to-date and clinical usefulness. *Ther. Adv. Musculoskelet. Dis.* **2017**, *9*, 75–85. [CrossRef]
152. Gordon, J.K.; Martyanov, V.; Franks, J.M.; Bernstein, E.J.; Szymonifka, J.; Magro, C.; Wildman, H.F.; Wood, T.A.; Whitfield, M.L.; Spiera, R.F. Belimumab for the Treatment of Early Diffuse Systemic Sclerosis: Results of a Randomized, Double-Blind, Placebo-Controlled, Pilot Trial. *Arthritis Rheumatol.* **2018**, *70*, 308–316. [CrossRef]
153. Abbasi, M.; Mousavi, M.J.; Jamalzehi, S.; Alimohammadi, R.; Bezvan, M.H.; Mohammadi, H.; Aslani, S. Strategies toward rheumatoid arthritis therapy; the old and the new. *J. Cell. Physiol.* **2019**, *234*, 10018–10031. [CrossRef]
154. Nie, J.; Li, Y.Y.; Zheng, S.G.; Tsun, A.; Li, B. FOXP3+ Treg Cells and Gender Bias in Autoimmune Diseases. *Front. Immunol.* **2015**, *6*, 493. [CrossRef]
155. Okazaki, T.; Nakao, A.; Nakano, H.; Takahashi, F.; Takahashi, K.; Shimozato, O.; Takeda, K.; Yagita, H.; Okumura, K. Impairment of Bleomycin-Induced Lung Fibrosis in CD28-Deficient Mice. *J. Immunol.* **2001**, *167*, 1977–1981. [CrossRef]
156. Chakravarty, E.F.; Martyanov, V.; Fiorentino, D.; Wood, T.A.; Haddon, D.J.; Jarrell, J.A.; Utz, P.J.; Genovese, M.C.; Whitfield, M.L.; Chung, L. Gene expression changes reflect clinical response in a placebo-controlled randomized trial of abatacept in patients with diffuse cutaneous systemic sclerosis. *Arthritis Res. Ther.* **2015**, *17*, 159. [CrossRef]
157. Khanna, D.; Spino, C.; Johnson, S.; Chung, L.; Whitfield, M.L.; Denton, C.P.; Berrocal, V.; Franks, J.; Mehta, B.; Molitor, J.; et al. Abatacept in Early Diffuse Cutaneous Systemic Sclerosis: Results of a Phase II Investigator-Initiated, Multicenter, Double-Blind, Randomized, Placebo-Controlled Trial. *Arthritis Rheumatol.* **2020**, *72*, 125–136. [CrossRef]
158. Elhai, M.; Meunier, M.; Matucci-Cerinic, M.; Maurer, B.; Riemekasten, G.; Leturcq, T.; Pellerito, R.; Von Mühlen, C.A.; Vacca, A.; Airo, P.; et al. Outcomes of patients with systemic sclerosis-associated polyarthritis and myopathy treated with tocilizumab or abatacept: A EUSTAR observational study. *Ann. Rheum. Dis.* **2013**, *72*, 1217–1220. [CrossRef]
159. Castellví, I.; Elhai, M.; Bruni, C.; Airò, P.; Jordan, S.; Beretta, L.; Codullo, V.; Montecucco, C.M.; Bokarewa, M.; Iannonne, F.; et al. Safety and effectiveness of abatacept in systemic sclerosis: The EUSTAR experience. *Semin. Arthritis Rheum.* **2020**, *50*, 1489–1493. [CrossRef]
160. Turnier, J.L.; Brunner, H.I. Tocilizumab for treating juvenile idiopathic arthritis. *Expert Opin. Biol. Ther.* **2016**, *16*, 559–566. [CrossRef]
161. Scott, L.J. Tocilizumab: A Review in Rheumatoid Arthritis. *Drugs* **2017**, *77*, 1865–1879. [CrossRef]
162. Khanna, D.; Denton, C.P.; Lin, C.J.F.; van Laar, J.M.; Frech, T.M.; Anderson, M.E.; Baron, M.; Chung, L.; Fierlbeck, G.; Lakshminarayanan, S.; et al. Safety and efficacy of subcutaneous tocilizumab in systemic sclerosis: Results from the open-label period of a phase II randomised controlled trial (faSScinate). *Ann. Rheum. Dis.* **2018**, *77*, 212–220. [CrossRef]
163. Denton, C.P.; Ong, V.H.; Xu, S.; Chen-Harris, H.; Modrusan, Z.; Lafyatis, R.; Khanna, D.; Jahreis, A.; Siegel, J.; Sornasse, T. Therapeutic interleukin-6 blockade reverses transforming growth factor-beta pathway activation in dermal fibroblasts: Insights from the faSScinate clinical trial in systemic sclerosis. *Ann. Rheum. Dis.* **2018**, *77*, 1362–1371. [CrossRef]
164. Kono, M.; Yasuda, S.; Kono, M.; Atsumi, T. Tocilizumab reduced production of systemic sclerosis-related autoantibodies and anti-cyclic citrullinated protein antibodies in two patients with overlapping systemic sclerosis and rheumatoid arthritis. *Scand. J. Rheumatol.* **2018**, *47*, 248–250. [CrossRef]
165. Shima, Y.; Hosen, N.; Hirano, T.; Arimitsu, J.; Nishida, S.; Hagihara, K.; Narazaki, M.; Ogata, A.; Tanaka, T.; Kishimoto, T.; et al. Expansion of range of joint motion following treatment of systemic sclerosis with tocilizumab. *Mod. Rheumatol.* **2013**, *25*, 134–137. [CrossRef]
166. Kondo, M.; Murakawa, Y.; Matsumura, T.; Matsumoto, O.; Taira, M.; Moriyama, M.; Sumita, Y.; Yamaguchi, S. A case of overlap syndrome successfully treated with tocilizumab: A hopeful treatment strategy for refractory dermatomyositis? *Rheumatology* **2014**, *53*, 1907–1908. [CrossRef]
167. Fernandes das Neves, M.; Oliveira, S.; Amaral, M.C.; Delgado Alves, J. Treatment of systemic sclerosis with tocilizumab. *Rheumatology* **2015**, *54*, 371–372. [CrossRef]
168. Saito, E.; Sato, S.; Nogi, S.; Sasaki, N.; Chinen, N.; Honda, K.; Wakabayashi, T.; Yamada, C.; Nakamura, N.; Suzuki, Y. A case of rheumatoid arthritis and limited systemic sclerosis overlap successfully treated with tocilizumab for arthritis and concomitant generalized lymphadenopathy and primary biliary cirrhosis. *Case Rep. Rheumatol.* **2014**, *2014*, 386328. [CrossRef]
169. Khanna, D.; Lin, C.J.F.; Furst, D.E.; Goldin, J.; Kim, G.; Kuwana, M.; Allanore, Y.; Matucci-Cerinic, M.; Distler, O.; Shima, Y.; et al. Tocilizumab in systemic sclerosis: A randomised, double-blind, placebo-controlled, phase 3 trial. *Lancet Respir. Med.* **2020**, *8*, 963–974. [CrossRef]
170. Roofeh, D.; Lin, C.J.F.; Goldin, J.; Kim, G.H.; Furst, D.E.; Denton, C.P.; Huang, S.; Khanna, D. Tocilizumab Prevents Progression of Early Systemic Sclerosis-Associated Interstitial Lung Disease. *Arthritis Rheumatol.* **2021**, *73*, 1301–1310. [CrossRef]
171. Hoffman, H.M. Rilonacept for the treatment of cryopyrin-associated periodic syndromes (CAPS). *Expert Opin. Biol. Ther.* **2009**, *9*, 519–531. [CrossRef]
172. Mantero, J.C.; Kishore, N.; Ziemek, J.; Stifano, G.; Zammitti, C.; Khanna, D.; Gordon, J.K.; Spiera, R.; Zhang, Y.; Simms, R.W.; et al. Randomised, double-blind, placebo-controlled trial of IL1-trap, rilonacept, in systemic sclerosis. A phase I/II biomarker trial. *Clin. Exp. Rheumatol.* **2018**, *36* (Suppl. S1), 146–149.
173. Tan, J.; Yang, S.; Wu, W. Basiliximab (Simulect) reduces acute rejection among sensitized kidney allograft recipients. *Transplant. Proc.* **2005**, *37*, 903–905. [CrossRef]

174. Scherer, H.U.; Burmester, G.-R.; Riemekasten, G. Targeting activated T cells: Successful use of anti-CD25 monoclonal antibody basiliximab in a patient with systemic sclerosis. *Ann. Rheum. Dis.* **2006**, *65*, 1245–1247. [CrossRef]
175. Becker, M.O.; Brückner, C.; Scherer, H.U.; Wassermann, N.; Humrich, J.Y.; Hanitsch, L.G.; Schneider, U.; Kawald, A.; Hanke, K.; Burmester, G.R.; et al. The monoclonal anti-CD25 antibody basiliximab for the treatment of progressive systemic sclerosis: An open-label study. *Ann. Rheum. Dis.* **2011**, *70*, 1340–1341. [CrossRef]
176. Rice, L.M.; Padilla, C.M.; McLaughlin, S.R.; Mathes, A.; Ziemek, J.; Goummih, S.; Nakerakanti, S.; York, M.; Farina, G.; Whitfield, M.L.; et al. Fresolimumab treatment decreases biomarkers and improves clinical symptoms in systemic sclerosis patients. *J. Clin. Investig.* **2015**, *125*, 2795–2807. [CrossRef]
177. Denton, C.P.; Merkel, P.A.; Furst, D.E.; Khanna, D.; Emery, P.; Hsu, V.M.; Silliman, N.; Streisand, J.; Powell, J.; Akesson, A.; et al. Recombinant human anti-transforming growth factor beta1 antibody therapy in systemic sclerosis: A multicenter, randomized, placebo-controlled phase I/II trial of CAT-192. *Arthritis Rheum.* **2007**, *56*, 323–333. [CrossRef]
178. Gerber, E.E.; Gallo, E.M.; Fontana, S.C.; Davis, E.C.; Wigley, F.M.; Huso, D.L.; Dietz, H.C. Integrin-modulating therapy prevents fibrosis and autoimmunity in mouse models of scleroderma. *Nature* **2013**, *503*, 126–130. [CrossRef]
179. Lehtonen, S.T.; Veijola, A.; Karvonen, H.; Lappi-Blanco, E.; Sormunen, R.; Korpela, S.; Zagai, U.; Sköld, M.C.; Kaarteenaho, R. Pirfenidone and nintedanib modulate properties of fibroblasts and myofibroblasts in idiopathic pulmonary fibrosis. *Respir. Res.* **2016**, *17*, 14. [CrossRef]
180. Khanna, D.; Albera, C.; Fischer, A.; Khalidi, N.; Raghu, G.; Chung, L.; Chen, D.; Schiopu, E.; Tagliaferri, M.; Seibold, J.R.; et al. An open-label, phase II study of the safety and tolerability of pirfenidone in patients with scleroderma-associated interstitial lung disease: The LOTUSS trial. *J. Rheumatol.* **2016**, *43*, 1672–1679. [CrossRef]
181. Huang, H.; Feng, R.E.; Li, S.; Xu, K.; Bi, Y.L.; Xu, Z.J. A case report: The efficacy of pirfenidone in a Chinese patient with progressive systemic sclerosis-associated interstitial lung disease: A CARE-compliant article. *Medicine* **2016**, *95*, e4113. [CrossRef]
182. Miura, Y.; Saito, T.; Fujita, K.; Tsunoda, Y.; Tanaka, T.; Takoi, H.; Yatagai, Y.; Rin, S.; Sekine, A.; Hayashihara, K.; et al. Clinical experience with pirfenidone in five patients with scleroderma-related interstitial lung disease. *Sarcoidosis Vasc. Diffus. Lung Dis.* **2014**, *31*, 235–238.
183. Zarrin, A.A.; Bao, K.; Lupardus, P.; Vucic, D. Kinase inhibition in autoimmunity and inflammation. *Nat. Rev. Drug Discov.* **2021**, *20*, 39–63. [CrossRef]
184. Pottier, C.; Fresnais, M.; Gilon, M.; Jérusalem, G.; Longuespée, R.; Sounni, N.E. Tyrosine Kinase Inhibitors in Cancer: Breakthrough and Challenges of Targeted Therapy. *Cancers* **2020**, *12*, 731. [CrossRef]
185. Tamura, T.; Koch, A. Tyrosine Kinases as Targets for Anti-Inflammatory Therapy. *Anti-Inflamm. Anti-Allergy Agents Med. Chem.* **2007**, *6*, 47–60. [CrossRef]
186. Iwamoto, N.; Distler, J.H.W.; Distler, O. Tyrosine kinase inhibitors in the treatment of systemic sclerosis: From animal models to clinical trials. *Curr. Rheumatol. Rep.* **2011**, *13*, 21–27. [CrossRef]
187. Gajski, G.; Gerić, M.; Domijan, A.-M.; Golubović, I.; Garaj-Vrhovac, V. Evaluation of oxidative stress responses in human circulating blood cells after imatinib mesylate treatment—Implications to its mechanism of action. *Saudi Pharm. J.* **2019**, *27*, 1216–1221. [CrossRef]
188. Haddon, D.J.; Wand, H.E.; Jarrell, J.A.; Spiera, R.F.; Utz, P.J.; Gordon, J.K.; Chung, L.S. Proteomic Analysis of Sera from Individuals with Diffuse Cutaneous Systemic Sclerosis Reveals a Multianalyte Signature Associated with Clinical Improvement during Imatinib Mesylate Treatment. *J. Rheumatol.* **2017**, *44*, 631–638. [CrossRef]
189. Spiera, R.F.; Gordon, J.K.; Mersten, J.N.; Magro, C.M.; Mehta, M.; Wildman, H.F.; Kloiber, S.; Kirou, K.A.; Lyman, S.; Crow, M.K. Imatinib mesylate (Gleevec) in the treatment of diffuse cutaneous systemic sclerosis: Results of a 1-year, phase IIa, single-arm, open-label clinical trial. *Ann. Rheum. Dis.* **2011**, *70*, 1003–1009. [CrossRef]
190. Divekar, A.A.; Khanna, D.; Abtin, F.; Maranian, P.; Saggar, R.; Saggar, R.; Furst, D.E.; Singh, R.R. Treatment with imatinib results in reduced IL-4-producing T cells, but increased CD4(+) T cells in the broncho-alveolar lavage of patients with systemic sclerosis. *Clin. Immunol.* **2011**, *141*, 293–303. [CrossRef]
191. Chung, L.; Fiorentino, D.F.; Benbarak, M.J.; Adler, A.S.; Mariano, M.M.; Paniagua, R.T.; Milano, A.; Connolly, M.K.; Ratiner, B.D.; Wiskocil, R.L.; et al. Molecular framework for response to imatinib mesylate in systemic sclerosis. *Arthritis Rheum.* **2009**, *60*, 584–591. [CrossRef]
192. Fraticelli, P.; Gabrielli, B.; Pomponio, G.; Valentini, G.; Bosello, S.; Riboldi, P.; Gerosa, M.; Faggioli, P.; Giacomelli, R.; Del Papa, N.; et al. Low-dose oral imatinib in the treatment of systemic sclerosis interstitial lung disease unresponsive to cyclophosphamide: A phase II pilot study. *Arthritis Res. Ther.* **2014**, *16*, R144. [CrossRef]
193. Khanna, D.; Saggar, R.; Mayes, M.D.; Abtin, F.; Clements, P.J.; Maranian, P.; Assassi, S.; Saggar, R.; Singh, R.R.; Furst, D.E. A one-year, phase I/IIa, open-label pilot trial of imatinib mesylate in the treatment of systemic sclerosis-associated active interstitial lung disease. *Arthritis Rheum.* **2011**, *63*, 3540–3546. [CrossRef]
194. Bradeen, H.A.; Eide, C.A.; O'Hare, T.; Johnson, K.J.; Willis, S.G.; Lee, F.Y.; Druker, B.J.; Deininger, M.W. Comparison of imatinib mesylate, dasatinib (BMS-354825), and nilotinib (AMN107) in an N-ethyl-N-nitrosourea (ENU)-based mutagenesis screen: High efficacy of drug combinations. *Blood* **2006**, *108*, 2332–2338. [CrossRef]
195. Gordon, J.K.; Martyanov, V.; Magro, C.; Wildman, H.F.; Wood, T.A.; Huang, W.-T.; Crow, M.K.; Whitfield, M.L.; Spiera, R.F. Nilotinib (Tasigna^{TM}) in the treatment of early diffuse systemic sclerosis: An open-label, pilot clinical trial. *Arthritis Res. Ther.* **2015**, *17*, 213. [CrossRef]

196. Wollin, L.; Distler, J.H.W.; Redente, E.F.; Riches, D.W.H.; Stowasser, S.; Schlenker-Herceg, R.; Maher, T.M.; Kolb, M. Potential of nintedanib in treatment of progressive fibrosing interstitial lung diseases. *Eur. Respir. J.* **2019**, *54*, 1900161. [CrossRef]
197. Sato, S.; Shinohara, S.; Hayashi, S.; Morizumi, S.; Abe, S.; Okazaki, H.; Chen, Y.; Goto, H.; Aono, Y.; Ogawa, H.; et al. Anti-fibrotic efficacy of nintedanib in pulmonary fibrosis via the inhibition of fibrocyte activity. *Respir. Res.* **2017**, *18*, 172. [CrossRef]
198. Cutolo, M.; Gotelli, E.; Montagna, P.; Tardito, S.; Paolino, S.; Pizzorni, C.; Sulli, A.; Smith, V.; Soldano, S. Nintedanib downregulates the transition of cultured systemic sclerosis fibrocytes into myofibroblasts and their pro-fibrotic activity. *Arthritis Res. Ther.* **2021**, *23*, 205. [CrossRef]
199. Distler, O.; Highland, K.B.; Gahlemann, M.; Azuma, A.; Fischer, A.; Mayes, M.D.; Raghu, G.; Sauter, W.; Girard, M.; Alves, M.; et al. Nintedanib for Systemic Sclerosis-Associated Interstitial Lung Disease. *N. Engl. J. Med.* **2019**, *380*, 2518–2528. [CrossRef]
200. Bruni, T.; Varone, F. The adoption of nintedanib in systemic sclerosis: The SENSCIS study. *Breathe* **2020**, *16*, 200005. [CrossRef]
201. Wang, W.; Bhattacharyya, S.; Marangoni, R.G.; Carns, M.; Dennis-Aren, K.; Yeldandi, A.; Wei, J.; Varga, J. The JAK/STAT pathway is activated in systemic sclerosis and is effectively targeted by tofacitinib. *J. Scleroderma Relat. Disord.* **2019**, *5*, 40–50. [CrossRef]
202. Kucharz, E.J.; Stajszczyk, M.; Kotulska-Kucharz, A.; Batko, B.; Brzosko, M.; Jeka, S.; Leszczyński, P.; Majdan, M.; Olesińska, M.; Samborski, W.; et al. Tofacitinib in the treatment of patients with rheumatoid arthritis: Position statement of experts of the Polish Society for Rheumatology. *Reumatologia* **2018**, *56*, 203–211. [CrossRef]
203. You, H.; Xu, D.; Hou, Y.; Zhou, J.; Wang, Q.; Li, M.; Zeng, X. Tofacitinib as a possible treatment for skin thickening in diffuse cutaneous systemic sclerosis. *Rheumatology* **2021**, *60*, 2472–2477. [CrossRef]
204. Karalilova, R.V.; Batalov, Z.A.; Sapundzhieva, T.L.; Matucci-Cerinic, M.; Batalov, A.Z. Tofacitinib in the treatment of skin and musculoskeletal involvement in patients with systemic sclerosis, evaluated by ultrasound. *Rheumatol. Int.* **2021**, *41*, 1743–1753. [CrossRef]
205. Vonk, M.C.; Smith, V.; Sfikakis, P.P.; Cutolo, M.; del Galdo, F.; Seibold, J.R. Pharmacological treatments for SSc-ILD: Systematic review and critical appraisal of the evidence. *Autoimmun. Rev.* **2021**, *20*, 102978. [CrossRef]
206. Lepri, G.; Hughes, M.; Bruni, C.; Cerinic, M.M.; Randone, S.B. Recent advances steer the future of systemic sclerosis toward precision medicine. *Clin. Rheumatol.* **2020**, *39*, 1–4. [CrossRef]

 *biomedicines* 

Review

Targeting Systemic Sclerosis from Pathogenic Mechanisms to Clinical Manifestations: Why IL-6?

Anca Cardoneanu [1,2,*], Alexandra Maria Burlui [1,2], Luana Andreea Macovei [1,2], Ioana Bratoiu [1,2], Patricia Richter [1,2] and Elena Rezus [1,2]

1 Department of Rheumatology, University of Medicine and Pharmacy "Grigore T Popa", 700115 Iasi, Romania; maria-alexandra.burlui@umfiasi.ro (A.M.B.); luana.macovei@umfiasi.ro (L.A.M.); ioana.bratoiu@umfiasi.ro (I.B.); patricia.richter@umfiasi.ro (P.R.); elena.rezus@umfiasi.ro (E.R.)
2 Rehabilitation Hospital, 700661 Iasi, Romania
* Correspondence: anca.cardoneanu@umfiasi.ro

Abstract: Systemic sclerosis (SS) is a chronic autoimmune disorder, which has both cutaneous and systemic clinical manifestations. The disease pathogenesis includes a triad of manifestations, such as vasculopathy, autoimmunity, and fibrosis. Interleukin-6 (IL-6) has a special role in SS development, both in vascular damage and in the development of fibrosis. In the early stages, IL-6 participates in vascular endothelial activation and apoptosis, leading to the release of damage-associated molecular patterns (DAMPs), which maintain inflammation and autoimmunity. Moreover, IL-6 plays an important role in the development of fibrotic changes by mediating the transformation of fibroblasts into myofibroblasts. All of these are associated with disabling clinical manifestations, such as skin thickening, pulmonary fibrosis, pulmonary arterial hypertension (PAH), heart failure, and dysphagia. Tocilizumab is a humanized monoclonal antibody that inhibits IL-6 by binding to the specific receptor, thus preventing its proinflammatory and fibrotic actions. Anti-IL-6 therapy with Tocilizumab is a new hope for SS patients, with data from clinical trials supporting the favorable effect, especially on skin and lung damage.

Keywords: systemic sclerosis; interleukin-6; Tocilizumab; pulmonary fibrosis

Citation: Cardoneanu, A.; Burlui, A.M.; Macovei, L.A.; Bratoiu, I.; Richter, P.; Rezus, E. Targeting Systemic Sclerosis from Pathogenic Mechanisms to Clinical Manifestations: Why IL-6? *Biomedicines* **2022**, *10*, 318. https://doi.org/10.3390/biomedicines10020318

Academic Editors: Yong-Gil Kim and Michal Tomcik

Received: 29 December 2021
Accepted: 26 January 2022
Published: 29 January 2022

1. Introduction

Systemic sclerosis (SS) is a chronic autoimmune disease that has both cutaneous and systemic clinical manifestations, with the latter being associated with high disease morbidity [1]. The pathogenic mechanisms of this condition are not yet fully understood. However, it is known that the pathogenesis of the disease is composed of a triad: vasculopathy, autoimmunity, and fibrosis [2]. All these mechanisms are interconnected, with each having a particular role in the occurrence of clinical manifestations and disease progression. Immunological mechanisms include both innate and adaptative immunity and involve many cells, such as: B and T lymphocytes, mast cells, macrophages, and dendritic cells [3]. Vasculopathy is characterized by damage and endothelial activation and participates in the initiation and perpetuation of autoimmunity [4]. Activation of fibroblasts and excess deposition of extracellular matrix causes the appearance of fibrosis, which is also related to autoimmunity and vasculopathy [3].

In the early stage of the disease, an inflammatory profile was highlighted, characterized by an increased secretion of proinflammatory cytokines, with the predominant immune response being Th1 and Th17 cells [5]. Moreover, innate immunity, by activating NK/NKT-like cells, plays a decisive role in this initial stage of the disease [6]. The late stage of SS has a predominant Th2-type immune response and is defined by the presence of fibrotic changes [5]. Knowing these important data involved in the pathogenesis of the disease, numerous studies have correlated the clinical manifestations with the excessive secretion of some molecules, with the latter being considered potential biomarkers of the disease [7] (Figure 1).

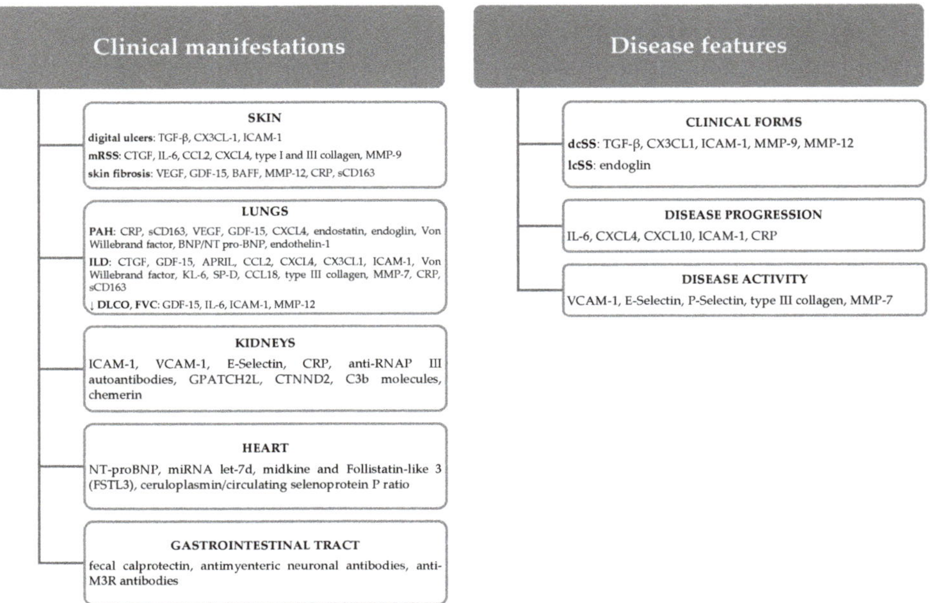

Figure 1. Potential biomarkers of organ involvement and disease features in SS. Key: TGF: transforming growth factor; GDF-15: growth differentiation factor 15; BAFF: B-cell-activating factor belonging to the tumor necrosis factor family; APRIL: a proliferation-inducing ligand; MMP: matrix metalloproteinases; BNP: brain natriuretic peptide; NT-proBNP: N-terminal-pro hormone BNP; CTG: connective tissue growth factor; mRSS: modified Rodnan total skin thickness score; ILD: interstitial lung disease; IL-6: interleukin 6; DLCO: diffusing capacity of carbon monoxide; PAH: pulmonary arterial hypertension; ICAM-1: intercellular adhesion molecule 1; dcSSc: diffuse cutaneous systemic sclerosis; VEGF: vascular endothelial growth factor; lcSSc: limited cutaneous systemic sclerosis; KL-6: krebs von den Lungen-6; SP-D: surfactant protein-D; CCL2: monocyte chemoattractant protein-1; CXCL4: platelet factor 4; VCAM-1: vascular cell adhesion molecule-1; CRP: C-reactive protein; sCD163: soluble CD163; anti-RNAP III antibodies: anti-RNA polymerase III antibodies; anti-M3R antibodies: anti-human muscarinic receptor M3 antibodies; miRNA let-7d: micro RNA let-7d; GPATCH2L: G-patch domain-containing protein 2-like; CTNND2: catenin delta 2.

Given the complex pathogenic mechanisms and the multitude of cells and cytokines, we will continue to focus on interleukin 6 (IL-6), which is considered to be one of the major fibrogenic cytokines along with interleukin 4 (IL-4) and transforming growth factor (TGF)-β [3].

2. IL-6: General Data

IL-6 is part of a large family of cytokines, having important roles in regulating immunity, hematopoiesis, inflammation, and oncogenesis. In the past, depending on its biological effect, IL-6 had different names: B lymphocyte-stimulating factor 2 (BSF-2), hepatocyte-stimulating factor (HSF), or interferon-β2 [8]. In 1989, it was found that all these molecules are identical, so it was renamed IL-6, a name used until today [9].

The biological effect of IL-6 can be achieved by two signaling pathways: the classical pathway, which uses a membrane receptor, and the trans-signaling pathway, which involves the presence of a soluble IL-6 receptor. The classical signaling pathway involves the binding of IL-6 to the membrane receptor-IL-6R [10]. Normally, the expression of this receptor is decreased, being found in several cells, especially hepatocytes and some immune cells [11,12]. IL-6R has no intrinsic signaling capacity. The IL-6/IL-6R interaction induces the dimerization of an IL-6R signaling-related cytokine called gp130 [13–15]. Unlike IL-6R, this gp130 molecule is ubiquitously expressed and has an intrinsic signaling capacity. The IL-6/IL-6R complex bound to gp130 initiates cellular signaling via several intracellular or extracellular pathways [10,12,16]. Gp130-mediated activation of Janus kinase (Jak) family members and subsequent phosphorylation of specific tyrosine residues cause phosphorylation and dimerization of STAT (signal transducer and activator of transcription proteins) [10,16]. The latter determines the transcriptional activation of IL-6-dependent genes [16].

The alternative IL-6 trans-signaling pathway involves the presence of a soluble receptor (sIL-6R) [16–19]. This receptor is formed after cleavage of IL-6R by activated ADAM17 (Disintegrin Metalloproteases 17) and ADAM10 (Disintegrin Metalloproteases 10) [20,21]. Binding of IL-6 to sIL-6R and subsequently to gp130 IL-6/sIL-6R determines intracellular signaling, which is very similar to classical IL-6R signaling. In addition, sIL-6R has the advantage of binding circulating IL-6, thereby prolonging the half-life of IL-6 [22]. Moreover, sIL-6R participates in the interaction between leukocytes and vascular endothelium, causing the endothelial production of MCP-1 (monocyte chemoattractant protein-1), a key chemokine that regulates monocyte/macrophage infiltration of the vascular wall [23].

To limit the proinflammatory effects of IL-6, IL-6 signaling is regulated by several inhibitory molecules and processes. Inhibition of IL-6 signaling is important in order to limit inflammation and its side effects, such as extensive tissue damage. Thus, STAT3-dependent transcription is rapidly inactivated by the cytokine 3 signaling suppressor (SOCS3) [24–27]. IL-6-induced SOCS3 expression acts as a major negative feedback loop for IL-6 signaling. Another inhibitory mechanism, a soluble form of gp130 (sgp130), has been identified in patients' plasma [28,29]. sgp130 serves as a functional antagonist of IL-6 trans-signaling by binding to the IL-6/sIL-6R complex and preventing IL-6/IL-6R complex binding to the gp130 membrane molecule [30]. A third mechanism of inhibition of IL-6 signaling involves a STAT-3 inhibitor (PIAS3-Protein Inhibitor of Activated STAT 3) that blocks the interaction between phosphorylated STAT3 and cellular DNA and finally genetic transcription [31].

Figure 2 schematically shows these two intracellular signaling pathways of IL-6, and the main negative feedback loops that block the production of its biological effects (Figure 2).

IL-6 is a key cytokine in inflammation, having a systemic polymorphic pathway. Thereby, after the initial stage of inflammation and local production, IL-6 is transported through the blood vessels in the liver. Here, it determines the formation and increased secretion of acute phase proteins, such as C-reactive protein (CRP), serum amyloid A, fibrinogen, and hepcidin. On the other hand, IL-6 has an inhibitory effect on the secretion of fibronectin, albumin, and transferrin [32,33]. Due to the stimulatory effect of hepcidin formation, the iron transporter is blocked in the gut (ferroportin 1) [34]. Thus, the IL-6-hepcidin axis is responsible for hyposideremia and anemia associated with chronic inflammation. In addition, IL-6 has the ability to promote zinc deposition in hepatocytes by increasing the expression of ZIP14-specific transporter. The regulation of ZIP14 by IL-6 contributes to hepatic accumulation of zinc and to the decrease in the serum zinc concentration highlighted in inflammation [32,35].

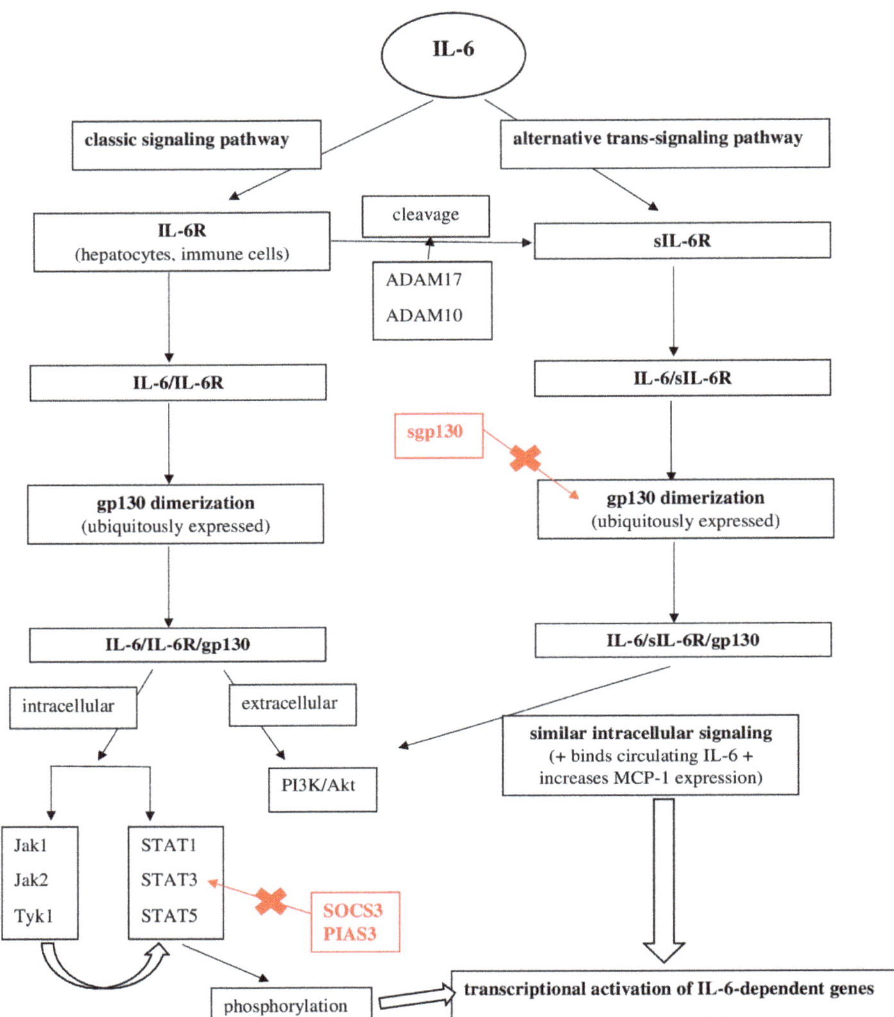

Figure 2. The intracellular signaling pathways of IL-6 and the main negative feedback loops that block the production of its biological effects. Key: IL-6: interleukin-6; IL-6R: interleukin-6 receptor; sIL-6R: soluble interleukin-6 receptor; gp130: glycoprotein 130; Jak: janus kinase; STAT: signal transducer and activator of transcription proteins; SOCS3: cytokine 3 signaling suppressor; PIAS3: protein inhibitor of activated STAT 3; ADAM: disintegrin metalloproteases; MCP: monocyte chemoattractant protein-1.

IL-6 plays an important role in the innate or adaptative immune response by stimulating CD4+ T cell differentiation [32]. Data from the literature support the indispensable role of the IL-6-TGF-β axis in differentiating Th17 cells from naive T CD4+ cells [36]. Moreover, inhibition of regulatory T cell differentiation (Treg) occurs [37]. This Th17/Treg imbalance is responsible for disrupting immune tolerance, leading to chronic inflammatory autoimmune disorders [38]. IL-6 has the ability to stimulate follicular T helper cell differentiation and the production of IL-21, which is responsible for the synthesis of immunoglobulins (Ig), especially IgG4 [39]. IL-6 can induce CD8+ cell differentiation into cytotoxic T cells [40].

IL-6 also acts on B lymphocytes, activating and transforming them into plasma cells, thus promoting the production of autoantibodies and hypergammaglobulinemia [32].

IL-6 can have a direct effect on osteoblast precursors, blocking their differentiation and maturation [41]. IL-6 induces the secretion of matrix metalloproteinases (MMP-1, MMP-3, MMP-13) from chondrocytes and synovial cells. MMP-1 and MMP-13 are able to break type II collagen, and MMP-3 exerts its action on the extracellular matrix, fibronectin, and laminin [42]. In addition, elevated levels of IL-6 in bone marrow stromal cells stimulate the activator receptor of nuclear factor kappa-B ligand (RANKL), leading to osteoclast differentiation and activation [43].

The action of IL-6 is systemic, being a key element in the development and perpetuation of inflammation. IL-6 participates in the formation of a cytokine storm by activating immune cells and stimulating the secretion of new inflammatory mediators. Figure 3 shows the main cells on which IL-6 acts and the systemic effects it can have.

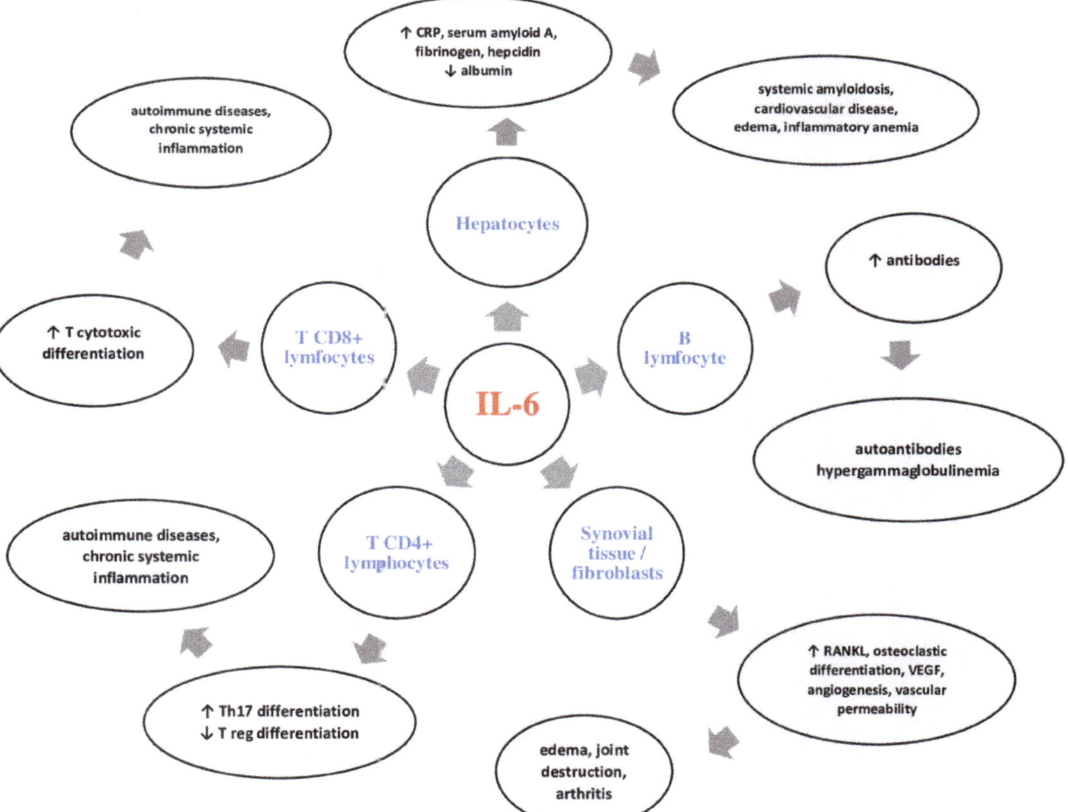

Figure 3. The action of IL-6 on various cells and its local and systemic manifestations. Key: IL-6: interleukin 6; CD: cluster of differentiation; CRP: C-reactive protein; RANKL: activator receptor of nuclear factor kappa-B ligand; VEGF: vascular endothelial growth factor; Treg: regulatory T cells.

3. IL-6 in Systemic Sclerosis

IL-6 has a particular role in the pathogenesis of SS, both in vascular damage and in the development of fibrosis. The increase in collagen production is achieved through different pathways, such as differentiation of myofibroblasts, inhibition of the secretion of matrix metalloproteinases with collagenolytic effect, and activation of fibroblasts [44]. The trans-signaling pathway is mandatory for collagen production and for the appearance of fibrotic changes due to the presence of sIL-6R [45]. This is supported by data from murine studies. The occurrence of an SS-like condition with pulmonary fibrosis and skin thickening was determined by IL-6 signaling [46]. On the other hand, a decrease in pulmonary fibrosis and inflammation was observed in IL-6-deficient mice [47].

Genetic susceptibility plays an important role in the pathogenesis of SS; therefore, the polymorphism of the IL-6 gene has been studied in various clinical trials. The most studied gene is *rs1800795* (also known as *−174 C/G*), with G alleles being associated with increased IL-6 levels [48,49]. An analysis of *−174 C/G* IL-6 polymorphism in a cohort of 102 SS patients revealed increased IL-6 gene expression in G-allele carriers. This genetic polymorphism of IL-6 was correlated only with the presence of gastrointestinal manifestations, not with skin and lung damage [50]. Other data obtained after studying 20 cases with SS support the link between IL-6 polymorphism, especially the homozygous *GG* form, and an active and disabling disease [51].

3.1. IL-6 and SS Features

Elevated serum and skin levels of IL-6 have been highlighted in both early and late stages of SS [52,53]. In the early stages, IL-6 participates in vascular endothelial activation and apoptosis [54], leading to the release of damage-associated molecular patterns (DAMPs), which maintain inflammation and autoimmunity [55].

Numerous data support this increase in IL-6 levels, both in the tissues and in the serum of patients with SS. Feghali's study showed elevated levels of IL-6 in the skin tissue supernatant cultures of SS patients, even 30 times higher than the control arm [56]. By culturing peripheral blood mononuclear cells isolated from cases of SS with type I collagen, Gurram highlighted an increased level of IL-6 in the supernatant, which shows that IL-6 can be highly secreted by other unaffected tissues [57]. Various methods for measuring serum IL-6 levels have been used over time, such as bioassays and ELISA (enzyme-linked immunosorbent assay). Needleman, using a bioassay method, found detectable concentrations of IL-6 in the serum of SS patients [58]. The ELISA method provided important data on elevated IL-6 levels in these patients, levels that correlated with the degree of skin damage quantified by the cutaneous score [59,60].

It seems that the concentration of IL-6 depends on the duration and form of SS. Thus, a study that included 55 SS patients analyzed the levels of IL-6, sIL-6R, and sgp130 according to the form of the disease. Patients were divided into 4 arms: 12 having an lcSS (limited cutaneous SS) and an early form of disease (less than 3 years from disease onset), 22 with lcSS and a late form (over 3 years from disease onset), 9 cases of an early dcSS form (diffuse cutaneous SS), and 12 cases of a late dcSS form [61]. Following the statistical analysis, the authors found significantly higher levels of IL-6 in the early SS form compared to the control arm, with high levels found especially in patients with pulmonary fibrosis. IL-6 was inversely correlated with vital lung capacity (VPC). IL-6 levels have also been associated with acute phase reactants and A and G immunoglobulin levels [61]. Regarding sIL-6R, elevated levels were observed in a limited SS form compared to the control. For sgp130, no statistically significant data were highlighted [61].

The fact that there is a different secretion of cytokines depending on the stage of the disease has been demonstrated by clinical trials. Thus, Matsushita showed a significant increase in IL-6 and IL-10 in the early stages of the disease, and a significant numerical decrease after 6 years of evolution. Opposite data have been described for IL-12, whose values were decreased in the early stages and more than 15 times increased in the disease with a 6-year evolution [62]. The level of IL-6 has been shown to be inversely proportional to that of IL-13, with early-stage patients having a high IL-6 and low IL-13 amount [63].

3.2. IL-6 and Skin Manifestations

IL-6 plays an important role in the development of fibrotic changes by mediating the transformation of fibroblasts into myofibroblasts, with the latter producing an excessive amount of collagen that infiltrates various organs and tissues, including the heart [64]. Myofibroblasts were found in the dermis in the early and progressive SS forms, with histological analysis indicating their disappearance in late (atrophic) forms of the disease [65]. Furthermore, a study that included a model of bleomycin SS model mice showed a numerical decrease in myofibroblasts after administration of anti-IL-6R antibodies [66]. The same data were highlighted by Kawaguchi, who performed cultures of cutaneous fibroblasts isolated from SS patients and found a high level of procollagen type I in the supernatant. After incubating these cultures with anti-IL-6R antibodies, a numerical decrease in collagen fibers was observed [67].

3.3. IL-6 and Lung Manifestations

The involvement of IL-6 in fibrotic lesions has been proven by clinical trials, especially in lung damage. Pulmonary fibrosis in SS has a multifactorial etiopathogenesis that includes both infiltration with inflammatory cells at the alveolar, peribronchiolar, and interstitium levels, and an intense and aberrant proliferation of fibroblasts that cause the deposition of fibronectin, type I and III collagen fibers, and tenascin [68,69].

Regarding IL-6, Crestani's data included the analysis of cultures of alveolar macrophages obtained after bronchoalveolar lavage in SS patients. In the supernatant, the researchers found an increased level of IL-6 compared to the control group [70]. Other recently published data point to the same increased level of IL-6 in cases of SS with advanced pulmonary fibrosis due to the perpetuation of chronic inflammation, inhibition of the secretion of metalloproteinases, and increased collagen fiber synthesis [71]. On the other hand, another important pathogenic mechanism involved in the development of pulmonary fibrosis is the enhancement of IL-6 trans-signaling via ADAM-17 by activated macrophages, which increases the extracellular matrix deposits and the proliferation of fibroblasts [72]. In this situation, it seems that inhibition of the IL-6 trans-signaling pathway may be a useful therapeutic approach [71,72].

IL-6 levels, especially in early SS with mild forms of pulmonary fibrosis, appear to have a prognostic value for impaired lung function and increased mortality. This is supported by the analysis of 74 cases of SS associated with pulmonary fibrosis. An IL-6 value of more than 7.67 pg/mL can be considered a predictor of decreased DLCO (diffusing capacity of lung for carbon monoxide) and FVC (forced vital capacity) in the first year. It also correlates with an increase in mortality in the first 30 months of the disease [73]. Other data obtained after analyzing a cohort of 68 SS patients support the prognostic value of IL-6, with increased values being correlated with the extension of skin involvement at the 3-year follow-up, the development of pulmonary fibrosis, and worse long-term survival. Therefore, IL-6 can be considered a predictive marker for disease progression [60,74]. Furthermore, IL-6 levels may be useful in stratifying patients regarding disease activity and survival outcome [75–77].

3.4. IL-6 and Cardiovascular Manifestations

IL-6 plays an important role in the cardiovascular involvement of SS. In this regard, an analysis of 20 cases with SS showed positive correlations between the level of IL-6 and the disease duration; EUSTAR (European Scleroderma Study Group) activity score; musculoskeletal, vascular, and respiratory status; and pressure peak in the pulmonary artery [78]. Other positive associations were noted with the pulmonary fibrosis score quantified by HRCT (high-resolution computer tomograph). Negative correlations were scored between IL-6, DLCO, the 6-min walk distance, and right ventricle systolic function parameters [78].

Furthermore, IL-6 may be considered a marker for pulmonary arterial hypertension (PAH) [79]. PAH mainly occurs in the limited form of the disease [80] and is closely related to the proliferation of endothelial cells and the formation of extracellular matrix deposits that cause intimate thickening of the capillaries and pulmonary arterioles [81]. Systemic inflammation, represented by IL-6, plays an important role in the development of PAH due to perivascular inflammatory cell infiltrates [81]. The risk of death is higher compared to idiopathic PAH [82,83], being linked to an accentuated intimal hyperplasia, fewer plexiform lesions, and the involvement of pulmonary veins [81,84].

3.5. IL-6 and Gastrointestinal Manifestations

In addition to lung damage, 90% of SS patients also have gastrointestinal involvement (GI) [85], which is the 3rd leading cause of death [86]. GI impairment is associated with a marked disability and a decreased quality of life in SS patients, leading to reduced survival [86,87]. IL-6 gene polymorphism increases disease susceptibility, being associated with different clinical manifestations [49]. Zekovic's study, which included 102 SS cases, analyzed the correlations between IL-6 expression and GI manifestations and the presence of a specific disease genotype /phenotype [50]. The results showed that the presence of the C-allele correlates with increased IL-6 gene expression and with GI manifestations, with the most common being abdominal distension and worsening of the GI score (UCLA GIT 2.0) [50].

The pathogenic mechanisms involved in the occurrence of GI manifestations are complex and include concentric intimate thickening and deposition of collagen and mucoid molecules [88], adventitia fibrosis [89], autonomic dysfunction affecting neuromuscular junction, smooth muscle atrophy [90,91], presence of antimyenteric neuronal antibodies, anti-U3 RNP and anti-muscarinic-3 receptor antibodies [92–95], and small intestinal bacterial overgrowth serologically expressed by an increased fecal calprotectin [96]. The clinical manifestations are polymorphic and may include the entire GI tract, with patients complaining of dysphagia, intestinal transit disorders, fecal incontinence, pseudo-obstruction, or having clinical signs of malabsorption [97].

3.6. IL-6 and Kidney Manifestations

SS kidney disease has an incidence of up to 10% [98] and is associated with increased long-term mortality [99], with scleroderma renal crisis (SRC) being a medical emergency. Genetic predisposition, autoimmune mechanisms, inflammatory cascade, and vasculopathy are the main mechanisms involved in the pathogenesis of SRC. An increased expression of two proteins: GPATCH2L and CTNND2, was highlighted in a recent study that analyzed kidney biopsy pieces in SS patients, suggesting the involvement of genetic susceptibility in the development of the disease [100]. Among autoantibodies, anti-RNA polymerase III antibodies are associated with renal impairment and have a prognostic value [101]. The main inflammatory markers involved in kidney damage are interleukins (IL-2r, IL-6, IL-10, IL-18) and chemokines, such as MCP-1 (monocyte chemoattractant protein-1) [102,103]. Vasculopathy refers to intimal proliferation and thickening of the renal arcuate and interlobular arteries, which causes activation of the renin-angiotensin system [104,105].

Clinically, patients present with rapidly progressive acute renal failure, malignant hypertension that may be complicated by hypertensive encephalopathy and retinopathy, pulmonary edema, or seizures. Microangiopathic hemolytic anemia, cardiac arrhythmias, myocarditis, and fever may also occur [106,107].

All of these systemic harmful effects of IL-6 are shown schematically in Figure 4.

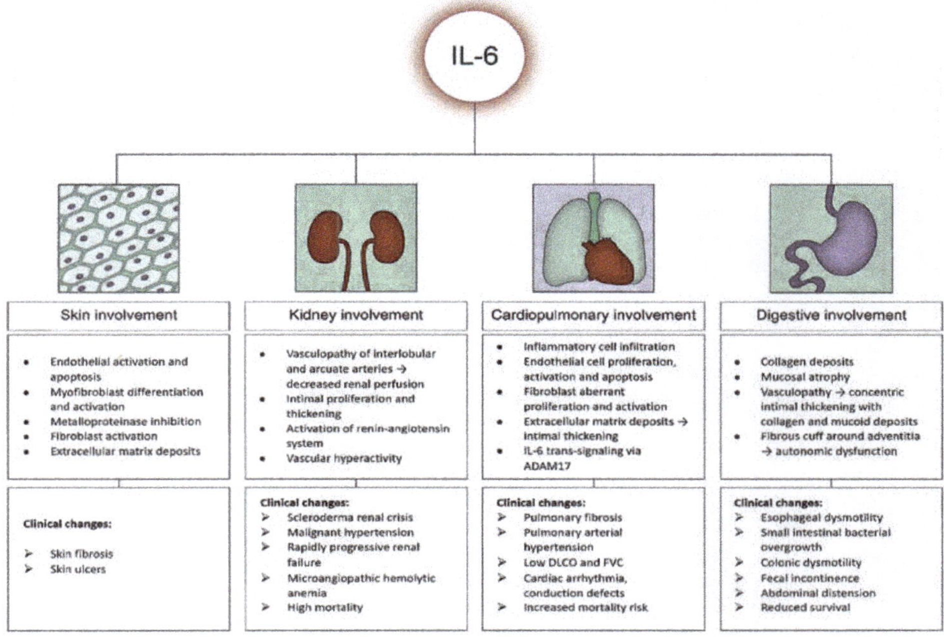

Figure 4. Systemic harmful effects of IL-6 on the skin, cardiopulmonary, gastrointestinal and renal system. Key: IL-6, interleukin-6; ADAM17: disintegrin metalloproteases 17; PAH: pulmonary arterial hypertension; DLCO: diffusing capacity of lung for carbon monoxide; FVC: forced vital capacity.

4. Targeting IL-6 in Systemic Sclerosis

Given the above data, we can sustain that IL-6 can be considered a key cytokine in the development and evolution of SS. For these reasons, anti-IL-6 therapy with Tocilizumab is a new hope for SS patients, with data from clinical trials supporting both cutaneous and pulmonary improvement. Tocilizumab is a humanized monoclonal antibody that inhibits IL-6 binding to mIL-6R (membrane IL-6R) and sIL-6R and prevents its pro-inflammatory and fibrotic effects [108]. In addition to SS, this beneficial effect has been shown in a variety of conditions, such as rheumatoid arthritis, juvenile idiopathic arthritis, Takayasu's arteritis, Still's adult disease, giant cell arteritis, and Castleman's disease [109–113].

The benefits of Tocilizumab in SS have been highlighted in phase 2 and 3 studies since 2016. The FaSScinated study is a randomized, double-blind, placebo-controlled phase 2 trial that included 87 SS patients from 35 sites [114]. Randomization was 1:1, with 43 cases in the SS group for whom the onset of the disease was less than 5 years and who received subcutaneous Tocilizumab 162 mg weekly versus a placebo arm, which included 44 patients. No patient had been on background immunosuppressive therapy. The primary endpoint was an improvement in the modified RODNAN skin score (mRSS) at 24 weeks, which was not achieved, but the data showed a cutaneous improvement after Tocilizumab administration (mRSS decrease with 3.92 for Tocilizumab vs. 1.22 for placebo, $p = 0.0915$). A more significant skin improvement was seen after 48 weeks with Tocilizumab, but this still did not reach statistical significance (mRSS decrease with 6.33 for Tocilizumab vs. 2.77 for placebo, $p = 0.0579$). Data regarding lung damage were promising, showing a significantly smaller decrease in FVC for Tocilizumab at week 48 compared with the placebo ($p = 0.0373$). Regarding fatigue, itching, clinician global disease severity, and disability, there were no differences between the two groups. The safety of Tocilizumab was comparable to the placebo (42 vs. 40), but more severe infections were reported in the active arm (7 vs. 2), with

1 death. The authors concluded that, although the skin primary endpoint was not achieved, the decrease in lung involvement was significant, thus opening new research perspectives.

Two years later, Khanna published the results of a phase 3 study (focuSSced study) that focused on Tocilizumab's effectiveness on skin and lung fibrosis, also including data regarding disability, treatment failure, and safety [115]. The randomized, double-blind, placebo-controlled trial included 210 patients with an early dcSS (<5 years) without background immunosuppressive treatment. Randomization was 1: 1, with 104 cases receiving subcutaneous Tocilizumab 162 mg weekly and 106 on placebo. The primary endpoint was an improvement in the mRRS skin score at week 48, which was not achieved ($p = 0.10$). Secondary objectives included the change in the predicted percentage of FVC, patient- and physician-reported outcomes, and time to treatment failure at week 48. Significant data were obtained for the predicted percentage of FVC, with Tocilizumab being effective in preserving lung functionality ($p = 0.002$). The results on disability and the overall physician and patient assessment of the disease were not significant. The Tocilizumab safety profile was good, with the most common side effects being infections.

Other recently published data support Tocilizumab's beneficial effect on skin fibrosis. A study on the effect of IL-6 blockade on the molecular, genomic, and functional characteristics of cultured fibroblasts from skin biopsies of SS patients was performed [116]. The analysis included 12 patients treated with Tocilizumab or placebo for 24 weeks. An activated molecular and functional phenotype was maintained after 24 weeks with placebo. In contrast, fibroblasts in Tocilizumab patient cultures underwent beneficial changes, such as decreased migration, proliferation and contraction, and low collagen fiber production. Furthermore, by analyzing the genetic profile dominated by genes that promote fibrosis (TGFβ-regulated genes, COL1A1), it was found that Tocilizumab can induce normalization of the genetic phenotype, thereby improving skin thickening [116]. The findings of this analysis were promising and highlighted the dual role of IL-6 blockade in inflammation and fibrosis.

Other data on the improvement of cutaneous fibrosis with Tocilizumab come from smaller studies, even from case reports and case series, considering the low incidence of the disease. Shima followed 2 patients with SS refractory to conventional therapies who received Tocilizumab 8 mg/kg/month for 6 months. Skin damage was assessed by both the mRRS score and with a Vesmeter device, which measures viscosity, elasticity, and skin thickening [117]. Tocilizumab therapy resulted in skin softening quantified by both scores. Another study by Shima included seven SS cases; its main purpose was to investigate the cutaneous efficacy of Tocilizumab quantified by the mRRS score and to analyze the factors that contribute to treatment response [63]. Skin improvement was evident after Tocilizumab therapy, but the data were not statistically consistent. Moreover, the decrease in the mRSS score was higher in cases with a significant inflammatory syndrome (quantified by CRP) and shorter disease duration. Negative correlations were found between the mRRS score and chemokine CCL-5 and IL-13 levels.

In addition to improving skin fibrosis, Tocilizumab appears to be effective in osteoarticular changes. Zacay et al. administered Tocilizumab in 16 cases of SS without clinical response to other immunosuppressants. All patients had musculoskeletal involvement. Twelve patients with arthritis/ arthralgia showed a significant improvement after treatment. One case of myalgia improved, and three out of four with myositis showed normalization of muscle enzymes after Tocilizumab [118]. In addition, the authors found a significant improvement in the mRSS skin score, which decreased by 11 points, and an improvement in lung capacity in 46% of cases. All these favorable responses were mainly recorded in patients with an early disease. The same positive effect on musculoskeletal manifestations is supported by EUSTAR (The European Scleroderma Trials and Research group), which analyzed the response to Tocilizumab and abatacept in a cohort of 27 SS cases. All patients with polyarthritis significantly improved after 5 months of therapy. However, the mRSS score did not show a statistically significant decrease [119]. In the two cases presented by Shima [117], one of the patients had a significantly limited range of joint motion, which improved after Tocilizumab therapy, thus strengthening the efficacy of the IL-6 inhibitor on joint involvement and supporting its inclusion in the SS treatment algorithm [120].

Regarding lung damage, one of the most important complications of the disease and the leading cause of death, Tocilizumab therapy is a recently approved solution linked to decreased lung worsening and maintenance of pulmonary function. The results of the extension of the phase 3 focuSSced study support this pulmonary improvement after switching from placebo to Tocilizumab [121]. Thus, 78 patients in the placebo arm switched to Tocilizumab. After 2 years of treatment, 50% of them maintained a constant level of FVC. Only 3% (2 cases) developed an increase in pulmonary fibrosis quantified by a decrease in FVC of over 10%. All of these patients showed not only maintenance of lung capacity, but even an improvement in FVC of 0.6% after 2 years. The results were similar to those in patients treated with Tocilizumab from the beginning. Moreover, the 48-week extension of the faSScinate study strengthened the effect of Tocilizumab therapy in stabilizing lung function. In addition, there was a further improvement in the mRSS score at week 96 [122].

Recently, in 2021, a post hoc analysis of the focuSSced trial was published, which aimed to analyze the preservation of lung function in 136 cases of SS and intestinal lung disease. As a novelty, stratification of patients was conducted according to the degree of pulmonary fibrosis: mild (5–10%), moderate (10–20%), and severe (>20%) [123]. The included patients had an early SS with progressive skin disease and pulmonary fibrosis, so they were in the immunoinflammatory phase of the disease. Moderate and severe fibrotic lung impairment had an increased incidence of 77%. The authors identified inverse correlations between the degree of pulmonary fibrosis and the predicted FVC percentage. FVC stabilization was similar in all study arms regardless of fibrosis and was significantly more important for Tocilizumab compared to the placebo (−0.1% Tocilizumab vs. −6.3% placebo, $p < 0.0001$). Because Tocilizumab's favorable pulmonary effects were observed in the early fibrotic phase of SS, the authors hypothesized that a window of therapeutic opportunity exists during this stage.

Tocilizumab may be considered as a rescue therapy in non-responsive cases. Narvaez included in an analysis nine cases of progressive SS with interstitial lung disease refractory to corticosteroid therapy (low and medium doses), rituximab, and other immunosuppressants (cyclophosphamide, azathioprine, mycophenolate mofetil). Patients received Tocilizumab (both intravenously and subcutaneously) in combination with mycophenolate mofetil for 6 months; prednisone below 5 mg/daily was used in 7 cases. The results regarding lung damage were promising, with 44% of patients experiencing both stabilization and an improvement in lung capacity quantified by FVC, DLCO, and the 6-min walking test [124].

Another recent analysis, reinforced by real-world data, indicated again the beneficial effect of Tocilizumab on skin and lung damage, its long-term safety, and an improvement in patient-reported outcomes. Thus, Khanna's study of 82 SS cases showed an improvement in mRSS and FVC and a good safety profile after 96 weeks of Tocilizumab administration [125]. Panopoulos' analysis of 21 SS patients receiving subcutaneous Tocilizumab for 1 year showed the same efficacy data on the mRSS score and polyarticular impairment, also providing stabilization of lung damage and an improvement in patient-reported outcomes [126].

The benefit of Tocilizumab therapy in SS heart disease has been less studied so far. Due to repeated lesions of ischemia and reperfusion, histological changes in band-like necrosis, and diffuse macular fibrosis, SS can be considered an independent risk factor for cardiovascular disease [127,128]. This year, the case of a young patient diagnosed with lcSS in 2015 who developed ventricular extrasystoles and changes in cardiac reperfusion highlighted on an interventricular septum biopsy test was published. Mycophenolate mofetil 2 g daily was administered for 3 months, but the heart condition worsened. Intravenous treatment with Tocilizumab 8 mg/kg/monthly was initiated with favorable results regarding clinical symptoms and reperfusion defects [129].

Finally, there are data indicating the favorable effect of Tocilizumab on SS juvenile forms (JSS). The first published study of the efficacy of Tocilizumab in JSS was published last year and included nine JSS cases on anti-IL-6R therapy in combination with methotrexate, mycophenolate mofetil, or corticosteroids. Patients mainly had pulmonary and gastrointestinal manifestations [130]. After a mean follow-up of 24 months, statistically significant data were recorded for skin damage (decreased mRSS), lung function (increased DLCO), global patient assessment (PGA), and Juvenile Systemic Sclerosis Severity (J4S), which improved.

All these data presented above regarding the clinical efficacy of Tocilizumab therapy in SS patients are summarized in Table 1.

Table 1. Data regarding the clinical efficacy of Tocilizumab therapy in SS patients.

Author	Number of Patients	Clinical and Paraclinical Parameters	Follow-Up Period	Results
Shima et al., 2010 [117]	2	mRSS	6 months	• skin thickening improvement (mRSS decrease)
Elhai et al., 2013 [119]	27	Polyarthritis Myopathy mRSS	5 months	• joint/muscle damage improvement • skin thickening improvement (mRSS decrease)
Khanna et al., 2016 [114]	87	mRSS FVC	6 months	• skin thickening improvement (mRSS decrease) • smaller decrease in FVC
Khanna et al., 2017 [122]	78	mRSS FVC	24 months	• skin thickening improvement (mRSS decrease) • lung function stabilization
Khanna et al., 2018 [115]	210	mRSS FVC	12 months	• skin thickening improvement (mRSS decrease) • lung function stabilization
Denton et al., 2018 [116]	12	mRSS	6 months	• skin thickening improvement (mRSS decrease) • decreased skin collagen fiber production
Zacay et al., 2018 [118]	16	Arthritis/arthralgia Myalgia/myositis FVC mRSS	8 months	• significant improvement in arthritis/arthralgia • muscle enzymes normalization • skin thickening improvement (mRSS decrease) • lung capacity improvement (FVC increase)
Shima et al., 2019 [63]	7	mRSS	6 months	• skin thickening improvement (mRSS decrease) especially in cases with significant inflammatory syndrome and short disease duration
Narvaez et al., 2019 [124]	9	FVC DLCO 6 min walking test	12 months	• lung function stabilization and improvement (FVC increase) in cases refractory to corticosteroids or other immunosuppressants
Khanna et al. 2020 [121]	78	FVC	24 months	• lung function stabilization/ improvement (FVC increase)
Roofeh et al., 2021 [123]	136	FVC	12 months	• lung function preservation according to the degree of pulmonary fibrosis
Khanna et al., 2021 [125]	82	mRSS FVC	24 months	• skin thickening improvement (mRSS decrease) • lung function improvement (FVC increase)
Adrovic et al., 2021 [130]	9	Juvenile SS mRSS DLCO	24 months	• skin thickening improvement (mRSS decrease) • lung function improvement (DLCO increase)
Panopoulos et al., 2022 [126]	21	mRSS Polyarthritis FVC PROs	12 months	• skin thickening improvement (mRSS decrease) • lung function stabilization • joint damage improvement (Disease Activity Score 28 decrease) • improving patients' quality of life

Key: mRSS: modified RODNAN skin score; DLCO: diffusing capacity of lung for carbon monoxide; FVC: forced vital capacity; PROs: patient-reported outcomes.

5. Conclusions

Although we have a rich therapeutic arsenal, SS remains a serious condition that involves a multidisciplinary approach and targeted therapies. The pathogenesis of the disease is complex and multifactorial, and is still incompletely known. IL-6 has an important role in both vasculopathy and fibrosis and is associated with various clinical manifestations. Blocking IL-6R with Tocilizumab results in many clinical and biological benefits due to its anti-inflammatory and anti-fibrotic effects. Recent studies have opened the way for Tocilizumab in SS, supporting its efficacy and safety.

Author Contributions: Conceptualization, A.C. and E.R.; methodology, A.C. and E.R.; software, A.M.B.; validation, E.R.; formal analysis, L.A.M. and E.R.; Investigation, I.B., P.R. and A.M.B.; resources, A.C., E.R., L.A.M. and A.M.B.; data curation, A.M.B. and E.R.; writing – original draft preparation, A.C.; writing – review and editing, A.C. and A.M.B.; visualization, E.R. and L.A.M.; supervision, E.R.; project administration, E.R.; funding acquisition, A.C. All authors conceived the article, co-wrote the final manuscript, and reviewed the literature. All authors have read and agreed to the published version of the manuscript.

Funding: This research received no external funding.

Institutional Review Board Statement: Not applicable.

Informed Consent Statement: Not applicable.

Data Availability Statement: Not applicable.

Conflicts of Interest: The authors declare no conflict of interests.

References

1. Allanore, Y.; Simms, R.; Distler, O.; Trojanowska, M.; Pope, J.; Denton, C.P.; Varga, J. Systemic sclerosis. *Nat. Rev. Dis. Prim.* **2015**, *1*, 15002. [CrossRef] [PubMed]
2. Denton, C.P.; Khanna, D. Systemic sclerosis. *Lancet* **2017**, *390*, 1685–1699. [CrossRef]
3. Brown, M.; O'Reilly, S. The immunopathogenesis of fibrosis in systemic sclerosis. *Clin. Exp. Immunol.* **2019**, *195*, 310–321. [CrossRef] [PubMed]
4. Asano, Y.; Sato, S. Vasculopathy in scleroderma. *Semin. Immunopathol.* **2015**, *37*, 489–500. [CrossRef] [PubMed]
5. Brembilla, N.C.; Chizzolini, C. T cell abnormalities in systemic sclerosis with a focus on Th17 cells. *Eur. Cytokine Netw.* **2012**, *23*, 128–139. [CrossRef] [PubMed]
6. Cossu, M.; van Bon, L.; Nierkens, S.; Bellocchi, C.; Santaniello, A.; Dolstra, H.; Beretta, L.; Radstake, T.R. The magnitude of cytokine production by stimulated CD56$^+$ cells is associated with early stages of systemic sclerosis. *Clin. Immunol.* **2016**, *173*, 76–80. [CrossRef]
7. Utsunomiya, A.; Oyama, N.; Hasegawa, M. Potential Biomarkers in Systemic Sclerosis: A Literature Review and Update. *J. Clin. Med.* **2020**, *9*, 3388. [CrossRef]
8. Kishimoto, T. Factors Affecting B-Cell Growth and Differentiation. *Annu. Rev. Immunol.* **1985**, *3*, 133–157. [CrossRef]
9. Kishimoto, T. The biology of interleukin-6. *Blood* **1989**, *74*, 1–10. [CrossRef]
10. Hunter, C.A.; Jones, S.A. IL-6 as a keystone cytokine in health and disease. *Nat. Immunol.* **2015**, *16*, 448–457. [CrossRef]
11. Chalaris, A.; Garbers, C.; Rabe, B.; Rose-John, S.; Scheller, J. The soluble Interleukin 6 receptor: Generation and role in inflammation and cancer. *Eur. J. Cell Biol.* **2011**, *90*, 484–494. [CrossRef] [PubMed]
12. Mihara, M.; Hashizume, M.; Yoshida, H.; Suzuki, M.; Shiina, M. IL-6/IL-6 receptor system and its role in physiological and pathological conditions. *Clin. Sci.* **2012**, *122*, 143–159. [CrossRef] [PubMed]
13. Saito, M.; Yoshida, K.; Hibi, M.; Taga, T.; Kishimoto, T. Molecular cloning of a murine IL-6 receptor-associated signal trans-ducer, gp130, and its regulated expression in vivo. *J. Immunol.* **1992**, *148*, 4066–4071.
14. Demyanets, S.; Huber, K.; Wojta, J. Vascular effects of glycoprotein130 ligands—Part II: Biomarkers and therapeutic targets. *Vasc. Pharmacol.* **2012**, *57*, 29–40. [CrossRef]
15. Klouche, M.; Bhakdi, S.; Hemmes, M.; Rose-John, S. Novel path to activation of vascular smooth muscle cells: Up-regulation of gp130 creates an autocrine activation loop by IL-6 and its soluble receptor. *J. Immunol.* **1999**, *163*, 4583–4589.
16. Ernst, M.; Jenkins, B.J. Acquiring signalling specificity from the cytokine receptor gp130. *Trends Genet.* **2004**, *20*, 23–32. [CrossRef]
17. Schöbitz, B.; Pezeshki, G.; Pohl, T.; Hemmann, U.; Heinrich, P.C.; Holsboer, F.; Reul, J.M. Soluble interleukin-6 (IL-6) receptor augments central effects of IL-6 in vivo. *FASEB J.* **1995**, *9*, 659–664. [CrossRef]
18. Garbers, C.; Aparicio-Siegmund, S.; Rose-John, S. The IL-6/gp130/STAT3 signaling axis: Recent advances towards specific inhibition. *Curr. Opin. Immunol.* **2015**, *34*, 75–82. [CrossRef]

19. Lacroix, M.; Rousseau, F.; Guilhot, F.; Malinge, P.; Magistrelli, G.; Herren, S.; Jones, S.A.; Jones, G.W.; Scheller, J.; Lissilaa, R.; et al. Novel Insights into Interleukin 6 (IL-6) Cis- and Trans-signaling Pathways by Differentially Manipulating the Assembly of the IL-6 Signaling Complex. *J. Biol. Chem.* **2015**, *290*, 26943–26953. [CrossRef]
20. Garbers, C.; Jänner, N.; Chalaris, A.; Moss, M.L.; Floss, D.M.; Meyer, D.; Koch-Nolte, F.; Rose-John, S.; Scheller, J. Species specificity of ADAM10 and ADAM17 proteins in interleukin-6 (IL-6) trans-signaling and novel role of ADAM10 in inducible IL-6 receptor shedding. *J. Biol. Chem.* **2011**, *286*, 14804–14811. [CrossRef]
21. Schumacher, N.; Meyer, D.; Mauermann, A.; Von Der Heyde, J.; Wolf, J.; Schwarz, J.; Knittler, K.; Murphy, G.; Michalek, M.; Garbers, C.; et al. Shedding of Endogenous Interleukin-6 Receptor (IL-6R) Is Governed by A Disintegrin and Metalloproteinase (ADAM) Proteases while a Full-length IL-6R Isoform Localizes to Circulating Microvesicles. *J. Biol. Chem.* **2015**, *290*, 26059–26071. [CrossRef]
22. Peters, M.; Jacobs, S.; Ehlers, M.; Vollmer, P.; Müllberg, J.; Wolf, E.; Brem, G.; Büschenfelde, K.H.M.Z.; Rose-John, S. The function of the soluble interleukin 6 (IL-6) receptor in vivo: Sensitization of human soluble IL-6 receptor transgenic mice towards IL-6 and prolongation of the plasma half-life of IL-6. *J. Exp. Med.* **1996**, *183*, 1399–1406. [CrossRef]
23. Marin, V.; Montero-Julian, F.A.; Grès, S.; Boulay, V.; Bongrand, P.; Farnarier, C.; Kaplanski, G. The IL-6-Soluble IL-6Rα Autocrine Loop of Endothelial Activation as an Intermediate Between Acute and Chronic Inflammation: An Experimental Model Involving Thrombin. *J. Immunol.* **2001**, *167*, 3435–3442. [CrossRef]
24. Croker, B.A.; Krebs, D.L.; Zhang, J.-G.; Wormald, S.; Willson, T.A.; Stanley, E.G.; Robb, L.; Greenhalgh, C.J.; Förster, I.; Clausen, B.E.; et al. SOCS3 negatively regulates IL-6 signaling in vivo. *Nat. Immunol.* **2003**, *4*, 540–545. [CrossRef]
25. Babon, J.J.; Varghese, L.N.; Nicola, N.A. Inhibition of IL-6 family cytokines by SOCS3. *Semin. Immunol.* **2014**, *26*, 13–19. [CrossRef]
26. Schmitz, J.; Weissenbach, M.; Haan, S.; Heinrich, P.C.; Schaper, F. SOCS3 Exerts Its Inhibitory Function on Interleukin-6 Signal Transduction through the SHP2 Recruitment Site of gp130. *J. Biol. Chem.* **2000**, *275*, 12848–12856. [CrossRef]
27. Zhang, L.; Badgwell, D.B.; Bevers, J.J., III; Schlessinger, K.; Murray, P.J.; Levy, D.E.; Watowich, S.S. IL-6 signaling via the STAT3/SOCS3 pathway: Functional Analysis of the Conserved STAT3 N-domain. *Mol. Cell. Biochem.* **2006**, *288*, 179–189. [CrossRef]
28. Narazaki, M.; Yasukawa, K.; Saito, T.; Ohsugi, Y.; Fukui, H.; Koishihara, Y.; Yancopoulos, G.D.; Taga, T.; Kishimoto, T. Soluble forms of the interleukin-6 signal-transducing receptor component gp130 in human serum possessing a potential to inhibit signals through membrane-anchored gp130. *Blood* **1993**, *82*, 1120–1126. [CrossRef]
29. Yasukawa, K.; Futatsugi, K.; Saito, T.; Yawata, H.; Narazaki, M.; Suzuki, H.; Taga, T.; Kishimoto, T. Association of recombinant soluble IL-6-signal transducer, gp130, with a complex of IL 6 and soluble IL-6 receptor, and establishment of an ELISA for soluble gp130. *Immunol. Lett.* **1992**, *31*, 123–130. [CrossRef]
30. Wolf, J.; Waetzig, G.H.; Chalaris, A.; Reinheimer, T.M.; Wege, H.; Rose-John, S.; Garbers, C. Different Soluble Forms of the Interleukin-6 Family Signal Transducer gp130 Fine-tune the Blockade of Interleukin-6 Trans-signaling. *J. Biol. Chem.* **2016**, *291*, 16186–16196. [CrossRef]
31. Heinrich, P.C.; Behrmann, I.; Haan, S.; Hermanns, H.M.; Müller-Newen, G.; Schaper, F. Principles of interleukin (IL)-6-type cytokine signalling and its regulation. *Biochem. J.* **2003**, *374*, 1–20. [CrossRef]
32. Tanaka, T.; Narazaki, M.; Kishimoto, T. IL-6 in inflammation, immunity, and disease. *Cold Spring Harb. Perspect. Biol.* **2014**, *6*, a016295. [CrossRef]
33. Heinrich, P.C.; Castell, J.V.; Andus, T. Interleukin-6 and the acute phase response. *Biochem. J.* **1990**, *265*, 621–636. [CrossRef]
34. Nemeth, E.; Rivera, S.; Gabayan, V.; Keller, C.; Taudorf, S.; Pedersen, B.K.; Ganz, T. IL-6 mediates hypoferremia of in-flammation by inducing the synthesis of the iron regulatory hormone hepcidin. *J. Clin. Investig.* **2004**, *113*, 1271–1276. [CrossRef]
35. Liuzzi, J.P.; Lichten, L.A.; Rivera, S.; Blanchard, R.K.; Aydemir, T.B.; Knutson, M.D.; Ganz, T.; Cousins, R.J. Interleukin-6 regulates the zinc transporter Zip14 in liver and contributes to the hypozincemia of the acute-phase response. *Proc. Natl. Acad. Sci. USA* **2005**, *102*, 6843–6848. [CrossRef]
36. Korn, T.; Bettelli, E.; Oukka, M.; Kuchroo, V.K. IL-17 and Th17 Cells. *Annu. Rev. Immunol.* **2009**, *27*, 485–517. [CrossRef]
37. Bettelli, E.; Carrier, Y.; Gao, W.; Korn, T.; Strom, T.B.; Oukka, M.; Weiner, H.L.; Kuchroo, V.K. Reciprocal developmental pathways for the generation of pathogenic effector TH17 and regulatory T cells. *Nature* **2006**, *441*, 235–238. [CrossRef]
38. Kimura, A.; Kishimoto, T. IL-6: Regulator of Treg/Th17 balance. *Eur. J. Immunol.* **2010**, *40*, 1830–1835. [CrossRef]
39. Ma, C.S.; Deenick, E.K.; Batten, M.; Tangye, S.G. The origins, function, and regulation of T follicular helper cells. *J. Exp. Med.* **2012**, *209*, 1241–1253. [CrossRef]
40. Okada, M.; Kitahara, M.; Kishimoto, S.; Matsuda, T.; Hirano, T.; Kishimoto, T. IL-6/BSF-2 functions as a killer helper factor in the in vitro induction of cytotoxic T cells. *J. Immunol.* **1988**, *141*, 1543–1549.
41. Kaneko, Y.; Takeuchi, T. An update on the pathogenic role of IL-6 in rheumatic diseases. *Cytokine* **2021**, *146*, 155645. [CrossRef] [PubMed]
42. Bondeson, J.; Wainwright, S.D.; Lauder, S.; Amos, N.; Hughes, C.E. The role of synovial macrophages and macrophage-produced cytokines in driving aggrecanases, matrix metalloproteinases, and other destructive and inflammatory responses in osteoarthritis. *Arthritis Res. Ther.* **2006**, *8*, R187. [CrossRef] [PubMed]
43. Yoshitake, F.; Itoh, S.; Narita, H.; Ishihara, K.; Ebisu, S. Interleukin-6 Directly Inhibits Osteoclast Differentiation by Suppressing Receptor Activator of NF-κB Signaling Pathways. *J. Biol. Chem.* **2008**, *283*, 11535–11540. [CrossRef] [PubMed]

44. Muangchan, C.; Pope, J.E. Interleukin 6 in Systemic Sclerosis and Potential Implications for Targeted Therapy. *J. Rheumatol.* **2012**, *39*, 1120–1124. [CrossRef] [PubMed]
45. Fuschiotti, P.; Medsger, T.A., Jr.; Morel, P.A. Effector CD8+ T cells in systemic sclerosis patients produce abnormally high levels of interleukin-13 associated with increased skin fibrosis. *Arthritis Care Res.* **2009**, *60*, 1119–1128. [CrossRef]
46. Yoshizaki, K. Pathogenic Role of IL-6 Combined with TNF-α or IL-1 in the Induction of Acute Phase Proteins SAA and CRP in Chronic Inflammatory Diseases. *Adv. Exp. Med. Biol.* **2010**, *691*, 141–150. [CrossRef]
47. Saito, F.; Tasaka, S.; Inoue, K.-I.; Miyamoto, K.; Nakano, Y.; Ogawa, Y.; Yamada, W.; Shiraishi, Y.; Hasegawa, N.; Fujishima, S.; et al. Role of Interleukin-6 in Bleomycin-Induced Lung Inflammatory Changes in Mice. *Am. J. Respir. Cell Mol. Biol.* **2008**, *38*, 566–571. [CrossRef]
48. Fishman, D.; Faulds, G.; Jeffery, R.; Mohamed-Ali, V.; Yudkin, J.S.; Humphries, S.; Woo, P. The effect of novel polymorphisms in the interleukin-6 (IL-6) gene on IL-6 transcription and plasma IL-6 levels, and an association with systemic-onset juvenile chronic arthritis. *J. Clin. Investig.* **1998**, *102*, 1369–1376. [CrossRef]
49. Steen, V.D.; Medsger, T.A., Jr. Severe organ involvement in systemic sclerosis with diffuse scleroderma. *Arthritis Rheum.* **2000**, *43*, 2437–2444. [CrossRef]
50. Khanna, D.; Hays, R.D.; Maranian, P.; Seibold, J.R.; Impens, A.; Mayes, M.D.; Clements, P.J.; Getzug, T.; Fathi, N.; Bechtel, A.; et al. Reliability and validity of the university of california, los angeles scleroderma clinical trial consortium gastrointestinal tract instrument. *Arthritis Care Res.* **2009**, *61*, 1257–1263. [CrossRef]
51. Sfrent-Cornateanu, R.; Mihai, C.; Balan, S.; Ionescu, R.; Moldoveanu, E. The IL—6 promoter polymorphism is associated with disease activity and disability in systemic sclerosis. *J. Cell. Mol. Med.* **2006**, *10*, 890–895. [CrossRef]
52. Needleman, B.W.; Wigley, F.M.; Stair, R.W. Interleukin-1, Interleukin-2, Interleukin-4, Interleukin-6, Tumor Necrosis Factor α, and Interferon-γ Levels in Sera from Patients with Scleroderma. *Arthritis Care Res.* **1992**, *35*, 67–72. [CrossRef]
53. Koch, A.E.; Kronfeld-Harrington, L.B.; Szekanecz, Z.; Cho, M.M.; Haines, K.; Harlow, L.A.; Strieter, R.M.; Kunkel, S.L.; Massa, M.C.; Barr, W.G.; et al. In situ Expression of Cytokines and Cellular Adhesion Molecules in the Skin of Patients with Systemic Sclerosis. Their role in early and late disease. *Pathobiology* **1993**, *61*, 239–246. [CrossRef]
54. Barnes, T.C.; Spiller, D.G.; Anderson, M.E.; Edwards, S.W.; Moots, R.J. Endothelial activation and apoptosis mediated by neutrophil-dependent interleukin 6 trans-signalling: A novel target for systemic sclerosis? *Ann. Rheum. Dis.* **2011**, *70*, 366–372. [CrossRef]
55. Henderson, J.; Bhattacharyya, S.; Varga, J.; O'Reilly, S. Targeting TLRs and the inflammasome in systemic sclerosis. *Pharmacol. Ther.* **2018**, *192*, 163–169. [CrossRef]
56. Feghali, C.A.; Bost, K.L.; Boulware, D.W.; Levy, L.S. Mechanisms of pathogenesis in scleroderma. I. Overproduction of in-terleukin 6 by fibroblasts cultured from affected skin sites of patients with scleroderma. *J. Rheumatol.* **1992**, *19*, 1207–1211.
57. Gurram, M.; Pahwa, S.; Frieri, M. Augmented interleukin-6 secretion in collagen-stimulated peripheral blood mononuclear cells from patients with systemic sclerosis. *Ann. Allergy* **1994**, *73*, 493–496.
58. Hasegawa, M.; Fujimoto, M.; Matsushita, T.; Hamaguchi, Y.; Takehara, K.; Sato, S. Serum chemokine and cytokine levels as indicators of disease activity in patients with systemic sclerosis. *Clin. Rheumatol.* **2011**, *30*, 231–237. [CrossRef]
59. Giacomelli, R.; Cipriani, P.; Danese, C.; Pizzuto, F.; Lattanzio, R.; Parzanese, I.; Passacantando, A.; Perego, M.A.; Tonietti, G. Peripheral blood mononuclear cells of patients with systemic sclerosis produce increased amounts of interleukin 6, but not transforming growth factor beta 1. *J. Rheumatol.* **1996**, *23*, 291–296.
60. Sato, S.; Hasegawa, M.; Takehara, K. Serum levels of interleukin-6 and interleukin-10 correlate with total skin thickness score in patients with systemic sclerosis. *J. Dermatol. Sci.* **2001**, *27*, 140–146. [CrossRef]
61. Hasegawa, M.; Sato, S.; Fujimoto, M.; Ihn, H.; Kikuchi, K.; Takehara, K. Serum levels of interleukin 6 (IL-6), oncostatin M, soluble IL-6 receptor, and soluble gp130 in patients with systemic sclerosis. *J. Rheumatol.* **1998**, *25*, 308–313. [PubMed]
62. Matsushita, T.; Hasegawa, M.; Hamaguchi, Y.; Takehara, K.; Sato, S. Longitudinal analysis of serum cytokine concentrations in systemic sclerosis: Association of interleukin 12 elevation with spontaneous regression of skin sclerosis. *J. Rheumatol.* **2006**, *33*, 275–284. [PubMed]
63. Shima, Y.; Kawaguchi, Y.; Kuwana, M. Add-on tocilizumab versus conventional treatment for systemic sclerosis, and cytokine analysis to identify an endotype to tocilizumab therapy. *Mod. Rheumatol.* **2019**, *29*, 134–139. [CrossRef] [PubMed]
64. Yamauchi-Takihara, K.; Kishimoto, T. Cytokines and their receptors in cardiovascular diseases—Role of gp130 signalling pathway in cardiac myocyte growth and maintenance. *Int. J. Exp. Pathol.* **2000**, *81*, 1–16. [CrossRef]
65. Rajkumar, V.S.; Howell, K.; Csiszar, K.; Denton, C.P.; Black, C.M.; Abraham, D.J. Shared expression of phenotypic markers in systemic sclerosis indicates a convergence of pericytes and fibroblasts to a myofibroblast lineage in fibrosis. *Arthritis Res. Ther.* **2005**, *7*, R1113–R1123. [CrossRef]
66. Kitaba, S.; Murota, H.; Terao, M.; Azukizawa, H.; Terabe, F.; Shima, Y.; Fujimoto, M.; Tanaka, T.; Naka, T.; Kishimoto, T.; et al. Blockade of Interleukin-6 Receptor Alleviates Disease in Mouse Model of Scleroderma. *Am. J. Pathol.* **2012**, *180*, 165–176. [CrossRef] [PubMed]
67. Kawaguchi, Y.; Hara, M.; Wright, T.M. Endogenous IL-1α from systemic sclerosis fibroblasts induces IL-6 and PDGF-A. *J. Clin. Investig.* **1999**, *103*, 1253–1260. [CrossRef]
68. Beon, M.; Harley, R.A.; Wessels, A.; Silver, R.M.; Ludwicka-Bradley, A. Myofibroblast induction and microvascular alteration in scleroderma lung fibrosis. *Clin. Exp. Rheumatol.* **2004**, *22*, 733–742.

69. Brissett, M.; Veraldi, K.L.; Pilewski, J.M.; Medsger, T.A., Jr.; Feghali-Bostwick, C.A. Localized expression of tenascin in systemic sclerosis-associated pulmonary fibrosis and its regulation by insulin-like growth factor binding protein 3. *Arthritis Care Res.* **2012**, *64*, 272–280. [CrossRef]
70. Crestani, B.; Seta, N.; De Bandt, M.; Soler, P.; Rolland, C.; Dehoux, M.; Boutten, A.; Dombret, M.C.; Palazzo, E.; Kahn, M.F. Interleukin 6 secretion by monocytes and alveolar macrophages in systemic sclerosis with lung involvement. *Am. J. Respir. Crit. Care Med.* **1994**, *149*, 1260–1265. [CrossRef] [PubMed]
71. Renaud, L.; da Silveira, W.A.; Takamura, N.; Hardiman, G.; Feghali-Bostwick, C. Prominence of IL6, IGF, TLR, and Bioenergetics Pathway Perturbation in Lung Tissues of Scleroderma Patients with Pulmonary Fibrosis. *Front. Immunol.* **2020**, *11*, 383. [CrossRef]
72. Le, T.-T.T.; Karmouty-Quintana, H.; Melicoff, E.; Le, T.-T.T.; Weng, T.; Chen, N.-Y.; Pedroza, M.; Zhou, Y.; Davies, J.; Philip, K.; et al. Blockade of IL-6 Trans Signaling Attenuates Pulmonary Fibrosis. *J. Immunol.* **2014**, *193*, 3755–3768. [CrossRef]
73. De Lauretis, A.; Sestini, P.; Pantelidis, P.; Hoyles, R.; Hansell, D.M.; Goh, N.S.; Zappala, C.J.; Visca, D.; Maher, T.M.; Denton, C.P.; et al. Serum Interleukin 6 Is Predictive of Early Functional Decline and Mortality in Interstitial Lung Disease Associated with Systemic Sclerosis. *J. Rheumatol.* **2013**, *40*, 435–446. [CrossRef]
74. Khan, K.; Xu, S.; Nihtyanova, S.; Derrett-Smith, E.; Abraham, D.; Denton, C.P.; Ong, V.H. Clinical and pathological significance of interleukin 6 overexpression in systemic sclerosis. *Ann. Rheum. Dis.* **2012**, *71*, 1235–1242. [CrossRef]
75. Giuggioli, D.; Lumetti, F.; Colaci, M.; Fallahi, P.; Antonelli, A.; Ferri, C. Rituximab in the treatment of patients with systemic sclerosis. Our experience and review of the literature. *Autoimmun. Rev.* **2015**, *14*, 1072–1078. [CrossRef]
76. Bosello, S.; De Santis, M.; Lama, G.; Spanò, C.; Angelucci, C.; Tolusso, B.; Sica, G.; Ferraccioli, G. B cell depletion in diffuse progressive systemic sclerosis: Safety, skin score modification and IL-6 modulation in an up to thirty-six months follow-up open-label trial. *Arthritis Res. Ther.* **2010**, *12*, R54. [CrossRef]
77. Abignano, G.; Del Galdo, F. Biomarkers as an opportunity to stratify for outcome in systemic sclerosis. *Eur. J. Rheumatol.* **2020**, *7*, 193–202. [CrossRef]
78. Abdel-Magied, R.A.; Kamel, S.; Said, A.F.; Ali, H.M.; Gawad, E.A.A.; Moussa, M.M. WITHDRAWN: Serum interleukin-6 in systemic sclerosis and its correlation with disease parameters and cardiopulmonary involvement. *Egypt. J. Chest Dis. Tuberc.* **2016**, *33*, 32. [CrossRef]
79. Pendergrass, S.A.; Hayes, E.; Farina, G.; Lemaire, R.; Farber, H.W.; Whitfield, M.L.; Lafyatis, R. Limited Systemic Sclerosis Patients with Pulmonary Arterial Hypertension Show Biomarkers of Inflammation and Vascular Injury. *PLoS ONE* **2010**, *5*, e12106. [CrossRef]
80. Steen, V.; Medsger, T.A., Jr. Predictors of isolated pulmonary hypertension in patients with systemic sclerosis and limited cutaneous involvement. *Arthritis Care Res.* **2003**, *48*, 516–522. [CrossRef]
81. Dorfmüller, P.; Humbert, M.; Perros, F.; Sanchez, O.; Simonneau, G.; Müller, K.-M.; Capron, F. Fibrous remodeling of the pulmonary venous system in pulmonary arterial hypertension associated with connective tissue diseases. *Hum. Pathol.* **2007**, *38*, 893–902. [CrossRef]
82. Kawut, S.M.; Taichman, D.B.; Archer-Chicko, C.L.; Palevsky, H.I.; Kimmel, S.E. Hemodynamics and Survival in Patients with Pulmonary Arterial Hypertension Related to Systemic Sclerosis. *Chest* **2003**, *123*, 344–350. [CrossRef]
83. Fisher, M.R.; Mathai, S.C.; Champion, H.C.; Girgis, R.E.; Housten-Harris, T.; Hummers, L.; Krishnan, J.A.; Wigley, F.; Hassoun, P.M. Clinical differences between idiopathic and scleroderma-related pulmonary hypertension. *Arthritis Care Res.* **2006**, *54*, 3043–3050. [CrossRef]
84. Cool, C.D.; Kennedy, D.; Voelkel, N.F.; Tuder, R.M. Pathogenesis and evolution of plexiform lesions in pulmonary hypertension associated with scleroderma and human immunodeficiency virus infection. *Hum. Pathol.* **1997**, *28*, 434–442. [CrossRef]
85. Sjogren, R.W. Gastrointestinal motility disorders in scleroderma. *Arthritis Care Res.* **1994**, *37*, 1265–1282. [CrossRef]
86. Cénit, M.C.; Simeón, C.P.; Vonk, M.C.; Callejas-Rubio, J.L.; Espinosa, G.; Carreira, P.; Blanco, F.J.; Narvaez, J.; Tolosa, C.; Román-Ivorra, J.A.; et al. Influence of the IL6 Gene in Susceptibility to Systemic Sclerosis. *J. Rheumatol.* **2012**, *39*, 2294–2302. [CrossRef]
87. Zekovic, A.; Vreća, M.; Spasovski, V.; Andjelković, M.; Pavlovic, S.; Damjanov, N. Association between the -174 C/G polymorphism in the interleukin-6 (IL-6) gene and gastrointestinal involvement in patients with systemic sclerosis. *Clin. Rheumatol.* **2018**, *37*, 2447–2454. [CrossRef]
88. Campbell, P.M.; LeRoy, E.C. Pathogenesis of systemic sclerosis: A vascular hypothesis. *Semin. Arthritis Rheum.* **1975**, *4*, 351–368. [CrossRef]
89. Steen, V.D.; Medsger, T.A., Jr. The value of the health assessment questionnaire and special patient-generated scales to demonstrate change in systemic sclerosis patients over time. *Arthritis Care Res.* **1997**, *40*, 1984–1991. [CrossRef]
90. Taroni, J.N.; Martyanov, V.; Huang, C.-C.; Mahoney, J.M.; Hirano, I.; Shetuni, B.; Yang, G.-Y.; Brenner, D.; Jung, B.; Wood, T.A.; et al. Molecular characterization of systemic sclerosis esophageal pathology identifies inflammatory and proliferative signatures. *Arthritis Res. Ther.* **2015**, *17*, 194. [CrossRef]
91. Thoua, N.M.; Abdel-Halim, M.; Forbes, A.; Denton, C.P.; Emmanuel, A.V. Fecal Incontinence in Systemic Sclerosis Is Secondary to Neuropathy. *Am. J. Gastroenterol.* **2012**, *107*, 597–603. [CrossRef]
92. Singh, J.; Mehendiratta, V.; Del Galdo, F.; Jimenez, S.; Cohen, S.; Dimarino, A.J.; Rattan, S. Immunoglobulins from scleroderma patients inhibit the muscarinic receptor activation in internal anal sphincter smooth muscle cells. *Am. J. Physiol. Liver Physiol.* **2009**, *297*, G1206–G1213. [CrossRef]

93. Hong, B.Y.; Giang, R.; Mbuagbaw, L.; Larche, M.; Thabane, L. Factors associated with development of gastrointestinal problems in patients with scleroderma: A systematic review. *Syst. Rev.* **2015**, *4*, 188. [CrossRef]
94. Fertig, N.; Domsic, R.T.; Rodriguez-Reyna, T.; Kuwana, M.; Lucas, M.; Medsger, T.A., Jr.; Feghali-Bostwick, C.A. Anti-U11/U12 RNP antibodies in systemic sclerosis: A new serologic marker associated with pulmonary fibrosis. *Arthritis Care Res.* **2009**, *61*, 958–965. [CrossRef]
95. Kawaguchi, Y.; Nakamura, Y.; Matsumoto, I.; Nishimagi, E.; Satoh, T.; Kuwana, M.; Sumida, T.; Hara, M. Muscarinic-3 acetylcholine receptor autoantibody in patients with systemic sclerosis: Contribution to severe gastrointestinal tract dysmotility. *Ann. Rheum. Dis.* **2009**, *68*, 710–714. [CrossRef]
96. Marie, I.; Leroi, A.-M.; Menard, J.-F.; Levesque, H.; Quillard, M.; Ducrotte, P. Fecal calprotectin in systemic sclerosis and review of the literature. *Autoimmun. Rev.* **2015**, *14*, 547–554. [CrossRef]
97. Miller, J.B.; Gandhi, N.; Clarke, J.; Mcmahan, Z. Gastrointestinal Involvement in Systemic Sclerosis. *JCR J. Clin. Rheumatol.* **2018**, *24*, 328–337. [CrossRef]
98. Denton, C.P.; Black, C.M. Scleroderma—Clinical and pathological advances. *Best Pr. Res. Clin. Rheumatol.* **2004**, *18*, 271–290. [CrossRef]
99. Penn, H.; Howie, A.J.; Kingdon, E.J.; Bunn, C.; Stratton, R.; Black, C.; Burns, A.; Denton, C. Scleroderma renal crisis: Patient characteristics and long-term outcomes. *QJM Int. J. Med.* **2007**, *100*, 485–494. [CrossRef]
100. Stern, E.P.; Guerra, S.G.; Chinque, H.; Acquaah, V.; González-Serna, D.; Ponticos, M.; Martin, J.; Ong, V.H.; Khan, K.; Nihtyanova, S.I.; et al. Analysis of Anti-RNA Polymerase III Antibody-positive Systemic Sclerosis and Altered GPATCH2L and CTNND2 Expression in Scleroderma Renal Crisis. *J. Rheumatol.* **2020**, *47*, 1668–1677. [CrossRef]
101. Liu, C.; Hou, Y.; Xu, D.; Li, L.; Zhang, Y.; Cheng, L.; Yan, S.; Zhang, F.; Li, Y. Analysis of anti-RNA polymerase III antibodies in Chinese Han systemic sclerosis patients. *Clin. Rheumatol.* **2020**, *39*, 1191–1197. [CrossRef]
102. Scala, E.; Pallotta, S.; Frezzolini, A.; Abeni, D.; Barbieri, C.; Sampogna, F.; DE Pità, O.; Puddu, P.; Paganelli, R.; Russo, G. Cytokine and chemokine levels in systemic sclerosis: Relationship with cutaneous and internal organ involvement. *Clin. Exp. Immunol.* **2004**, *138*, 540–546. [CrossRef]
103. Doran, J.P.; Veale, D.J. Biomarkers in systemic sclerosis. *Rheumatology* **2008**, *47*, v36–v38. [CrossRef]
104. Lee, S.; Lee, S.; Sharma, K. The pathogenesis of fibrosis and renal disease in scleroderma: Recent insights from glomerulosclerosis. *Curr. Rheumatol. Rep.* **2004**, *6*, 141–148. [CrossRef]
105. Charles, C.; Clements, P.; Furst, D.F. Systemic sclerosis: Hypothesis-driven treatment strategies. *Lancet* **2006**, *367*, 1683–1691. [CrossRef]
106. Steen, V.D. Scleroderma renal crisis. *Rheum. Dis. Clin. North Am.* **2003**, *29*, 315–333. [CrossRef]
107. Steen, V.D.; Medsger, T.A., Jr. Long-Term Outcomes of Scleroderma Renal Crisis. *Ann. Intern. Med.* **2000**, *133*, 600–603. [CrossRef]
108. Mihara, M.; Kasutani, K.; Okazaki, M.; Nakamura, A.; Kawai, S.; Sugimoto, M.; Matsumoto, Y.; Ohsugi, Y. Tocilizumab inhibits signal transduction mediated by both mIL-6R and sIL-6R, but not by the receptors of other members of IL-6 cytokine family. *Int. Immunopharmacol.* **2005**, *5*, 1731–1740. [CrossRef]
109. Burmester, G.R.; Feist, E.; Kellner, H.; Braun, J.; Iking-Konert, C.; Rubbert-Roth, A. Effectiveness and safety of the interleukin 6-receptor antagonist tocilizumab after 4 and 24 weeks in patients with active rheumatoid arthritis: The first phase IIIb real-life study (TAMARA). *Ann. Rheum. Dis.* **2011**, *70*, 755–759. [CrossRef]
110. Ohsugi, Y. The immunobiology of humanized Anti-IL6 receptor antibody: From basic research to breakthrough medicine. *J. Transl. Autoimmun.* **2020**, *3*, 100030. [CrossRef]
111. Castañeda, S.; Martínez-Quintanilla, D.; Martín-Varillas, J.L.; García-Castañeda, N.; Atienza-Mateo, B.; González-Gay, M.A. Tocilizumab for the treatment of adult-onset Still's disease. *Expert Opin. Biol. Ther.* **2019**, *19*, 273–286. [CrossRef]
112. Stone, J.H.; Tuckwell, K.; Dimonaco, S.; Klearman, M.; Aringer, M.; Blockmans, D.E.; Brouwer, E.; Cid, M.C.; Dasgupta, B.; Rech, J.; et al. Trial of Tocilizumab in Giant-Cell Arteritis. *N. Engl. J. Med.* **2017**, *377*, 317–328. [CrossRef]
113. Abid, M.B.; Peck, R.; Abid, M.A.; Al-Sakkaf, W.; Zhang, Y.; Dunnill, G.S.; Staines, K.; Sequeiros, I.-M.; Lowry, L. Is tocilizumab a potential therapeutic option for refractory unicentric Castleman disease? *Hematol. Oncol.* **2017**, *36*, 320–323. [CrossRef]
114. Khanna, D.; Denton, C.P.; Jahreis, A.; van Laar, J.M.; Frech, T.M.; Anderson, M.E.; Baron, M.; Chung, L.; Fierlbeck, G.; Lakshminarayanan, S.; et al. Safety and efficacy of subcutaneous tocilizumab in adults with systemic sclerosis (faSScinate): A phase 2, randomised, controlled trial. *Lancet* **2016**, *387*, 2630–2640. [CrossRef]
115. Khanna, D.; Lin, C.J.F.; Kuwana, M.; Allanore, Y.; Batalov, A.; Butrimiene, I.; Carreira, P.; Matucci Cerinic, M.; Distler, O.; Kaliterna, D.M.; et al. Efficacy and Safety of Tocilizumab for the Treatment of Systemic Sclerosis: Results from a Phase 3 Randomized Controlled Trial [abstract]. *Arthritis Rheumatol.* **2018**, *70*, 1000–1002.
116. Denton, C.P.; Ong, V.H.; Xu, S.; Chen-Harris, H.; Modrusan, Z.; Lafyatis, R.; Khanna, D.; Jahreis, A.; Siegel, J.; Sornasse, T. Therapeutic interleukin-6 blockade reverses transforming growth factor-beta pathway activation in dermal fibroblasts: Insights from the faSScinate clinical trial in systemic sclerosis. *Ann. Rheum. Dis.* **2018**, *77*, 1362–1371. [CrossRef]
117. Shima, Y.; Kuwahara, Y.; Murota, H.; Kitaba, S.; Kawai, M.; Hirano, T.; Arimitsu, J.; Narazaki, M.; Hagihara, K.; Ogata, A.; et al. The skin of patients with systemic sclerosis softened during the treatment with anti-IL-6 receptor antibody tocilizumab. *Rheumatology* **2010**, *49*, 2408–2412. [CrossRef]
118. Zacay, G.; Levy, Y. Outcomes of patients with systemic sclerosis treated with tocilizumab: Case series and review of the literature. *Best Pr. Res. Clin. Rheumatol.* **2018**, *32*, 563–571. [CrossRef]

119. Elhai, M.; Meunier, M.; Matucci-Cerinic, M.; Maurer, B.; Riemekasten, G.; Leturcq, T.; Pellerito, R.; Von Mühlen, C.A.; Vacca, A.; Airo, P.; et al. Outcomes of patients with systemic sclerosis-associated polyarthritis and myopathy treated with tocilizumab or abatacept: A EUSTAR observational study. *Ann. Rheum. Dis.* **2013**, *72*, 1217–1220. [CrossRef]
120. Fernández-Codina, A.; Walker, K.M.; Pope, J.E. Treatment Algorithms for Systemic Sclerosis According to Experts. *Arthritis Rheumatol.* **2018**, *70*, 1820–1828. [CrossRef]
121. Khanna, D.; Lin, C.J.F.; Furst, D.E.; Goldin, J.; Kim, G.; Kuwana, M.; Allanore, Y.; Matucci-Cerinic, M.; Distler, O.; Shima, Y.; et al. Tocilizumab in systemic sclerosis: A randomised, double-blind, placebo-controlled, phase 3 trial. *Lancet Respir. Med.* **2020**, *8*, 963–974. [CrossRef]
122. Khanna, D.; Jahreis, A.; Furst, D.E. Tocilizumab Treatment of Patients with Systemic Sclerosis: Clinical Data. *J. Scleroderma Relat. Disord.* **2017**, *2*, S29–S35. [CrossRef]
123. Roofeh, D.; Lin, C.J.F.; Goldin, J.; Kim, G.H.; Furst, D.E.; Denton, C.P.; Huang, S.; Khanna, D. Tocilizumab Prevents Progression of Early Systemic Sclerosis–Associated Interstitial Lung Disease. *Arthritis Rheumatol.* **2021**, *73*, 1301–1310. [CrossRef]
124. Narváez, J.; Lluch, J.; Alegre Sancho, J.J.A.; Molina-Molina, M.; Nolla, J.M.; Castellvi, I. Effectiveness and safety of tocilizumab for the treatment of refractory systemic sclerosis associated interstitial lung disease: A case series. *Ann. Rheum. Dis.* **2019**, *78*, e123. [CrossRef]
125. Khanna, D.; Lin, C.J.F.; Furst, D.E.; Wagner, B.; Zucchetto, M.; Raghu, G.; Martinez, F.J.; Goldin, J.; Siegel, J.; Denton, C.P.; et al. Long-Term Safety and Efficacy of Tocilizumab in Early Systemic Sclerosis–Interstitial Lung Disease: Open Label Extension of a Phase 3 Randomized Controlled Trial. *Am. J. Respir. Crit. Care Med.* **2021**, *204*, 1245–1252. [CrossRef]
126. Panopoulos, S.T.; Tektonidou, M.G.; Bournia, V.-K.; Arida, A.; Sfikakis, P.P. Anti-interleukin 6 Therapy Effect for Refractory Joint and Skin Involvement in Systemic Sclerosis: A Real-world, Single-center Experience. *J. Rheumatol.* **2022**, *49*, 68–73. [CrossRef]
127. Kahan, A.; Allanore, Y. Primary myocardial involvement in systemic sclerosis. *Rheumatology* **2006**, *45*, iv14–iv17. [CrossRef]
128. Meléndez, G.C.; McLarty, J.L.; Levick, S.P.; Du, Y.; Janicki, J.S.; Brower, G.L. Interleukin 6 Mediates Myocardial Fibrosis, Concentric Hypertrophy, and Diastolic Dysfunction in Rats. *Hypertension* **2010**, *56*, 225–231. [CrossRef]
129. Ishizaki, Y.; Ooka, S.; Doi, S.; Kawasaki, T.; Sakurai, K.; Mizushima, M.; Kiyokawa, T.; Takakuwa, Y.; Tonooka, K.; Kawahata, K. Treatment of myocardial fibrosis in systemic sclerosis with tocilizumab. *Rheumatology* **2020**, *60*, e205–e206. [CrossRef]
130. Adrovic, A.; Yildiz, M.; Haslak, F.; Koker, O.; Aliyeva, A.; Sahin, S.; Barut, K.; Kasapcopur, O. Tocilizumab therapy in juvenile systemic sclerosis: A retrospective single centre pilot study. *Rheumatol. Int.* **2021**, *41*, 121–128. [CrossRef]

Review

Novel Concepts in Systemic Sclerosis Pathogenesis: Role for miRNAs

Iulia Szabo [1], Laura Muntean [1,2,*], Tania Crisan [3,4], Voicu Rednic [5,6], Claudia Sirbe [1] and Simona Rednic [1,2]

1 Department of Rheumatology, "Iuliu Hațieganu" University of Medicine and Pharmacy Cluj-Napoca, 400000 Cluj-Napoca, Romania; rotaru.iulia@umfcluj.ro (I.S.); claudia.sirbe@yahoo.com (C.S.); srednic.umfcluj@gmail.com (S.R.)
2 Department of Rheumatology, County Emergency Hospital Cluj-Napoca, 400000 Cluj-Napoca, Romania
3 Department of Medical Genetics, "Iuliu Hațieganu" University of Medicine and Pharmacy Cluj-Napoca, 400000 Cluj-Napoca, Romania; tania.crisan@gmail.com
4 Department of Internal Medicine and Radboud Institute for Molecular Life Sciences (RIMLS), Radboud University Medical Center, 6525 GA Nijmegen, The Netherlands
5 Department of Gastroenterology, "Iuliu Hațieganu" University of Medicine and Pharmacy Cluj-Napoca, 400000 Cluj-Napoca, Romania; rednicvoicu@gmail.com
6 Department of Gastroenterology II, "Prof. Dr. Octavian Fodor" Regional Institute of Gastroenterology and Hepatology, 400000 Cluj-Napoca, Romania
* Correspondence: lmuntean13@yahoo.com

Abstract: Systemic sclerosis (SSc) is a rare connective tissue disease with heterogeneous clinical phenotypes. It is characterized by the pathogenic triad: microangiopathy, immune dysfunction, and fibrosis. Epigenetic mechanisms modulate gene expression without interfering with the DNA sequence. Epigenetic marks may be reversible and their differential response to external stimuli could explain the protean clinical manifestations of SSc while offering the opportunity of targeted drug development. Small, non-coding RNA sequences (miRNAs) have demonstrated complex interactions between vasculature, immune activation, and extracellular matrices. Distinct miRNA profiles were identified in SSc skin specimens and blood samples containing a wide variety of dysregulated miRNAs. Their target genes are mainly involved in profibrotic pathways, but new lines of evidence also confirm their participation in impaired angiogenesis and aberrant immune responses. Research approaches focusing on earlier stages of the disease and on differential miRNA expression in various tissues could bring novel insights into SSc pathogenesis and validate the clinical utility of miRNAs as biomarkers and therapeutic targets.

Keywords: systemic sclerosis; pathogenesis; epigenetic mechanisms; miRNAs

Citation: Szabo, I.; Muntean, L.; Crisan, T.; Rednic, V.; Sirbe, C.; Rednic, S. Novel Concepts in Systemic Sclerosis Pathogenesis: Role for miRNAs. *Biomedicines* **2021**, *10*, 1471. https://doi.org/10.3390/biomedicines9101471

Academic Editor: Estefania Nuñez

Received: 12 September 2021
Accepted: 8 October 2021
Published: 14 October 2021

1. Introduction

Systemic sclerosis (SSc) is a rare autoimmune disease with miscellaneous clinical manifestations and a distinct autoantibody profile [1,2]. It is characterized by high morbidity and mortality related to the extent of fibrosis and obliterative vasculopathy of the internal organs [3–5]. The etiology of SSc is not fully unraveled, but evidence supports a complex interaction between genetic variants, environmental exposures, and epigenetic modifications [6]. The modest effect size of SSc-associated genetic risk loci shifted the interest of the scientific community toward the contribution of epigenetics to disease predisposition and its complex pathogenesis [7–9].

SSc pathophysiology is distinguished by the interaction between three main altered pathways: microangiopathy, immune dysfunction, and fibrosis [10–12]. An inaugural vascular injury [13–15] leads to activation of cell-mediated and humoral immune responses [16,17], subsequently resulting in fibroblast to myofibroblast differentiation [18] with production and deposition of collagen and other extracellular matrix (ECM) components into the vascular walls, skin, and internal organs [19–21].

Epigenetics refers to the modulation of gene expression through heritable and reversible alterations of the chromatin structure without interfering with the DNA sequence. Epigenetic mechanisms have previously been linked to the pathogenesis of SSc, extensively reviewed elsewhere [22–26]. Evidence of association to SSc has previously been reported for all major epigenetic alterations, including DNA methylation [27–31], histone modifications [32–34], non-coding small (miRNA), and long (lncRNA) RNA transcripts [35–37]. Epigenetic mechanisms regulate diverse physiological processes such as cell division and differentiation, growth, and development, being responsible at least in part for the variable phenotypic traits in both health and disease [7,9,37]. The epigenome is susceptible to change and can be influenced by various environmental factors, including air pollution, infection, diet, drugs, metals, and chemicals [38–40].

DNA methylation is an enzyme-mediated process occurring mostly at the CpG sites where cytosine is located in the vicinity of guanidine in the nucleotide sequence of the DNA structure. DNA methyltransferases (DNMTs) catalyze the addition of a methyl (CH3) group to the 5-carbon of the cytosine ring, generating 5-methylcytosine (5-mC). The methylation status (5-mC content) of a CpG island (cluster of CpG sites) in the promoter region of a gene modulates gene transcription. This translates into either gene-silencing if highly methylated or active gene transcription in low methylated states [24,36,41]. Histone modifications refer to post-translational alterations (such as methylation, acetylation, phosphorylation, ubiquitylation, or sumoylation) of the histone proteins, which alter their interaction with the DNA strand. The subsequent conformational changes in the chromatin architecture make the DNA more or less accessible to transcriptional factors, resulting in activation or repression of gene transcription [42,43]. Non-coding RNAs, miRNAs (<30 nucleotides) [44,45], and lncRNAs (>200 nucleotides) [46,47] are functional regulators of gene expression at the transcriptional and post-transcriptional level. These RNA fragments are transcribed from the DNA but are not translated into proteins [22,23,26,48].

MiRNAs, the focus of this review, bind post-transcriptionally to a complementary sequence from a target mRNA and induce gene silencing. This can be achieved by blocking mRNA translation or promoting mRNA cleavage based on the degree of complementarity [24,37]. MiRNAs have the ability of regulating multiple mRNA targets, whereas translation of one mRNA transcript into protein can be modulated by various miRNAs [49,50] (Figure 1).

Upregulation or downregulation of diverse miRNAs has been identified in blood samples and tissue biopsies from patients with SSc [34,51]. MiRNAs involved in fibrosis received particular attention compared to the scarce data on immune disfunction and vasculopathy [52]. An even more attractive aspect besides a better understanding of the contribution to disease pathogenesis is their potential use as diagnostic and prognostic markers as well as the possibility of developing targeted therapies [53].

The purpose of this review is to illustrate the current knowledge on the role of miRNAs in modulating the three main pathogenic pathways in SSc as well as depicting their clinical utility as biomarkers and therapeutic targets.

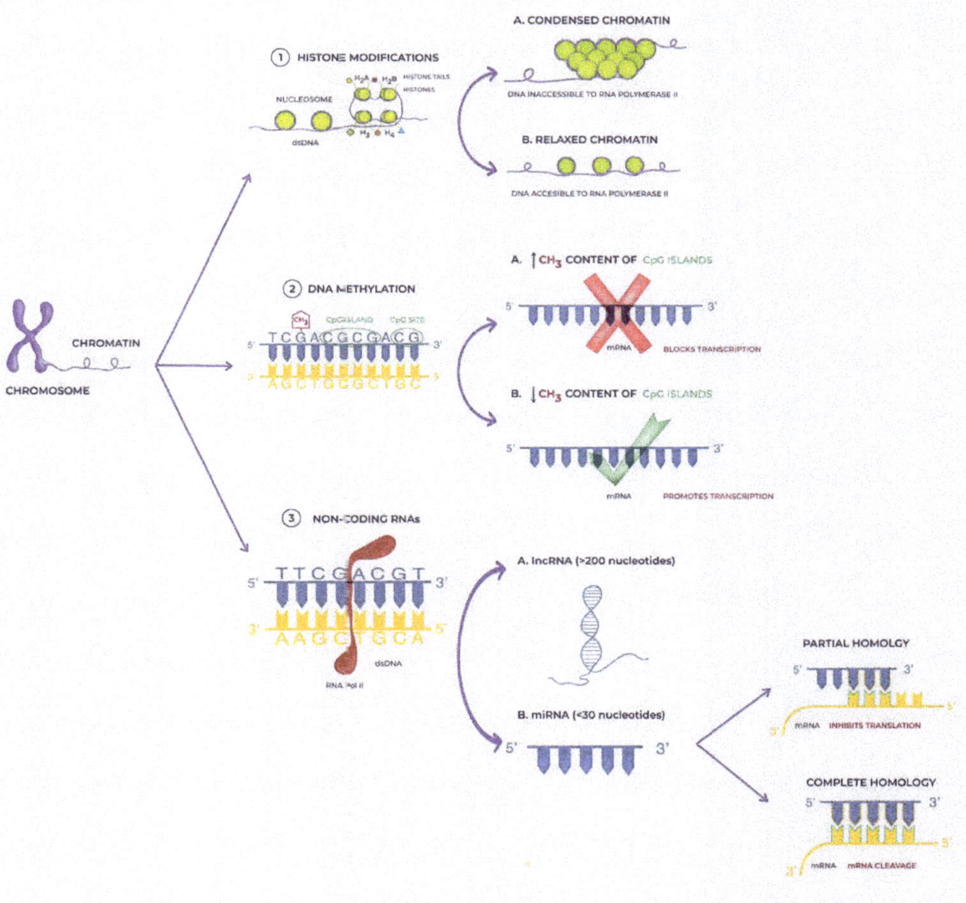

Figure 1. Illustration of epigenetic mechanisms. This figure is a schematical representation of the epigenetic mechanisms that modulate gene expression: (1) *Histone modifications* refer to post-translational modifications of the histone proteins leading to conformational changes that make DNA more or less accessible to RNA polymerase II (RNA POL II); (2) *DNA methylation* is an enzyme-mediated process consisting of the addition of a methyl (CH3) group to the 5-carbon of the cytosine ring from a CpG site. Clusters of CpG sites form a CpG island. The methylation status of a CpG island located in the promotor region of a gene can either lead to gene silencing if highly methylated or active gene transcription if slightly methylated; (3) *Non-coding RNAs* (lncRNAs and miRNAs) are functional RNA fragments transcribed from the DNA by RNA Pol II but unable to be translated into proteins. LncRNAs possess diverse functions, such as the capacity of altering mRNA splicing or recruiting chromatin remodeling proteins and transcription factors. MiRNAs have the ability to bind post-transcriptionally to a complementary sequence from a target mRNA and induce gene silencing. Depending on the degree of homology they can either inhibit transcription or induce mRNA cleavage.

2. Serum- and Tissue-Specific miRNA Signatures in SSc

Multiple studies have aimed at identifying miRNAs involved in the pathogenesis of SSc and their potential as diagnostic or prognostic biomarkers, as well as therapeutic targets. In this regard, Zhu H (2012) identified a plethora of miRNAs differentially expressed in SSc skin biopsies compared to healthy controls (HC). The miRNA profiles differed between the limited (lcSSc) and diffuse (dcSSc) clinical subtypes. Twenty-one miRNAs overlapped between the two SSc subgroups, out of which six (miR-21, miR-31, miR-503, miR-146, miR-29b, miR-145) were predicted to target mRNAs involved in fibrosis. Further, the analysis was restricted to the TGF-β-associated genes and the miRNAs that regulate their expression levels in both skin specimens and SSc fibroblasts: SMAD7 (miR-21 predicted target), SMAD3 (miR-145 predicted target), and COL1A1 (miR-29b predicted target). In these samples, miR-21 increased levels were mirrored by SMAD7 downregulation, whereas miR-145 and miR29b decreased levels were associated with SMAD3 and COL1A1 upregulation. Stimulation of healthy dermal fibroblasts with recombinant TGF-β resulted in increased miR-21/decreased SMAD7, increased miR-145/decreased SMAD3, and decreased miR-29b/increased COL1A1 levels, suggesting that these miRNAs do not directly control their target mRNAs [34].

Interestingly, these miRNAs were not reproduced in the study conducted by Li (2012). By means of miRNA microarray analysis, 24 miRNAs were identified as being differentially expressed in SSc skin samples. Results were confirmed by real-time PCR. Target genes with a known role in SSc pathogenesis were identified for six miRNAs (hsa-miR-206, hsa-miR-133a, hsa-miR-125b, hsa-miR-140-5p, hsa-miR-23b, hsa-let-7g) using bioinformatics analysis. Hsa-miR-206 received particular attention as it regulates an impressive number of genes, 15 of them being correlated with SSc pathogenesis [54].

As expected, miRNAs identified in SSc serum samples differ from tissue miRNAs. Steen (2015) proposed a circulating miRNA signature in a large cohort of 189 patients. The study included 120 SSc patients, 29 systemic lupus erythematosus (SLE) patients, and 40 HC. From the 37 identified miRNAs, 19 were significantly dysregulated (14 miRNAs decreased and five miRNAs increased). Quantitative PCR reflected the main differences between SSc patients and HC with respect to the expression of the miRNA 17~92 cluster, as well as miR-16, miR-223, and miR-638. Predicted targets of these miRNAs are mRNAs involved in different fibrotic pathways, including TGF-β [51].

A different miRNA circulating profile was demonstrated by means of microarray analysis in the serum of 10 SSc patients compared to six HC in a study by Rusek et al. (2019). Out of the 15 miRNAs differentially expressed, miR-4484 was remarkably increased (18-fold). Bioinformatics analysis suggested miR-4484 as a potential regulator of fibrosis through the identification of a wide range of target genes involved in the TGF-β/SMAD and Wnt/β-catenin signaling pathways, as well as collagen synthesis and extracellular matrix (ECM) homeostasis [45]. Matrix metalloproteinase-21 (MMP-21), even though not a direct target gene according to computational analysis, was hypothesized to be up-regulated by miR-4484 due to their close chromosomal vicinity and the increased MMP-21 serum levels. These findings further enabled the authors to suggest that miR-4484 and MMP-21 might play a role in SSc pathogenesis and proposed them as serum biomarkers [45].

Another relevant aspect is that circulating miRNA profiles are able to discriminate between SSc clinical subtypes (lcSSc versus dcSSc) and autoantibody specificities, as shown by Wuttge (2015). Out of 45 selected miRNAs, four miRNAs (miR-223, miR-181b, miR-342-3p, miR-184) consistently exhibited different expression levels in the lcSSc and dcSSc subgroups. In the autoantibody subgroups, five miRNAs (miR-409, miR-184, miR-92a, miR-29a, miR-101) showed statistically different expression levels [55].

A distinct miRNA signature in SSc and idiopathic pulmonary fibrosis (IPF) lung fibroblasts was expressed in the experiment led by Mullenbrock (2018). The author proved that various miRNAs were differentially expressed compared to controls. To validate their function, transfection of miR-29b-3p, miR-138-5p, and miR-146b-5p mimics was performed and their effects on gene expression were quantified using a Nanostring fibrosis panel. One hundred seventy-five pro-fibrotic target genes were consequently downregulated in the SSc and IPF lung fibroblasts, supporting a role for miR-29b-3p, miR-138-5p, and miR-146b-5p in fibrosis in these disease models [56].

3. MiRNAs: Culprits in SSc Pathogenesis

3.1. Profibrotic miRNA Transcripts

Zhu H (2012) identified altered expression levels of miR-21, miR-145, and miR-29b in SSc skin and cultured fibroblasts [34]. A further study from the same group explored the expression levels of miR-21 and its target gene Smad7 in SSc and bleomycin-treated mice skin biopsies. They validated miR-21 as an important regulator of the TGF-β signaling pathway through the manipulation of its direct target, SMAD7. On the one hand, TGF-β fibroblast stimulation induced upregulation of miR-21 and downregulation of Smad7, and on the other hand, transfection of small interfering RNA (siRNA) decreased Smad7 protein levels. Smad7 is a negative regulatory component of the TGF-β signaling pathway. Therefore, decreased levels of Smad7 will have the opposite effect by stimulating fibrosis. Similar results were obtained in the bleomycin-treated mice with upregulation of miR-21 and downregulation of Smad7. After treatment with bortezomib, miR-21 decreased, Smad7 levels were restored, and skin fibrosis improved [57]. The same profibrotic phenotype of miR-21 was demonstrated by S. Jafarinejad-Farsangi (2019). Upregulation of miR-21 was observed in both diffuse cutaneous SSc (dcSSc) and TGF-β-stimulated fibroblasts, leading to increased type I collagen production [58].

Ly (2020) has recently proven that miR-145 mediates α-smooth muscle actin (α-SMA) myofibroblast differentiation through downregulation of transcription factor Kruppel-like factor 4 (KLF4) in TGF-β1-stimulated dermal fibroblasts and SSc fibroblasts. KLF4 has a prohibitory effect on the XYLT1 gene. XYLT1 encodes xylosyltransferase-1 (XT-1), a proteoglycan synthesis biomarker. Experiments revealed that exogenous delivery of KLF4 siRNA into normal human fibroblasts led to downregulation of KLF4 mRNA levels and upregulation of XYLT1 expression levels in a dose-dependent manner in response to TGF-β1. The same trend was identified in SSc fibroblasts, therefore leading to the identification of a new miR-145/KLF4 profibrotic pathway [59].

Another validated profibrotic miRNA is miR-92a. Transfection of miR-92a mimics in normal fibroblasts resulted in decreased expression levels of matrix metalloproteinase-1 (MMP-1). MiR-92a upregulation in SSc fibroblasts and serum from SSc patients might be a consequence of TGF-β endogenous activation as increased miR-92a levels were evidenced in normal dermal fibroblasts stimulated with TGF-β and decreased expression levels were shown after inhibition of TGF-β with siRNA [60]. MMP-1 is also the target gene for another profibrotic miRNA, miR-202-3p, as shown by Zhou (2017). In SSc skin samples and cultured fibroblasts, miR-202-3p was upregulated and MMP-1 was downregulated. Luciferase reporter assays identified MMP-1 as the target gene for miR-202-3p and gain and loss of function assays showed that in SSc fibroblasts MMP-1 was regulated by miR-202-3p [61].

Nakayama (2017) showed that miR-4458 plays a decisive role in type I collagen production via the IL-23 immune pathway, therefore indicating IL-23 as an important factor in SSc fibrogenesis and a possible therapeutic target. In normal fibroblasts, IL-23 stimulation leads to increased miR-4458 levels and downregulation of type I collagen production. Conversely, IL-23 stimulation of SSc fibroblasts also prompts miR-4458 upregulation, but the effect at the protein level is enhanced type I collagen synthesis [62].

MiR-155 also proved to play a role in SSc fibrogenesis by regulating Wnt/β catenin and Akt profibrotic pathways. This finding was illustrated after transfection of mouse fibroblasts with miR-155 inhibitor, which resulted in increased degradation of β-catenin, decreased phosphorylation of Akt, and, subsequently, decreased type I collagen production. In bleomycin-treated miR-155 knockout mice and after topical administration of antagomiR-155 in bleomycin-induced fibrosis mouse models, decreased protein levels of β-catenin and pAkt were evidenced. Additionally, improvement of skin fibrosis was noted, therefore supporting the therapeutic potential of miR-155 inhibition [63]. Christmann (2016) further suggested miR-155 as a potential therapeutic target since miR-155 knockout mice exhibited less aggressive lung involvement and better survival rates after bleomycin administration compared to wild-type controls [64]. Additionally, the same group suggested a promising role for miR-155 as a prognostic biomarker in SSc-ILD due to its correlation with higher high-resolution computed tomography (HRCT) fibrosis scores and lower performances on pulmonary function tests (PFTs) [64].

Data from Artlett (2017) showed that miR-155 expression levels depend upon inflammasome activation. The study depicted the strong link between inflammasome activation, miR-155 expression, and collagen synthesis in SSc fibroblasts and bleomycin mouse models. Inflammasome inhibition in SSc fibroblasts via caspase-1 inhibitor determined downregulation of miR-155 and decreased collagen production. Fibroblasts from NLRP3 knockout mice did not exhibit enhanced miR-155 expression levels after stimulation with bleomycin, showing that miR-155 expression cannot be achieved without inflammasome activation [65].

The study by Henderson (2021) validated miR27a-3p as a profibrotic epigenetic direct regulator of the sFRP-1 protein, a Wnt pathway antagonist. Transfection of miR27a-3p mimic in TGF-β1-stimulated normal dermal fibroblasts induced COL1A1 and Axin-2 upregulation, as well as downregulation of the antifibrotic PPARγ mRNA and decline in MMP-1 protein levels. The authors also revealed decreased sFRP-1 protein levels in the serum and skin biopsies of early dcSSc patients and increased miR27a-3p expression levels in SSc dermal fibroblasts. A 33% drop in collagen synthesis resulted following exogenous delivery of antagomiR27a-3p in sFRP-1-depleted SSc dermal fibroblasts. These results suggest a role for miR27a-3p in SSc fibrosis [66]. Another study from the same group confirmed miR33a-3p as an additional epigenetic regulator of the Wnt pathway through direct repression of Dickkopf-1 (DKK-1) mRNA translation. MiR33a-3p was increased in SSc fibroblasts, whereas DKK-1 was decreased. AntagomiR33a-3p transfection into SSc fibroblasts led to a significant reduction in collagen 1 synthesis, again supporting a profibrotic role for this miRNA in SSc pathogenesis [67].

MiR-483-5p displayed a profibrotic phenotype in SSc. Serum levels of miR-483-5p are elevated in such patients. Transfection of miR-483-5p mimics in primary human fibroblasts and pulmonary endothelial cells caused increased synthesis of type IV collagen via modulation of COL4A1 and COL4A2 target genes. Transfection of miR-483-5p in endothelial cells also increased the expression levels of αSMA and SM22A mRNA, suggesting that miR-483-5p orchestrates the myofibroblast differentiation of endothelial cells [44].

Table 1 summarizes the main profibrotic miRNAs identified so far along with their targeted genes.

Table 1. Profibrotic miRNAs involved in SSc pathogenesis.

miRNA	Expression	Tissue Specimen(s)	Target Gene(s)	Reference(s)
miR-21	Upregulated	Fibroblasts Skin Bleomycin-treated mice skin samples	SMAD7	Zhu et al. [34,57] Jafarinejad-Farsangi et al. [58]
miR-145	Upregulated	Fibroblasts TGF-β1-stimulated fibroblasts	KLF4	Ly et al. [59]
miR-92a	Upregulated	Fibroblasts Serum TGF-β-stimulated fibroblasts	MMP1	Sing et al. [60]
miR-202-3p	Upregulated	Fibroblasts Skin	MMP1	Zhou et al. [61]
miR-4458	Upregulated	Fibroblasts	Unknown	Nakayama et al. [62]
miR-155	Upregulated	Fibroblasts Skin Serum	CSNK1A1 SHIP1	Yan et al. [63] Christmann et al. [64] Artlett et al. [65]

Table 1. *Cont.*

miRNA	Expression	Tissue Specimen(s)	Target Gene(s)	Reference(s)
miR-27a-3p	Upregulated	Fibroblasts Skin Serum	sFRP-1	Henderson et al. [66]
miR-33a-3p	Upregulated	Fibroblasts	DKK-1	Henderson et al. [67]

KLF4: Kruppel-like factor 4; MMP1: matrix metalloproteinase 1; sFRP-1: secreted frizzled-related protein-1; DKK-1: Dickkopf-1.

3.2. Antifibrotic miRNA Transcripts

TGF-β, a promoter of collagen synthesis and fibroblast proliferation and differentiation, plays a central role in SSc pathogenesis [68]. TGF-β signaling is mediated through its receptors, TGF-β receptor type 1 (TGFBR1) and type 2 (TGFBR2) [69]. Numerous in vitro and in vivo experiments have shown that TGFBR2 is involved in dermal and internal organ fibrosis [70–72]. From that perspective, Shi (2018) has demonstrated that TGFBR2 upregulation in SSc dermal fibroblasts and in dermal biopsies is a direct consequence of miR-3606-3p downregulation. Additionally, transfection of miR-3606-3p mimics in SSc dermal fibroblasts resulted in a reduction of TGFBR2 expression, as well as reduced p-SMAD2/3 and type I collagen protein levels [73]. MiR-3606-3p silencing of the TGFBR2 mRNA could represent a new therapeutic strategy in SSc.

Besides the profibrotic phenotype displayed by miR-4458, Nakayama (2017) likewise showed that miR-18a influences type I collagen production. In normal fibroblasts, IL-23 stimulation led to decreased miR-18a expression levels and downregulation of type I collagen synthesis. In contrast, IL-23 stimulation of SSc fibroblasts caused miR-18a downregulation and increased type I collagen synthesis. This paradox is explained by strong downregulation of miR-18a, a potent antifibrotic miRNA, due to intrinsic activation of TGF-β in SSc fibroblasts. The profibrotic activity of IL-23 was subsequently demonstrated by accelerated skin fibrosis after IL-23 injection of bleomycin-treated mice [62].

Five members of the let-7 family were dysregulated in SSc and localized scleroderma (LSc) skin samples compared to normal controls and keloid skin specimens. Let-7a was significantly downregulated in scleroderma tissues, with lower levels in the LSc group compared to the SSc group. TGF-β1 stimulation of normal fibroblasts resulted in decreased expression levels of Let-7a and increased production of type I collagen, suggesting that downregulation of Let-7a might mitigate the overexpression of extracellular matrices, mainly the secretion of type I collagen. This hypothesis was validated by transfection of the Let-7a inhibitor in the normal fibroblasts, which led to increased production of type I collagen. Serum levels of Let-7a were also downregulated and the same trend of lower levels in the LSc subset compared to the SSc subset was maintained. Injection of Let-7a in bleomycin-induced fibrosis mouse models resulted in improvement of skin fibrosis [74].

Maurer (2010) demonstrated significant downregulation of miR-29a in SSc-cultured fibroblasts, SSc skin biopsies, and bleomycin-induced fibrosis mouse models. In order to validate its function and role in SSc fibrogenesis, transfection of pre-miR-29a/29b/29c in SSc fibroblasts was conducted. This manipulation led to downregulation of type I collagen and markedly decreased expression levels of type III collagen being observed after pre-miR-29a transfection. Conversely, transfection of anti-miR-29a in normal fibroblasts determined upregulation of type I and type III collagens. COL3A1 proved to be a direct target of miR-29a after cotransfection of HEK 293 cells with pre-miR-29a and pGL3 luciferase reporter containing the 3′-UTR of COL3A1. Cotransfection resulted in reduced relative luciferase activity, whereas cotransfection with anti-miR-29a and pGL3 luciferase reporter led to enhanced relative luciferase activity. The group subsequently analyzed the influence of several profibrotic cytokines, namely TGF-β, PDGF-B, and IL-4, on miR-29a expression. They demonstrated that stimulation of normal fibroblasts with these molecules resulted in downregulation of miR-29a similar to levels seen in SSc fibroblasts, whereas inhibition of TGF-β and PDGF-B pathways with imatinib restored miR-29a levels in SSc fibroblasts as well as bleomycin-induced skin fibrosis. Given the direct regulation of collagen genes by miR-29a, this miRNA could be a potential antifibrotic therapeutic target [75]. In a recent study by Jafarinejad-Farsangi (2019), transfection of miR-29a mimics significantly reduced collagen type I expression levels in SSc and TGF-β-stimulated fibroblasts, further supporting the antifibrotic role of miR-29a [58].

Similarly, Ciechomska (2014) validated TAB1 as another target gene for miR-29a, demonstrating an important role for this miRNA in SSc fibrosis. Transfection of miR-29a in normal fibroblasts led to downregulation of TIMP-1 and upregulation of MMP-1, resulting in decreased extracellular matrix deposition. Bioinformatics analysis identified TAB1 as a possible target gene for miR-29a. Validation of TAB1 was performed through cotransfection of HeLa cells with pre-miR-29a and TAB1 3′UTR luciferase reporter. Luciferase analysis showed a 20% reduction in luciferase activity after cotransfection. Subsequently, pharmacological inhibition of TBA1 or transfection of anti-TAB1 siRNA in normal fibroblasts resulted in TIMP-1 reduction, demonstrating that TAB1 plays a key role in the regulation of TIMP-1 expression levels [76].

Honda (2013) depicted miR-150 as an antifibrotic miRNA that mediates its effects via integrin β3 inhibition. Integrin β3 is an adhesion molecule that is supposed to play an important role in the endogenous TGF-β activation in SSc fibroblasts. In SSc skin and cultured fibroblasts, low miR-150 levels and high integrin β3 levels were identified. Transfection of miR-150 mimics in SSc fibroblasts resulted in decreased integrin β3, phosphorylated SMAD3, and type I collagen, while on the contrary miR-150 antisense inhibition in normal fibroblasts caused enhanced expression of the aforementioned molecules [77].

PDGF receptor β is the target gene for miR-30b. Tanaka (2013) demonstrated that miR-30b was repressed in SSc serum samples. Decreased levels were also seen in SSc skin specimens and experimental mouse models, whereas PDGFR-β was highly expressed in SSc fibroblasts compared to controls. Hence, downregulation of miR-30b leads to a profibrotic phenotype via enhanced expression of the PDGFR-β [78].

MiR-135b and miR-196a are validated antifibrotic miRNAs. O'Reilly (2016) proved that IL-13 signaling leads to increased extracellular matrix deposition in SSc fibroblasts through regulation of the signal transducer and activator of transcription-6 (STAT6). IL-13-induced downregulation of miR-135b results in upregulation of STAT6 and increased collagen synthesis [79]. The involvement of epigenetics in SSc fibrosis is also illustrated by Makino (2013). Regulation of discoidin domain receptor 2 (DDR2) mRNA and protein level is accomplished through negative feedback: decreased DDR2 stimulates miR-196a expression and decreased collagen synthesis in normal fibroblasts. In SSc fibroblasts, this feedback is incompetent due to downregulation of miR-196a by endogenous activation and downstream signaling of TGF-β, generating enhanced collagen production [80].

MiR-125b modulates both the activation of fibroblasts into myofibroblasts and fibroblast apoptosis. It exerts a tissue dependent effect as seen with cancer and cardiac fibrosis [81,82]. Kozlova (2019) demonstrated that miR-125b is downregulated in SSc dermal fibroblasts and skin samples. This leads to enhanced fibroblast apoptosis through induction of apoptosis genes BAK1, BMF, and BBC3, but also reduces fibroblast proliferation and differentiation as shown by decreased αSMA mRNA expression and protein levels. Hence, miR-125b plays a protective, antifibrotic role in SSc pathogenesis [83].

MiR-16-5p inhibits tissue fibrosis by repressing myofibroblast activation through direct inhibition of NOTCH2 expression. Yao (2020) revealed that transfection of antogomiR-16-5p in cultured skin fibroblasts led to a rise in the levels of several profibrotic markers, such as COL1A1, COL1A2, connective tissue growth factor (CTGF), as well as α-SMA, a marker of myofibroblast differentiation. On the contrary, MMP-1 and matrix metalloproteinase-8 (MMP-8) levels were decreased in response to miR-16-5p inhibition. Additional exogenous delivery of siNOTCH2 partially reversed the expression of the abovementioned biomarkers. Decreased miR-16-5p and increased NOTCH2 expression levels were identified in SSc serum samples, suggesting that miR-16-5p interferes in SSc pathogenesis by modulating fibroblast to myofibroblast differentiation [84].

Table 2 outlines the main characteristics of the antifibrotic miRNAs identified in patients with SSc.

Table 2. Antifibrotic miRNAs involved in SSc pathogenesis.

miRNA	Expression	Tissue Specimen(s)	Target Gene(s)	Reference(s)
miR-145	Downregulated	Fibroblasts Skin	SMAD3	Zhu et al. [34]
miR-29b	Downregulated	Fibroblasts Skin	COL1A1	Zhu et al. [34]
miR-let-7a	Downregulated	Fibroblasts Skin Serum	Unknown	Makino et al. [74]
miR-29a	Downregulated	Fibroblasts Skin Bleomycin-treated mice skin samples	COL1A1 COL3A1 TAB1	Maurer et al. [75] Jafarinejad-Farsangi et al. [58] Ciechomska et al. [76]
miR-3606-3p	Downregulated	Fibroblasts Skin	TGFBR2	Shi et al. [73]
miR-18a	Downregulated	Fibroblasts	Unknown	Nakayama et al. [62]
miR-150	Downregulated	Fibroblasts Skin	ITGB3	Honda et al. [77]
miR-30b	Downregulated	Serum Experimental mouse model	Unknown	Tanaka et al. [78]
miR-135b	Downregulated	Fibroblasts Serum Monocytes	STAT6	O'Reilly et al. [79]
miR-16-5p	Downregulated	Fibroblasts Serum	NOTCH2	Yao et al. [84]

COL1A1: collagen type 1 alpha 1 chain; COL3A1: collagen type 3 alpha 1 chain; TAB1: transforming growth factor beta activated protein kinase 1; TGFBR2: transforming growth factor beta receptor 2.

Figure 2 is an illustration of the regulatory effects of various profibrotic and antifibrotic miRNAs involved in SSc tissue fibrosis.

3.3. Apoptosis and miRNAs

SSc fibroblasts are resistant to apoptosis. This dysfunctional programmed cell death further contributes to increased extracellular matrix deposition [85]. Two members of the Bcl-2 family, namely Bax and Bcl-2, control apoptosis. Mir-29a and miR-21 regulate the expression levels of Bax and Bcl-2 [86,87].

Accordingly, Jafarinejad-Farsangi (2015) demonstrated the proapoptotic role of miR-29a in SSc and TGF-β-stimulated fibroblasts through regulation of the expression levels of the Bcl-2 family members. MiR-29a downregulates the antiapoptotic Bcl-2 and Bcl-XL proteins, therefore increasing the Bax:Bcl-2 ratio. An elevated Bax:Bcl-2 ratio translates into enhanced apoptosis. The dual properties of mir-29a, both antifibrotic and proapoptotic, make it an excellent therapeutic target [86].

In a subsequent study, Jafarinejad-Farsangi (2016) confirmed the increased Bcl-2 levels in SSc fibroblasts and the resulting decreased Bax:Bcl-2 ratio. This confers resistance to apoptosis, as previously demonstrated [86]. Transfection of SSc fibroblasts with miR-21 mimics additionally upregulated Bcl-2 levels and lowered the Bax:Bcl-2 ratio, supporting miR-21 as an antiapoptotic factor. On the contrary, transfection of miR-21 inhibitor increased Bax expression levels and consequently enhanced apoptosis. MiR-21 inhibition is an attractive therapeutic target in inducing apoptosis and reversing fibrosis in SSc [87].

Table 3 is a schematic representation of the miRNAs that modulate apoptosis in SSc.

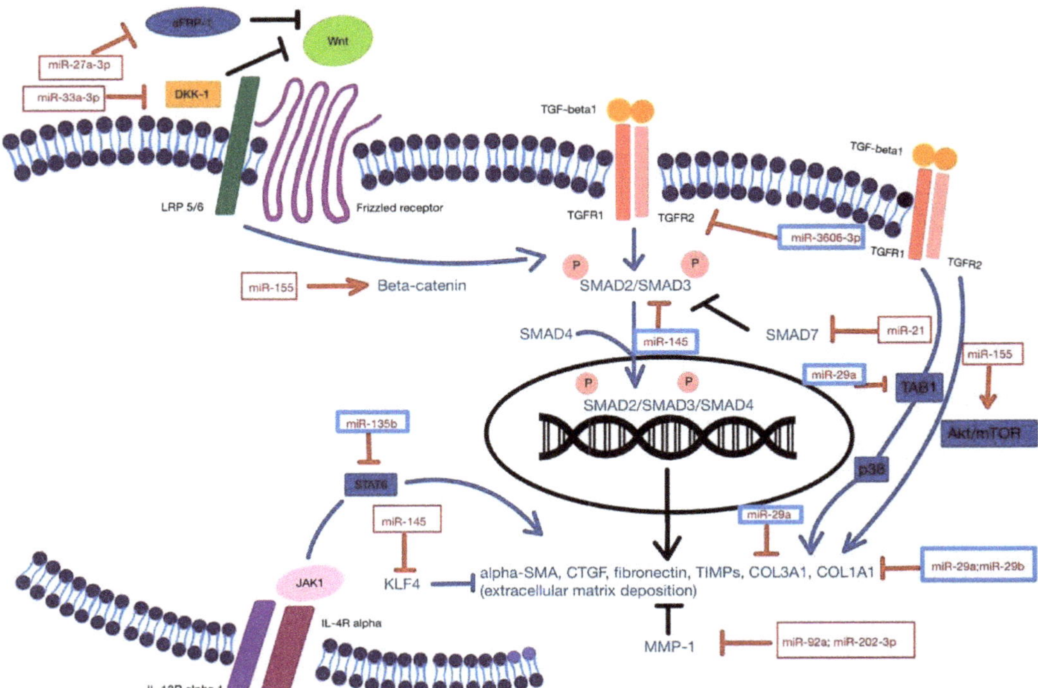

Figure 2. TGF-β1, the main regulator of fibrosis, plays a central role in SSc pathogenesis. SMAD and non-SMAD TGF-β signaling pathways lead to transcription of fibrosis-related genes responsible for fibroblast proliferation, myofibroblast differentiation, and extracellular matrix deposition. Upregulation (red squares) or downregulation (blue squares) of diverse miRNAs interfere with these mechanisms and promote tissue fibrosis. KLF4: Kruppel-like factor 4; MMP1: matrix metalloproteinase 1; sFRP-1: secreted frizzled-related protein-1; DKK-1: Dickkopf-1; LRP 5/6: lipoprotein receptor-related proteins (LRP) 5 and 6; COL1A1: collagen type 1 alpha 1 chain; COL3A1: collagen type 3 alpha 1 chain; TAB1: transforming growth factor beta activated protein kinase 1; TGFR1: transforming growth factor beta receptor 1; TGFR2: transforming growth factor beta receptor 2; Wnt: Wnt signaling pathway; alpha-SMA: alpha-smooth muscle actin; CTGF: connective tissue growth factor; TIMPs: tissue inhibitors of metalloproteinases; IL-13R alpha 1: interleukin-13 receptor alpha 1; IL-4R alpha: interleukin-4 receptor alpha; STAT6: signal transducer and activator of transcription 6.

Table 3. Apoptosis and miRNAs.

miRNAs	Expression	Tissue Sample(s)	Regulatory Effect	Consequence	Reference
miR-29a	Downregulated	Fibroblasts TGF-β-stimulated fibroblasts	Increased Bax:Bcl2 ratio	Proapoptotic	Jafarinejad-Farsangi et al. [86]
miR-21	Upregulated	Fibroblasts	Decreased Bax:Bcl2 ratio	Antiapoptotic	Jafarinejad-Farsangi et al. [87]

3.4. Microangiopathy and miRNAs

Proliferative microangiopathy is responsible for severe manifestations such as digital ulcers and pulmonary arterial hypertension. Iwamoto (2016) investigated the potential role of epigenetics in mediating SSc vasculopathy. The study revealed that SSc fibroblasts and dermal biopsies exhibited lower levels of miR-193b compared to normal controls. By means of computational analysis, several genes were identified as potential targets and urokinase-type plasminogen activator (uPA) was the most significantly dysregulated by miR-193b stimulation or inhibition. Accordingly, transfection of miR-193b mimics lowered uPA levels and transfection of miR-193b inhibitors upregulated uPA expression. These findings were also shown in HC and primary human pulmonary artery smooth muscle cell (HPASMCs) cultures and validated at the protein level. Immunohistochemistry and double-staining of SSc skin samples with uPA and α-SMA proved that uPA is highly expressed in vascular structures, especially by vascular smooth muscle cells (VSMCs), and modestly expressed by skin fibroblasts. Obtained data suggest a possible role of uPA in SSc vasculopathy that was further validated on HPASMCs cultures where uPA enhanced the expression of the PCNA proliferation marker and decreased apoptosis detected by flow cytometry Upregulation of MiR-193b represents a potential treatment for targeting vasculopathy in SSc [88].

MiR-126 is a negative regulator of epidermal growth factor like-domain 7 (EGFL7), a modulator of angiogenesis that exhibits pro-angiogenic properties. According to the publication of Liakouli (2019), EGFL7 expression levels are increased in early onset dcSSc skin specimens but decreased in long-standing dcSSc skin biopsies. The authors further showed that exogenous delivery of human recombinant (rh)EGFL7 suppressed the impaired angiogenesis in cocultures of early-onset and long-standing dcSSc fibroblasts with HUVECs. Moreover, (rh)EGFL7 suppressed COL1A1 expression levels in early-onset SSc fibroblasts, whereas EGFL7 small interfering (si)RNA increased COL1A1 mRNA levels. These results emphasize the dual role of EGFL7 in SSc pathogenesis, modulating both angiogenesis and fibrosis [89].

Table 4 represents a summary of the modulatory effects of miRNAs in SSc vasculopathy.

Table 4. Microangiopathy and miRNAs.

miRNA	Tissue Samples	Regulatory Effect	Reference
miR-193b	Fibroblasts Skin HPASMCs cultures	uPA expression	Iwamoto et al. [88]
miR-126	Fibroblasts Skin HUVECs	EGFL7 expression	Liakouli et al. [89]

uPA: urokinase-type plasminogen activator; EGFL7: epidermal growth factor like-domain.

3.5. Immune Dysfunction and miRNAs

B cell-activating factor (BAFF), a TNF superfamily member, revealed its important role in the pathogenesis of several autoimmune diseases by modulating the activity and survival of B cells. In SSc, stimulation of dermal fibroblasts with either Poly(I:C) or IFN-γ (known upregulators of BAFF) resulted in decreased expression levels of miR-30a-3p. Conversely, transfection of miR-30a-3p mimics in these cells lowered BAFF expression levels and consequently determineed decreased B cell survival. To a further extent, transfection of normal fibroblasts with miR-30a-3p inhibitor enhanced BAFF levels, demonstrating that miR-30a-3p is an important regulator of BAFF production and secretion [90].

The interplay between miRNA dysregulation and interferon (IFN) signatures in SSc was explored by Ciechomska (2020) through mRNA–miRNA sequencing and functional studies on monocytes. Accordingly, miR-26a-2-3p was significantly downregulated in SSc monocytes compared to controls, while expression of selected IFN-stimulated genes was increased in SSc monocytes but not in controls or rheumatoid arthritis samples. Transfection of miR-26a-2-3p mimics to TLR-stimulated THP-1 cells proved that this miRNA is a negative regulator of IFN-stimulated genes. These findings suggest that miR-26a-2-3p downregulation might be responsible, at least in part, for the increased IFN production in SSc [91].

Table 5 illustrates the characteristics of miRNAs involved in the dysregulation of the immune system in SSc.

Table 5. Immune dysfunction and miRNAs.

miRNA	Tissue Sample	Regulatory Effect(s)	Reference
miR-30a-3p	Fibroblasts	BAFF production and secretion	Alsaleh et al. [90]
miR-26a-2-3p	Monocytes	Regulation of IFN-stimulated genes	Ciechomska et al. [91]

4. MiRNAs: Diagnostic and Prognostic Biomarkers

An interesting finding was that of Makino (2012), who found highly expressed levels of miR-142-3p in the serum of SSc patients. These levels were significantly dysregulated compared to the scleroderma spectrum disorder (SSD), systemic lupus erythematosus (SLE), and dermatomyositis (DM) cohorts, suggesting that miR-142-3p could be a potential diagnostic marker in distinguishing SSc from SSD [92].

Izumiya (2015) illustrated the relationship between five let-7 family miRNA members and the severity of pulmonary hypertension in SSc patients. Microarray analysis of skin biopsies from six patients without pulmonary hypertension (PH) and nine patients with PH identified 32 miRNAs that were upregulated and 14 miRNAs that were downregulated. After validation by quantitative real-time PCR, the expression levels of let-7a, let-7d, let-7e, let-7f, and let-7g were significantly dysregulated in the PH group. Furthermore, let-7d and let-7b were correlated with an increased pulmonary arterial pressure measured by echocardiography, making them possible candidates as biomarkers of PH severity in SSc patients [93].

The association between cancer and SSc is another troubling aspect in the management of these patients. SSc patients have a higher risk of developing certain types of cancer, mainly breast, lung, and hematological malignancies [94]. Dolcino (2018) investigated the potential role of epigenetics in promoting carcinogenesis in SSc. The expression levels of 5 MiRNAs (miR-21-5p, miR-92a-3p, miR-155-5p, miR-16-5p, miR-126) with proven implication in these types of malignancies were detected by real-time PCR in the serum of 30 SSc patients and 10 HC. MiR-21-5p, miR-92a-3p, miR-155-5p, and miR-16-5p were significantly dysregulated in the SSc group compared to controls. Mir-126 levels were not statistically different between SSc patients and controls. The upregulation of miR-21-5p, miR-92a-3p, and miR-155-5p in both SSc and cancer specimens, with implications in fibrosis as well as angiogenesis and proliferation, suggests that there might be a defining role for epigenetic mechanisms in cancer predisposition in SSc [95].

5. Role of miRNAs in SSc Interstitial Lung Disease (SSc-ILD) Pathogenesis

Pulmonary involvement in SSc is associated with increased morbidity and mortality and therefore warrants special attention with respect to the role miRNAs might play in lung fibrosis.

Wu (2021) analyzed one miRNA and three mRNA datasets retrieved from the Gene Expression Omnibus (GEO) database and identified nine differentially expressed miRNAs in SSc-ILD lung samples compared to controls. These miRNAs regulate various fibrosis-related signaling pathways, such as the integrin family, TNF-related apoptosis inducing ligand (TRAIL) protein, and vascular endothelial growth factor (VEGF)/VEGF receptor (VEGFR) signaling networks [96].

Compared to idiopathic pulmonary fibrosis (IPF), an organ-specific fibrotic disease, Mullenbrock (2018) identified a similar miRNA profile in SSc lung fibroblasts. Transfection of miR-29b-3p, miR-138-5p, and miR-146b-5p in both IPF and SSc pulmonary fibroblasts resulted in downregulation of several profibrotic genes, COL1A1 (miR-29b-3p target gene), connective tissue growth factor (CTGF; miR138-5p target gene), and actin alpha 2 (ACTA2; miR-146b-5p target gene) [97].

As previously mentioned, data from Christmann (2016) promoted miR-155 as an attractive therapeutic target and also a promising prognostic biomarker in SSc-ILD [64]. Another proposed prognostic biomarker is miR-200c identified in peripheral blood mononuclear cells (PBMCs) among patients with ILD and different connective tissue diseases (CTDs). Higher miR-200c levels were detected among patients with SSc-ILD compared to other

CTDs and in patients with more severe forms of lung fibrosis (defined by the decline of FVC and FEV1) [98]. Results from a different study showed that transfection of miR-30c in experimental mouse models resulted in decreased dermal thickness and collagen production as well as improved vascular dysfunction and lung fibrosis scores, promoting miR-30c as a versatile therapeutic target in SSc [99].

MiR-320a is downregulated in serum and PBMCs of SSc-ILD patients and lung samples of bleomycin-induced ILD. Through its target genes, TGFR2 and insulin-like growth factor receptor 1 (IGF1R), miR-320a modulates the expression of type I collagen in normal human pulmonary fibroblasts cell lines. Further stimulation of these cells with TGF-β upregulated both miR-320a and collagen genes, again pointing toward the central role of the TGF-β signaling pathway in tissue fibrosis [100].

Pulmonary endothelial myofibroblast differentiation and type IV collagen synthesis are induced in vitro by miR-483-5p, a profibrotic miRNA that was detected in high levels in SSc serum samples [44].

6. Future Directions

Research in the field of miRNAs in SSc has mostly focused on miRNAs exhibiting an antifibrotic or profibrotic effect in the hope of identifying and developing more targeted therapies. Some of the earliest promising results came from Montgomery (2014). In this study, bleomycin-induced pulmonary fibrosis improved after intravenous administration of double-stranded miR-29b mimics. These chemically modified miRNA transcripts were able to restore COL1A1 and COL3A1 expression levels and even decrease total collagen amount in lung biopsies. These findings suggest that miR-29b therapeutic delivery may not only stop progression of pulmonary fibrosis but also reverse already established lung fibrosis [101]. In this respect, miRagen Therapeutics has a phase 2, double-blind, placebo-controlled clinical trial investigating the potential use of Remlarsen/MRG-201 (miR-29 mimic) for the treatment of keloid scars (www.clinicaltrials.gov, accessed on 27 March 2021). MRG-229, a second-generation miR-29 mimic designed for treatment of idiopathic pulmonary fibrosis (IPF), has recently shown favorable efficacy and safety profiles in preclinical studies (www.miragen.com, accessed on 27 March 2021). Another attractive target could be the inhibition of miR-155. Several lines of research showed that bleomycin-treated miR-155 knockout mice achieved improved skin and pulmonary fibrosis scores [63,64]. Therefore, silencing profibrotic miRNAs with synthetic antagomiRs could also represent an approach in SSc therapy. Besides miRNA mimics and antagomiRs, several other methods of delivery have been developed, but the most important questions remain their stability, cell and tissue specificity, and subsequent immune response [24,101,102].

Exosomes, small membrane vesicles containing genetic information, are emerging as a new direction in the study of SSc pathogenesis. They mediate intercellular interactions within the same tissue but also modulate cell phenotypes away from their origin in distant organs. Thus, free-circulating exosomes could explain the progression of fibrosis from skin to different organs [103,104]. Current evidence even implies that exosomes might represent the link between the three disrupted mechanistic pathways in SSc: microangiopathy, immune disfunction, and fibrosis [105]. The serum exosome content of 28 miRNAs previously shown to mediate various fibrotic pathways in SSc was evaluated by means of semiquantitative real-time PCR (RT-PCR) in three lcSSc patients, three dcSSc patients, and HC. The expression levels of six profibrotic miRNAs were increased and 10 antifibrotic miRNAs were decreased in both SSc subsets compared to normal controls. A significant difference was also observed in the expression levels of eight antifibrotic miRNAs (miR-let-7a, miR-29), miR-92a, miR-1250, miR-133, miR-140, miR-146a, miR-200a) that were markedly downregulated in the dcSSc subgroup compared to levels observed in the lcSSc subgroup. Furthermore, normal dermal fibroblasts were exposed in vitro to three different concentrations of exosomes isolated from both SSc subgroups in order to validate their involvement in fibrosis. RT-PCR evidenced dose-dependent upregulation of COL1A1, COL3A1, and fibronectin 1 (FN1), genes encoding type I collagen, type III collagen, and fibronectin. Their corresponding protein levels were also increased after exosome exposure. Other genes that were induced after exosome treatment were genes involved in myofibroblast activation as well as genes encoding TGF-β and CTGF 56. As a result, exosomes act as potent biological tools that could be used as diagnostic and even prognostic biomarkers, whereas their manipulation as therapy delivery carriers is an exciting perspective in many diseases, including SSc [105].

7. Limitations

This is a narrative literature review that focused on the current state of research in the field of epigenetics, particularly miRNAs, and their role in SSc pathogenesis. "Systemic sclerosis", "pathogenesis", "epigenetic mechanisms", and "miRNAs" were the MeSH terms used to select and retrieve information from the National Library of Medicine (PubMed.gov). Original articles, narrative reviews, systematic reviews, and meta-analysis were considered and included in the present study after applying text availability (only full-text articles) and language (only English) filters. This review is therefore prone to the inherited limitations of this research methodology such as selection bias, difficulty in determining complex interactions, and drawing conclusions.

8. Conclusions

MiRNAs are involved in various physiological and pathological processes. These molecules have validated their role in modulating vasculopathy, immune responses, and fibrosis in SSc and represent promising therapeutic targets. Even though advances in the field are continuously expanding, certain limitations remain to be addressed in future studies. SSc heterogeneity, small cohorts, permissive inclusion criteria without a clear distinction in terms of disease severity, and status, together with scarce data regarding current treatments, are just some of the culprits responsible for the discordance between reports. MiRNAs are cell- and tissue-specific, therefore their expression is expected to differ between body compartments and internal organs. In this respect, a more balanced research agenda should also be considered given the fact that most protocols investigated miRNA expression levels in skin biopsies and dermal fibroblasts with an emphasis on profibrotic and antifibrotic transcripts.

Author Contributions: Conceptualization, I.S. and L.M.; methodology, I.S. and T.C.; software, V.R.; validation, I.S., L.M., T.C. and S.R.; formal analysis, I.S.; investigation, I.S. and C.S.; resources, I.S., V.R. and C.S.; data curation, I.S.; writing - original draft preparation, I.S.; writing - review and editing, I.S., L.M. and S.R.; visualization, I.S.; supervision, I.S.; project administration, I.S.; funding acquisition, I.S. All authors have read and agreed to the published version of the manuscript.

Funding: This research was funded by "Iuliu Hatieganu" University of Medicine and Pharmacy Cluj-Napoca through doctoral research projects no. 2461/64/17 January 2020 and 1033/57/13 January 2021 as part of the doctoral studies no. 4335/01.10.2018.

Institutional Review Board Statement: Not applicable.

Informed Consent Statement: Not applicable.

Data Availability Statement: Not applicable.

Conflicts of Interest: The authors declare no conflict of interest.

References

1. Leroy, E.C.; Black, C.; Fleischmajer, R.; Jablonska, S.; Krieg, T.; Medsger, T.A.; Rowell, N.; Wollheim, F. Scleroderma (systemic sclerosis): Classification, subsets and pathogenesis. *J. Rheumatol.* **1988**, *15*, 202–205. [PubMed]
2. Steen, V.D. Autoantibodies in Systemic Sclerosis. *Semin. Arthritis Rheum.* **2005**, *35*, 35–42. [CrossRef] [PubMed]
3. Allanore, Y.; Simms, R.; Distler, O.; Trojanowska, M.; Pope, J.; Denton, C.P.; Varga, J. Systemic sclerosis. *Nat. Rev. Dis. Prim.* **2015**, *1*, 15002. [CrossRef] [PubMed]
4. Denton, C.P.; Khanna, D. Systemic sclerosis. *Lancet* **2017**, *390*, 1685–1699. [CrossRef]
5. Hoffmann-Vold, A.-M.; Molberg, Ø.; Midtvedt, Ø.; Garen, T.; Gran, J.T. Survival and Causes of Death in an Unselected and Complete Cohort of Norwegian Patients with Systemic Sclerosis. *J. Rheumatol.* **2013**, *40*, 1127–1133. [CrossRef] [PubMed]

6. Varga, J.; Trojanowska, M.; Kuwana, M. Pathogenesis of systemic sclerosis: Recent insights of molecular and cellular mechanisms and therapeutic opportunities. *J. Scleroderma Relat. Disord.* **2017**, *2*, 137–152. [CrossRef]
7. Salazar, G.; Mayes, M.D. Genetics, Epigenetics, and Genomics of Systemic Sclerosis. *Rheum. Dis. Clin. North Am.* **2015**, *41*, 345–366. [CrossRef] [PubMed]
8. Aslani, S.; Sobhani, S.; Gharibdoost, F.; Jamshidi, A.; Mahmoudi, M. Epigenetics and pathogenesis of systemic sclerosis; the ins and outs. *Hum. Immunol.* **2018**, *79*, 178–187. [CrossRef]
9. Tsou, P.-S.; Sawalha, A.H. Unfolding the pathogenesis of scleroderma through genomics and epigenomics. *J. Autoimmun.* **2017**, *83*, 73–94. [CrossRef] [PubMed]
10. Cutolo, M.; Soldano, S.; Smith, V. Pathophysiology of systemic sclerosis: Current understanding and new insights. *Expert Rev. Clin. Immunol.* **2019**, *15*, 753–764. [CrossRef] [PubMed]
11. Asano, Y. Systemic sclerosis. *J. Dermatol.* **2018**, *45*, 128–138. [CrossRef] [PubMed]
12. Epattanaik, D.; Ebrown, M.; Postlethwaite, B.C.; Postlethwaite, A.E. Pathogenesis of Systemic Sclerosis. *Front. Immunol.* **2015**, *6*, 272. [CrossRef]
13. Matucci-Cerinic, M.; Kahaleh, B.; Wigley, F.M. Review: Evidence That Systemic Sclerosis Is a Vascular Disease. *Arthritis Rheum.* **2013**, *65*, 1953–1962. [CrossRef] [PubMed]
14. Cipriani, P.; Di Benedetto, P.; Ruscitti, P.; Capece, D.; Zazzeroni, F.; Liakouli, V.; Pantano, I.; Berardicurti, O.; Carubbi, F.; Pecetti, G.; et al. The Endothelial-mesenchymal Transition in Systemic Sclerosis Is Induced by Endothelin-1 and Transforming Growth Factor-β and May Be Blocked by Macitentan, a Dual Endothelin-1 Receptor Antagonist. *J. Rheumatol.* **2015**, *42*, 1808–1816. [CrossRef] [PubMed]
15. Rajkumar, V.S.; Sundberg, C.; Abraham, D.J.; Rubin, K.; Black, C.M. Activation of microvascular pericytes in autoimmune Raynaud's phenomenon and systemic sclerosis. *Arthritis Rheum.* **1999**, *42*, 930–941. [CrossRef]
16. Dowson, C.; Simpson, N.; Duffy, L.; O'Reilly, S. Innate Immunity in Systemic Sclerosis. *Curr. Rheumatol. Rep.* **2017**, *19*, 2. [CrossRef]
17. Chizzolini, C.; Boin, F. The role of the acquired immune response in systemic sclerosis. *Semin. Immunopathol.* **2015**, *37*, 519–528. [CrossRef] [PubMed]
18. Cipriani, P.; Di Benedetto, P.; Ruscitti, P.; Liakouli, V.; Berardicurti, O.; Carubbi, F.; Ciccia, F.; Guggino, G.; Zazzeroni, F.; Alesse, E.; et al. Perivascular Cells in Diffuse Cutaneous Systemic Sclerosis Overexpress Activated ADAM12 and Are Involved in Myofibroblast Transdifferentiation and Development of Fibrosis. *J. Rheumatol.* **2016**, *43*, 1340–1349. [CrossRef] [PubMed]
19. Brown, M.; O'Reilly, S. The immunopathogenesis of fibrosis in systemic sclerosis. *Clin. Exp. Immunol.* **2018**, *195*, 310–321. [CrossRef]
20. Wynn, T.A. Cellular and molecular mechanisms of fibrosis. *J. Pathol.* **2007**, *214*, 199–210. [CrossRef] [PubMed]
21. Ho, Y.Y.; Lagares, D.; Tager, A.M.; Kapoor, M. Fibrosis—A lethal component of systemic sclerosis. *Nat. Rev. Rheumatol.* **2014**, *10*, 390–402. [CrossRef] [PubMed]
22. Altorok, N.; Almeshal, N.; Wang, Y.; Kahaleh, B. Epigenetics, the holy grail in the pathogenesis of systemic sclerosis. *Rheumatology* **2015**, *54*, 1759–1770. [CrossRef] [PubMed]
23. Broen, J.C.A.; Radstake, T.R.D.J.; Rossato, M. The role of genetics and epigenetics in the pathogenesis of systemic sclerosis. *Nat. Rev. Rheumatol.* **2014**, *10*, 671–681. [CrossRef] [PubMed]
24. Henderson, J.; Distler, J.; O'Reilly, S. The Role of Epigenetic Modifications in Systemic Sclerosis: A Druggable Target. *Trends Mol. Med.* **2019**, *25*, 395–411. [CrossRef]
25. Bergmann, C.; Distler, J.H. Epigenetic factors as drivers of fibrosis in systemic sclerosis. *Epigenomics* **2017**, *9*, 463–477. [CrossRef]
26. Altorok, N.; Kahaleh, B. Epigenetics and systemic sclerosis. *Semin. Immunopathol.* **2015**, *37*, 453–462. [CrossRef] [PubMed]
27. Hattori, M.; Yokoyama, Y.; Hattori, T.; Motegi, S.-I.; Amano, H.; Hatada, I.; Ishikawa, O. Global DNA hypomethylation and hypoxia-induced expression of the ten eleven translocation (TET) family, TET1, in scleroderma fibroblasts. *Exp. Dermatol.* **2015**, *24*, 841–846. [CrossRef]
28. Ding, W.; Pu, W.; Wang, L.; Jiang, S.; Zhou, X.; Tu, W.; Yu, L.; Zhang, J.; Guo, S.; Liu, Q.; et al. Genome-Wide DNA Methylation Analysis in Systemic Sclerosis Reveals Hypomethylation of IFN-Associated Genes in CD4+ and CD8+ T Cells. *J. Investig. Dermatol.* **2018**, *138*, 1069–1077. [CrossRef] [PubMed]
29. Wang, Y.; Shu, Y.; Xiao, Y.; Wang, Q.; Kanekura, T.; Li, Y.; Wang, J.; Zhao, M.; Lu, Q.; Xiao, R. Hypomethylation and overexpression of ITGAL (CD11a) in CD4+ T cells in systemic sclerosis. *Clin. Epigenet.* **2014**, *6*, 25. [CrossRef] [PubMed]
30. Lei, W.; Luo, Y.; Yan, K.; Zhao, S.; Li, Y.; Qiu, X.; Zhou, Y.; Long, H.; Zhao, M.; Liang, Y.; et al. Abnormal DNA methylation in CD4+ T cells from patients with systemic lupus erythematosus, systemic sclerosis, and dermatomyositis. *Scand. J. Rheumatol.* **2009**, *38*, 369–374. [CrossRef]
31. Dees, C.; Schlottmann, I.; Funke, R.; Distler, A.; Palumbo-Zerr, K.; Zerr, P.; Lin, N.-Y.; Beyer, C.; Distler, O.; Schett, G.; et al. The Wnt antagonists DKK1 and SFRP1 are downregulated by promoter hypermethylation in systemic sclerosis. *Ann. Rheum. Dis.* **2013**, *73*, 1232–1239. [CrossRef]
32. Krämer, M.; Dees, C.; Huang, J.; Schlottmann, I.; Palumbo-Zerr, K.; Zerr, P.; Gelse, K.; Beyer, C.; Distler, A.; Marquez, V.E.; et al. Inhibition of H3K27 histone trimethylation activates fibroblasts and induces fibrosis. *Ann. Rheum. Dis.* **2012**, *72*, 614–620. [CrossRef]

33. Deng, Q.; Luo, Y.; Chang, C.; Wu, H.; Ding, Y.; Xiao, R. The Emerging Epigenetic Role of CD8+T Cells in Autoimmune Diseases: A Systematic Review. *Front. Immunol.* **2019**, *10*, 856. [CrossRef] [PubMed]
34. Zhu, H.; Li, Y.; Qu, S.; Luo, H.; Zhou, Y.; Wang, Y.; Zhao, H.; You, Y.; Xiao, X.; Zuo, X. MicroRNA Expression Abnormalities in Limited Cutaneous Scleroderma and Diffuse Cutaneous Scleroderma. *J. Clin. Immunol.* **2012**, *32*, 514–522. [CrossRef]
35. Henry, T.W.; Mendoza, F.A.; Jimenez, S.A. Role of microRNA in the pathogenesis of systemic sclerosis tissue fibrosis and vasculopathy. *Autoimmun. Rev.* **2019**, *18*, 102396. [CrossRef] [PubMed]
36. Fioretto, B.S.; Rosa, I.; Romano, E.; Wang, Y.; Guiducci, S.; Zhang, G.; Manetti, M.; Matucci-Cerinic, M. The contribution of epigenetics to the pathogenesis and gender dimorphism of systemic sclerosis: A comprehensive overview. *Ther. Adv. Musculoskelet. Dis.* **2020**, *12*, 1759720–20918456. [CrossRef]
37. Ramahi, A.; Altorok, N.; Kahaleh, B. Epigenetics and systemic sclerosis: An answer to disease onset and evolution? *Eur. J. Rheumatol.*. **2020**, *7*, 147–156. [CrossRef] [PubMed]
38. Perera, B.; Faulk, C.; Svoboda, L.K.; Goodrich, J.M.; Dolinoy, D.C. The role of environmental exposures and the epigenome in health and disease. *Environ. Mol. Mutagen.* **2020**, *61*, 176–192. [CrossRef]
39. Toraño, E.G.; García, M.G.; Fernández-Morera, J.L.; Niño-García, P.; Fernandez, A. The Impact of External Factors on the Epigenome:In Uteroand over Lifetime. *BioMed Res. Int.* **2016**, *2016*, 1–17. [CrossRef] [PubMed]
40. Aguilera, O.; Fernandez, A.; Munoz, A.; Fraga, M. Epigenetics and environment: A complex relationship. *J. Appl. Physiol.* **2010**, *109*, 243–251. [CrossRef] [PubMed]
41. Moore, L.D.; Le, T.; Fan, G. DNA Methylation and Its Basic Function. *Neuropsychopharmacology* **2012**, *38*, 23–38. [CrossRef] [PubMed]
42. Tsou, P.-S.; Campbell, P.; Amin, M.A.; Coit, P.; Miller, S.; Fox, D.A.; Khanna, D.; Sawalha, A.H. Inhibition of EZH2 prevents fibrosis and restores normal angiogenesis in scleroderma. *Proc. Natl. Acad. Sci. USA* **2019**, *116*, 3695–3702. [CrossRef]
43. Wang, Q.; Xiao, Y.; Shi, Y.; Luo, Y.; Li, Y.; Zhao, M.; Lu, Q.; Xiao, R. Overexpression of JMJD3 may contribute to demethylation of H3K27me3 in CD4 + T cells from patients with systemic sclerosis. *Clin. Immunol.* **2015**, *161*, 396–399. [CrossRef] [PubMed]
44. Chouri, E.; Servaas, N.H.; Bekker, C.P.; Affandi, A.J.; Cossu, M.; Hillen, M.; Angiolilli, C.; Mertens, J.S.; Hoogen, L.L.V.D.; Silva-Cardoso, S.; et al. Serum microRNA screening and functional studies reveal miR-483-5p as a potential driver of fibrosis in systemic sclerosis. *J. Autoimmun.* **2018**, *89*, 162–170. [CrossRef]
45. Rusek, M.; Michalska-Jakubus, M.; Kowal, M.; Bełtowski, J.; Krasowska, D. A novel miRNA-4484 is up-regulated on microarray and associated with increased MMP-21 expression in serum of systemic sclerosis patients. *Sci. Rep.* **2019**, *9*, 1–13. [CrossRef] [PubMed]
46. Dolcino, M.; Tinazzi, E.; Puccetti, A.; Lunardi, C. In Systemic Sclerosis, a Unique Long Non Coding RNA Regulates Genes and Pathways Involved in the Three Main Features of the Disease (Vasculopathy, Fibrosis and Autoimmunity) and in Carcinogenesis. *J. Clin. Med.* **2019**, *8*, 320. [CrossRef]
47. Messemaker, T.C.; Chadli, L.; Cai, G.; Goelela, V.S.; Boonstra, M.; Dorjée, A.L.; Andersen, S.N.; Mikkers, H.M.; Hof, P.V.T.; Mei, H.; et al. Antisense Long Non-Coding RNAs Are Deregulated in Skin Tissue of Patients with Systemic Sclerosis. *J. Investig. Dermatol.* **2018**, *138*, 826–835. [CrossRef] [PubMed]
48. Mazzone, R.; Zwergel, C.; Artico, M.; Taurone, S.; Ralli, M.; Greco, A.; Mai, A. The emerging role of epigenetics in human autoimmune disorders. *Clin. Epigenet.* **2019**, *11*, 1–15. [CrossRef]
49. Treiber, T.; Treiber, N.; Meister, G. Regulation of microRNA biogenesis and function. *Thromb. Haemost.* **2012**, *107*, 605–610. [CrossRef] [PubMed]
50. O'Brien, J.; Hayder, H.; Zayed, Y.; Peng, C. Overview of MicroRNA Biogenesis, Mechanisms of Actions, and Circulation. *Front. Endocrinol.* **2018**, *9*, 402. [CrossRef]
51. Steen, S.O.; Iversen, L.V.; Carlsen, A.L.; Burton, M.; Nielsen, C.T.; Jacobsen, S.; Heegaard, N.H. The Circulating Cell-free microRNA Profile in Systemic Sclerosis Is Distinct from Both Healthy Controls and Systemic Lupus Erythematosus. *J. Rheumatol.* **2015**, *42*, 214–221. [CrossRef] [PubMed]
52. Zhang, L.; Wu, H.; Zhao, M.; Lu, Q. Meta-analysis of differentially expressed microRNAs in systemic sclerosis. *Int. J. Rheum. Dis.* **2020**, *23*, 1297–1304. [CrossRef]
53. Tsou, P.-S. Epigenetic Control of Scleroderma: Current Knowledge and Future Perspectives. *Curr. Rheumatol. Rep.* **2019**, *21*, 69. [CrossRef] [PubMed]
54. Li, H.; Yang, R.; Fan, X.; Gu, T.; Zhao, Z.; Chang, D.; Wang, W.; Wang, C. MicroRNA array analysis of microRNAs related to systemic scleroderma. *Rheumatol. Int.* **2012**, *32*, 307–313. [CrossRef] [PubMed]
55. Wuttge, D.M.; Carlsen, A.L.; Teku, G.; Steen, S.O.; Wildt, M.; Vihinen, M.; Hesselstrand, R.; Heegaard, N.H.H. Specific autoantibody profiles and disease subgroups correlate with circulating micro-RNA in systemic sclerosis. *Rheumatology* **2015**, *54*, 2100–2107. [CrossRef] [PubMed]
56. Wermuth, P.; Piera-Velazquez, S.; Jimenez, S.A. Exosomes isolated from serum of systemic sclerosis patients display alterations in their content of profibrotic and antifibrotic microRNA and induce a profibrotic phenotype in cultured normal dermal fibroblasts. *Clin. Exp. Rheumatol.* **2017**, *35*, 21–30.
57. Zhu, H.; Luo, H.; Li, Y.; Zhou, Y.; Jiang, Y.; Chai, J.; Xiao, X.; You, Y.; Zuo, X. MicroRNA-21 in Scleroderma Fibrosis and its Function in TGF-β- Regulated Fibrosis-Related Genes Expression. *J. Clin. Immunol.* **2013**, *33*, 1100–1109. [CrossRef] [PubMed]

58. Jafarinejad-Farsangi, S.; Gharibdoost, F.; Farazmand, A.; Kavosi, H.; Jamshidi, A.; Karimizadeh, E.; Noorbakhsh, F.; Mahmoudi, M. MicroRNA-21 and microRNA-29a modulate the expression of collagen in dermal fibroblasts of patients with systemic sclerosis. *Autoimmunity* **2019**, *52*, 108–116. [CrossRef] [PubMed]
59. Ly, T.-D.; Riedel, L.; Fischer, B.; Schmidt, V.; Hendig, D.; Distler, J.; Kuhn, J.; Knabbe, C.; Faust, I. microRNA-145 mediates xylosyltransferase-I induction in myofibroblasts via suppression of transcription factor KLF4. *Biochem. Biophys. Res. Commun.* **2020**, *523*, 1001–1006. [CrossRef] [PubMed]
60. Sing, T.; Jinnin, M.; Yamane, K.; Honda, N.; Makino, K.; Kajihara, I.; Makino, T.; Sakai, K.; Masuguchi, S.; Fukushima, S.; et al. microRNA-92a expression in the sera and dermal fibroblasts increases in patients with scleroderma. *Rheumatology* **2012**, *51*, 1550–1556. [CrossRef]
61. Zhou, B.; Zhu, H.; Luo, H.; Gao, S.; Dai, X.; Li, Y.; Zuo, X. MicroRNA-202-3p regulates scleroderma fibrosis by targeting matrix metalloproteinase 1. *Biomed. Pharmacother.* **2017**, *87*, 412–418. [CrossRef] [PubMed]
62. Nakayama, W.; Jinnin, M.; Tomizawa, Y.; Nakamura, K.; Kudo, H.; Inoue, K.; Makino, K.; Honda, N.; Kajihara, I.; Fukushima, S.; et al. Dysregulated interleukin-23 signalling contributes to the increased collagen production in scleroderma fibroblasts via balancing microRNA expression. *Rheumatology* **2017**, *56*, 145–155. [CrossRef]
63. Yan, Q.; Chen, J.; Li, W.; Bao, C.; Fu, Q. Targeting miR-155 to Treat Experimental Scleroderma. *Sci. Rep.* **2016**, *6*, 20314. [CrossRef]
64. Christmann, R.B.; Wooten, A.; Sampaio-Barros, P.; Borges, C.L.; Carvalho, C.R.R.; Kairalla, R.A.; Feghali-Bostwick, C.; Ziemek, J.; Mei, Y.; Goummih, S.; et al. miR-155 in the progression of lung fibrosis in systemic sclerosis. *Arthritis Res.* **2016**, *18*, 1–13. [CrossRef] [PubMed]
65. Artlett, C.M.; Sassi-Gaha, S.; Hope, J.L.; Feghali-Bostwick, C.A.; Katsikis, P.D. Mir-155 is overexpressed in systemic sclerosis fibroblasts and is required for NLRP3 inflammasome-mediated collagen synthesis during fibrosis. *Arthritis Res.* **2017**, *19*, 1–8. [CrossRef]
66. Henderson, J.; Wilkinson, S.; Przyborski, S.; Stratton, R.; O'Reilly, S. microRNA27a-3p mediates reduction of the Wnt antagonist sFRP-1 in systemic sclerosis. *Epigenetics* **2021**, *16*, 808–817. [CrossRef] [PubMed]
67. Henderson, J.; Pryzborski, S.; Stratton, R.; O'Reilly, S. Wnt antagonist DKK-1 levels in systemic sclerosis are lower in skin but not in blood and are regulated by microRNA33a-3p. *Exp. Dermatol.* **2021**, *30*, 162–168. [CrossRef] [PubMed]
68. Derk, C.T. Transforming Growth Factor-β (TGF-β) and its Role in the Pathogenesis of Systemic Sclerosis: A Novel Target for Therapy? *Recent Patents Inflamm. Allergy Drug Discov.* **2007**, *1*, 142–145. [CrossRef] [PubMed]
69. La, J.; Reed, E.; Chan, L.; Smolyaninova, L.V.; Akomova, O.A.; Mutlu, G.M.; Orlov, S.N.; Dulin, N.O. Downregulation of TGF-β Receptor-2 Expression and Signaling through Inhibition of Na/K-ATPase. *PLoS ONE* **2016**, *11*, e0168363. [CrossRef]
70. Meng, J.; Li, L.; Zhao, Y.; Zhou, Z.; Zhang, M.; Li, D.; Zhang, C.-Y.; Zen, K.; Liu, Z. MicroRNA-196a/b Mitigate Renal Fibrosis by Targeting TGF-β Receptor 2. *J. Am. Soc. Nephrol.* **2016**, *27*, 3006–3021. [CrossRef] [PubMed]
71. Li, M.; Krishnaveni, M.S.; Li, C.; Zhou, B.; Xing, Y.; Banfalvi, A.; Li, A.; Lombardi, V.; Akbari, O.; Borok, Z.; et al. Epithelium-specific deletion of TGF-β receptor type II protects mice from bleomycin-induced pulmonary fibrosis. *J. Clin. Investig.* **2011**, *121*, 277–287. [CrossRef] [PubMed]
72. Yu, F.; Chen, B.; Fan, X.; Li, G.; Dong, P.; Zheng, J. Epigenetically-Regulated MicroRNA-9-5p Suppresses the Activation of Hepatic Stellate Cells via TGFBR1 and TGFBR2. *Cell. Physiol. Biochem.* **2017**, *43*, 2242–2252. [CrossRef] [PubMed]
73. Shi, X.; Liu, Q.; Li, N.; Tu, W.; Luo, R.; Mei, X.; Ma, Y.; Xu, W.; Chu, H.; Jiang, S.; et al. MiR-3606-3p inhibits systemic sclerosis through targeting TGF-β type II receptor. *Cell Cycle* **2018**, *17*, 1967–1978. [CrossRef] [PubMed]
74. Makino, K.; Jinnin, M.; Hirano, A.; Yamane, K.; Eto, M.; Kusano, T.; Honda, N.; Kajihara, I.; Makino, T.; Sakai, K.; et al. The Downregulation of microRNA let-7a Contributes to the Excessive Expression of Type I Collagen in Systemic and Localized Scleroderma. *J. Immunol.* **2013**, *190*, 3905–3915. [CrossRef] [PubMed]
75. Maurer, B.; Stanczyk, J.; Jüngel, A.; Akhmetshina, A.; Trenkmann, M.; Brock, M.; Kowal-Bielecka, O.; Gay, R.E.; Michel, B.A.; Distler, J.H.W.; et al. MicroRNA-29, a key regulator of collagen expression in systemic sclerosis. *Arthritis Rheumatol.* **2010**, *62*, 1733–1743. [CrossRef] [PubMed]
76. Ciechomska, M.; O'Reilly, S.; Suwara, M.; Bogunia-Kubik, K.; Van Laar, J.M. MiR-29a Reduces TIMP-1 Production by Dermal Fibroblasts via Targeting TGF-β Activated Kinase 1 Binding Protein 1, Implications for Systemic Sclerosis. *PLoS ONE* **2014**, *9*, e115596. [CrossRef]
77. Honda, N.; Jinnin, M.; Kira-Etoh, T.; Makino, K.; Kajihara, I.; Makino, T.; Fukushima, S.; Inoue, Y.; Okamoto, Y.; Hasegawa, M.; et al. miR-150 Down-Regulation Contributes to the Constitutive Type I Collagen Overexpression in Scleroderma Dermal Fibroblasts via the Induction of Integrin β3. *Am. J. Pathol.* **2013**, *182*, 206–216. [CrossRef] [PubMed]
78. Tanaka, S.; Suto, A.; Ikeda, K.; Sanayama, Y.; Nakagomi, D.; Iwamoto, T.; Suzuki, K.; Kambe, N.; Matsue, H.; Matsumura, R.; et al. Alteration of circulating miRNAs in SSc: miR-30b regulates the expression of PDGF receptor β. *Rheumatol.* **2013**, *52*, 1963–1972. [CrossRef]
79. O'Reilly, S.; Ciechomska, M.; Fullard, N.; Przyborski, S.; Van Laar, J.M. IL-13 mediates collagen deposition via STAT6 and microRNA-135b: A role for epigenetics. *Sci. Rep.* **2016**, *6*, 25066. [CrossRef] [PubMed]
80. Makino, K.; Jinnin, M.; Aoi, J.; Hirano, A.; Kajihara, I.; Makino, T.; Sakai, K.; Fukushima, S.; Inoue, Y.; Ihn, H. Discoidin Domain Receptor 2–microRNA 196a—Mediated Negative Feedback against Excess Type I Collagen Expression Is Impaired in Scleroderma Dermal Fibroblasts. *J. Investig. Dermatol.* **2013**, *133*, 110–119. [CrossRef] [PubMed]

81. Nagpal, V.; Rai, R.; Place, A.T.; Murphy, S.B.; Verma, S.K.; Ghosh, A.K.; Vaughan, D.E. MiR-125b Is Critical for Fibroblast-to-Myofibroblast Transition and Cardiac Fibrosis. *Circulation* **2016**, *133*, 291–301. [CrossRef]
82. Banzhaf-Strathmann, J.; Edbauer, D. Good guy or bad guy: The opposing roles of microRNA 125b in cancer. *Cell Commun. Signal.* **2014**, *12*, 30. [CrossRef] [PubMed]
83. Kozlova, A.; Pachera, E.; Maurer, B.; Jüngel, A.; Distler, J.H.W.; Kania, G.; Distler, O. Regulation of Fibroblast Apoptosis and Proliferation by Micro RNA -125b in Systemic Sclerosis. *Arthritis Rheumatol.* **2019**, *71*, 2068–2080. [CrossRef]
84. Yao, Q.; Xing, Y.; Wang, Z.; Liang, J.; Lin, Q.; Huang, M.; Chen, Y.; Lin, B.; Xu, X.; Chen, W. MiR-16-5p suppresses myofibroblast activation in systemic sclerosis by inhibiting NOTCH signaling. *Aging* **2021**, *13*, 2640–2654. [CrossRef] [PubMed]
85. Kissin, E.; Korn, J.H. Apoptosis and myofibroblasts in the pathogenesis of systemic sclerosis. *Curr. Rheumatol. Rep.* **2002**, *4*, 129–135. [CrossRef]
86. Jafarinejad-Farsangi, S.; Farazmand, A.; Mahmoudi, M.; Gharibdoost, F.; Karimizadeh, E.; Noorbakhsh, F.; Faridani, H.; Jamshidi, A.R. MicroRNA-29a induces apoptosis via increasing the Bax:Bcl-2 ratio in dermal fibroblasts of patients with systemic sclerosis. *Autoimmunity* **2015**, *48*, 369–378. [CrossRef]
87. Jafarinejad-Farsangi, S.; Farazmand, A.; Gharibdoost, F.; Karimizadeh, E.; Noorbakhsh, F.; Faridani, H.; Mahmoudi, M.; Jamshidi, A.R. Inhibition of MicroRNA-21 induces apoptosis in dermal fibroblasts of patients with systemic sclerosis. *Int. J. Dermatol.* **2016**, *55*, 1259–1267. [CrossRef]
88. Iwamoto, N.; Vettori, S.; Maurer, B.; Brock, M.; Pachera, E.; Jüngel, A.; Calcagni, M.; Gay, R.E.; Whitfield, M.L.; Distler, J.H.; et al. Downregulation of miR-193b in systemic sclerosis regulates the proliferative vasculopathy by urokinase-type plasminogen activator expression. *Ann. Rheum. Dis.* **2016**, *75*, 303–310. [CrossRef]
89. Liakouli, V.; Cipriani, P.; Di Benedetto, P.; Panzera, N.; Ruscitti, P.; Pantano, I.; Berardicurti, O.; Carubbi, F.; Esteves, F.; Mavria, G.; et al. Epidermal Growth Factor Like-domain 7 and miR-126 are abnormally expressed in diffuse Systemic Sclerosis fibroblasts. *Sci. Rep.* **2019**, *9*, 4589. [CrossRef] [PubMed]
90. Alsaleh, G.; Francois, A.; Philippe, L.; Gong, Y.-Z.; Bahram, S.; Cetin, S.; Pfeffer, S.; Gottenberg, J.-E.; Wachsmann, D.; Georgel, P.; et al. MiR-30a-3p Negatively Regulates BAFF Synthesis in Systemic Sclerosis and Rheumatoid Arthritis Fibroblasts. *PLoS ONE* **2014**, *9*, e111266. [CrossRef] [PubMed]
91. Ciechomska, M.; Wojtas, B.; Swacha, M.; Olesinska, M.; Benes, V.; Maslinski, W. Global miRNA and mRNA expression profiles identify miRNA-26a-2-3p-dependent repression of IFN signature in systemic sclerosis human monocytes. *Eur. J. Immunol.* **2020**, *50*, 1057–1066. [CrossRef] [PubMed]
92. Makino, K.; Jinnin, M.; Kajihara, I.; Honda, N.; Sakai, K.; Masuguchi, S.; Fukushima, S.; Inoue, Y.; Ihn, H. Circulating miR-142-3p levels in patients with systemic sclerosis. *Clin. Exp. Dermatol.* **2012**, *37*, 34–39. [CrossRef] [PubMed]
93. Izumiya, Y.; Jinnn, M.; Kimura, Y.; Wang, Z.; Onoue, Y.; Hanatani, S.; Araki, S.; Ihn, H.; Ogawa, H. Expression of Let-7 family microRNAs in skin correlates negatively with severity of pulmonary hypertension in patients with systemic scleroderma. *Int. J. Cardiol. Heart Vasc.* **2015**, *8*, 98–102. [CrossRef] [PubMed]
94. Hashimoto, Y.; Akiyama, Y.; Yuasa, Y. Multiple-to-Multiple Relationships between MicroRNAs and Target Genes in Gastric Cancer. *PLoS ONE* **2013**, *8*, e62589. [CrossRef] [PubMed]
95. Dolcino, M.; Pelosi, A.; Fiore, P.F.; Patuzzo, G.; Tinazzi, E.; Lunardi, C.; Puccetti, A. Gene Profiling in Patients with Systemic Sclerosis Reveals the Presence of Oncogenic Gene Signatures. *Front. Immunol.* **2018**, *9*, 449. [CrossRef] [PubMed]
96. Wu, Q.; Liu, Y.; Xie, Y.; Wei, S.; Liu, Y. Identification of Potential ceRNA Network and Patterns of Immune Cell Infiltration in Systemic Sclerosis-Associated Interstitial Lung Disease. *Front. Cell Dev. Biol.* **2021**, *9*, 622021. [CrossRef]
97. Mullenbrock, S.; Liu, F.; Szak, S.; Hronowski, X.; Gao, B.; Juhasz, P.; Sun, C.; Liu, M.; McLaughlin, H.; Xiao, Q.; et al. Systems Analysis of Transcriptomic and Proteomic Profiles Identifies Novel Regulation of Fibrotic Programs by miRNAs in Pulmonary Fibrosis Fibroblasts. *Genes* **2018**, *9*, 588. [CrossRef] [PubMed]
98. Jiang, Z.; Tao, J.-H.; Zuo, T.; Li, X.-M.; Wang, G.-S.; Fang, X.; Xu, X.-L. The correlation between miR-200c and the severity of interstitial lung disease associated with different connective tissue diseases. *Scand. J. Rheumatol.* **2017**, *46*, 122–129. [CrossRef] [PubMed]
99. Kanno, Y.; Shu, E.; Niwa, H.; Seishima, M.; Ozaki, K.-I. MicroRNA-30c attenuates fibrosis progression and vascular dysfunction in systemic sclerosis model mice. *Mol. Biol. Rep.* **2021**, *48*, 3431–3437. [CrossRef]
100. Li, Y.; Huang, J.; Hu, C.; Zhou, J.; Xu, D.; Hou, Y.; Wu, C.; Zhao, J.; Li, M.; Zeng, X.; et al. MicroRNA-320a: An important regulator in the fibrotic process in interstitial lung disease of systemic sclerosis. *Arthritis Res.* **2021**, *23*, 1–11. [CrossRef]
101. Montgomery, R.L.; Yu, G.; Latimer, P.A.; Stack, C.; Robinson, K.; Dalby, C.M.; Kaminski, N.; Van Rooij, E. Micro RNA mimicry blocks pulmonary fibrosis. *EMBO Mol. Med.* **2014**, *6*, 1347–1356. [CrossRef] [PubMed]
102. O'Reilly, S. Epigenetic modulation as a therapy in systemic sclerosis. *Rheumatology* **2018**, *58*, 191–196. [CrossRef] [PubMed]
103. Raposo, G.; Stoorvogel, W. Extracellular vesicles: Exosomes, microvesicles, and friends. *J. Cell Biol.* **2013**, *200*, 373–383. [CrossRef]
104. Yuana, Y.; Sturk, A.; Nieuwland, R. Extracellular vesicles in physiological and pathological conditions. *Blood Rev.* **2013**, *27*, 31–39. [CrossRef]
105. Colletti, M.; Galardi, A.; De Santis, M.; Guidelli, G.M.; Di Giannatale, A.; Di Luigi, L.; Antinozzi, C. Exosomes in Systemic Sclerosis: Messengers Between Immune, Vascular and Fibrotic Components? *Int. J. Mol. Sci.* **2019**, *20*, 4337. [CrossRef] [PubMed]

Article

MiRNAs in Systemic Sclerosis Patients with Pulmonary Arterial Hypertension: Markers and Effectors

Mor Zaaroor Levy [1,2], Noa Rabinowicz [1,2], Maia Yamila Kohon [1,2], Avshalom Shalom [1,3], Ariel Berl [1,3], Tzipi Hornik-Lurie [4], Liat Drucker [1,5], Shelly Tartakover Matalon [1,2,*,†] and Yair Levy [1,2,6,*,†]

1 Sackler Faculty of Medicine, Tel Aviv University, Tel Aviv 6997801, Israel; mor.zaaroor@yahoo.com (M.Z.L.); noa_uziel@hotmail.com (N.R.); maia.kohon@clalit.org.il (M.Y.K.); avshalom.shalom@clalit.org.il (A.S.); arielberl23@gmail.com (A.B.); druckerl@clalit.org.il (L.D.)
2 Autoimmune Research Laboratory, Meir Medical Center, Kfar Saba 4428164, Israel
3 Department of Plastic Surgery, Meir Medical Center, Kfar Saba 4428164, Israel
4 Research Authority, Meir Medical Center, Kfar Saba 4428164, Israel; tzipi.hornik@clalit.org.il
5 Oncogenetic Laboratory, Meir Medical Center, Kfar Saba 4428164, Israel
6 Department of Internal Medicine E, Meir Medical Center, Kfar Saba 4428164, Israel
* Correspondence: matalon.shelly@clalit.org.il (S.T.M.); levy.yair@clalit.org.il (Y.L.); Tel./Fax: +972-9-74721992 (S.T.M.)
† These authors contributed equally to this work.

Citation: Zaaroor Levy, M.; Rabinowicz, N.; Yamila Kohon, M.; Shalom, A.; Berl, A.; Hornik-Lurie, T.; Drucker, L.; Tartakover Matalon, S.; Levy, Y. MiRNAs in Systemic Sclerosis Patients with Pulmonary Arterial Hypertension: Markers and Effectors. *Biomedicines* **2022**, *10*, 629. https://doi.org/10.3390/biomedicines10030629

Academic Editor: Michal Tomcik

Received: 20 December 2021
Accepted: 3 March 2022
Published: 8 March 2022

Abstract: Background: Pulmonary arterial hypertension (PAH) is a major cause of death in systemic sclerosis (SSc). Early detection may improve patient outcomes. Methods: We searched for circulating miRNAs that would constitute biomarkers in SSc patients with PAH (SSc-PAH). We compared miRNA levels and laboratory parameters while evaluating miRNA levels in white blood cells (WBCs) and myofibroblasts. Results: Our study found: 1) miR-26 and miR-let-7d levels were significantly lower in SSc-PAH (n = 12) versus SSc without PAH (SSc-noPAH) patients (n = 25); 2) a positive correlation between miR-26 and miR-let-7d and complement-C3; 3) GO-annotations of genes that are miR-26/miR-let-7d targets and that are expressed in myofibroblast cells, suggesting that these miRNAs regulate the TGF-β-pathway; 4) reduced levels of both miRNAs accompanied fibroblast differentiation to myofibroblasts, while macitentan (endothelin receptor-antagonist) increased the levels. WBCs of SSc-noPAH and SSc-PAH patients contained equal amounts of miR-26/miR-let-7d. During the study, an echocardiograph that predicted PAH development, showed increased pulmonary artery pressure in three SSc-noPAH patients. At study initiation, those patients and an additional SSc-noPAH patient, who eventually developed PAH, had miR-let-7d/miR-26 levels similar to those of SSc-PAH patients. This implies that reduced miR-let-7d/miR-26 levels might be an early indication of PAH. Conclusions: miR-26 and miR-let-7d may be serological markers for SSc-PAH. The results of our study suggest their involvement in myofibroblast differentiation and complement pathway activation, both of which are active in PAH development.

Keywords: systemic sclerosis; pulmonary arterial hypertension (PAH); biomarkers; miRNA; complement; myofibroblasts; macitentan

1. Introduction

Systemic sclerosis (SSc) is a rare connective tissue disease that primarily affects women. Its main manifestations are skin and visceral fibrosis, vascular hyperactivity and musculoskeletal changes [1]. It is generally accepted that an interplay between genetic and environmental factors induce SSc-specific gene programs in several cell populations, including immune cells, endothelial cells and myofibroblasts [2,3]. SSc involves three main pathophysiologic pathways: (1) An autoimmune attack, which is the first marker of the disease, (2) vascular injury followed by defective neovascularization and impaired remodeling (3) and extensive tissue fibrosis [3].

Pulmonary arterial hypertension (PAH) is a major cause of death in approximately 12% of SSc patients [3]. Older age at scleroderma onset is a risk factor for PAH [4,5]. SSc-associated PAH (SSc-PAH) is caused by functional alterations and structural fibroproliferative vasculopathy affecting the small- and medium-size pulmonary arteries, leading to an increase in mean pulmonary artery pressure of >20 mmHg in echocardiography (echo) at rest [6,7]. SSc-PAH has a poor prognosis, with a five-year overall survival rate of 50% [8].

Early detection of PAH in patients with SSc is essential because an asymptomatic phase is common and early treatment can significantly improve clinical outcomes and positively affect survival [5]. Accurate diagnosis of PAH is clinically challenging and relies on right heart catheterization. This invasive procedure is not suitable for screening and is typically performed only for patients with a high index of suspicion based on echo and other criteria that are less sensitive [8]. In recent years, great efforts have been directed to finding serum-based diagnostic biomarkers for SSc-PAH that would allow for rapid, noninvasive screening, and early diagnosis.

As the pathogenesis of SSc is influenced by environmental factors, it is reasonable to assume that they induce epigenetic regulatory changes that alter disease outcomes [9]. RNA interference via microRNAs (miRNA) is a leading mechanism for initiating and maintaining epigenetic changes. Previous publications showed that SSc patients have different levels of certain miRNAs in their circulation and skin fibroblasts compared to healthy individuals [9,10]. Yet, only a few reports described differences in miRNA expression between patients with SSc that do not have PAH (SSc-noPAH) vs. SSc-PAH.

The aim of this study was to find markers that would allow diagnosis of PAH in the early stages. We searched for a marker that would not require significant invasive intervention. Obtaining a blood sample does not require a complex invasive process; therefore, we chose to perform our analysis on blood samples from our biobank. We selected plasma samples of 25 SSc patients without PAH (SSc-noPAH) and 12 SSc-PAH patients. We searched for a molecule that affects signal transmission processes and the development of PAH. As miRNAs are often used to identify pathologic conditions such as cancers and autoimmune diseases, in the current study, we searched for circulating miRNAs that could serve as biomarkers for SSc patients with PAH (SSc-PAH). To learn more about the effects of these miRNAs in PAH, we correlated patients' miRNA levels and laboratory parameters and performed a bioinformatics analysis to find biological annotations of genes common to the miRNA targets and to the biology of immune, endothelial and myofibroblast cells. Since plasma miRNAs represent the total miRNAs secreted from various cells in the body, we also aimed to determine which cell population(s) is responsible for the changes in plasma miRNAs by evaluating levels of the biomarker miRNAs in white blood cells (WBCs) of SSc-noPAH and SSc-PAH patients, as well as in activated myofibroblasts.

2. Materials and Methods

2.1. Patient Data

The current study included the entire population of SSc-PAH patients at Meir Medical Center (12 patients) and 25 SSc-noPAH patients, whose blood samples were consecutively collected. SSc patients were diagnosed according to 2013 EULAR/ACR classification, by using Doppler echocardiography (EPIQCVx Release 5.0.2, PHILIPS)) and right heart catheterization. Doppler echocardiography estimated PASP (pulmonary artery systolic pressure, where PASP = 4 (TRV)2) exceeding 35 to 40 mmHg was considered elevated. When PAH was suspected based on this screening, we proceed to right-heart catheterization (RHC). Cut-off values for establishing SSc-PAH in right heart catheterization (RHC) is mPAP >20 mmHg. All samples were from patients at least 2 years after disease onset. Their clinical features are described in Table 1. The clinical data that were collected for each patient included C3 and C4 fragments of the complement, total IgG, IgG1–4 levels, autoantibodies positivity (anti-centromere, anti-RNA-polymerase III and anti-DNA-topo-isomerase I) and CRP.

Table 1. Demographic and clinical characteristics of the patients.

Demographic Parameters	SSc-noPAH (n = 25)		SSc-PAH (n = 12)	
	Valid N	Median (25–75 Percentile)/%	Valid N	Median (25–75 Percentile)/%
Female (a)	25	88%	12	83%
Age at diagnosis (years) (b)	25	46.0 (39.0–57.0) *	12	65.0 (61.0–73.0)
Duration of illness (years) (b)	25	6.0 (4.0–12.0)	12	3.5 (2.0–9.0)
Clinical parameters				
Diffuse vs. limited SSc (a)	25	92% *	12	42%
Echo (mPAP, mmHg) (b)	25	28.0 (25.0–30.0) *	12	65.0 (49.5–82.0)
FVC (%) (b)	18	82.0 (72.0–101.0)	10	70.5 (58.0–84.0)
FEV1 (%) (b)	18	82.5 (71.0–95.8)	10	64.0 (53.0–75.0)
DLCO (%) (b)	18	59.3 (51.3–68.0) *	9	43.0 (31.0–48.3)
6 MW (meter) (b)	17	513.0 (456.0–528.0) *	8	259.5 (151.0–430.0)
C-reactive protein CRP mg/dl (b)	22	0.8 (0.3–1.5)	12	0.8 (0.4–1.8)
Anti-centromere (a)	8	12% *	6	83%
Anti-DNA-topo-isomerase I (a)	24	S137%	12	17%
Anti-RNA-polymerase III (a)	24	33%	11	27%

Median and 25–75 percentile/% were used to demonstrate the distribution of non-categorical variables and % for the categorical variables. (a) Chi-square test was used to test relationships between categorical variables and the two independent groups. (b) Mann–Whitney U test was used to compare differences between the two independent groups: SSc-noPAH and SSc-PAH. DLCO: Diffusing capacity of lung for carbon monoxide; Echo: Echocardiogram; FEV1: Forced expiratory volume in one second; FVC: Forced vital capacity; 6 MW: Six-minute walk; * Results significantly different from SSc-PAH, $p < 0.05$.

2.1.1. C3, C4 and IgG Evaluation

C3, C4 and IgG levels were evaluated with NAS IGG, NAS C3 and NAS C4 kits (Siemens, Marburg, Germany), according to manufacturer's instructions. Briefly, the proteins in the plasma form immune complexes with specific antibodies, which scatter a beam of light passed through the sample. The intensity of the scattered light is proportional to the concentration of the relevant protein in the sample. The results were evaluated by comparison with standard values of known concentration.

2.1.2. IgG1-4 Evaluation

IgG1-4 levels were evaluated using Optilite IgG1-4 kits (Optilite, Birmingham, UK). The method for determining the protein concentration is similar to that of C3 (Section 2.1.1), using the Binding Site analyzer (Optilite, Birmingham, UK).

2.1.3. CRP Evaluation

CRP was evaluated using CRP Latex reagent (Beckman Coulter, Brea, CA, USA) and the Beckman Coulter System. During evaluation, CRP reacts specifically with anti-human CRP antibodies coated on the latex particles to yield insoluble aggregates. The absorbance of these aggregates is proportional to the CRP concentration in the sample.

2.1.4. Evaluation of Autoantibodies

Screening for autoantibodies was done using the BioPlex 2200 ANA Screen Pack (Bio-Rad, Hercules, CA, USA). Detection was done in a multiplex flow immunoassay with the BioPlex 2200 System (Bio-Rad, Hercules, CA, USA).

2.2. *Biobank*

The Autoimmune Laboratory at Meir Medical Center in Israel manages a rheumatologic bio-bank that contains plasma, serum and blood cells collected from rheumatology patients since April 2018. Blood samples are collected from participants who previously consented and authorized the use of their samples for research purposes. Blood samples were separated into serum, plasma and cells a few minutes after blood draw and immediately placed in a −80 °C freezer.

2.3. Blood Collection and Separation of Plasma and WBCs

Blood samples were taken from participants who consented to donate blood. Twenty ml of blood was drawn into a tube with EDTA and centrifuged (3000× *g*, 10 min, room temperature). The plasma was collected into a tube and immediately frozen at −80 °C. Then, the WBCs cells that lay over the RBC along with the top layer of the RBC were collected into a separate test tube and immediately frozen at −80 °C. Prior to RNA extraction, the cells were incubated for 10 min at room temperature with a buffer lysis (Lysing Solution (X10), Beckman Coulter, Villepinte, France), diluted with molecular biology grade water, Dnase and RNase free (Satorius, Beit Haemek, Israel) and centrifuged (1400 rpm, 5 min). The top fluid was discarded and RNA extraction was performed on the remaining cells at the bottom of the test tubes.

2.4. RNA Extraction and RT cDNA Synthesis

RNA was extracted from the SSc-noPAH and SSc-PAH patients' plasma using miRNeasy Serum/Plasma Advanced Kit (QIAGEN, Hilden, Germany) and from the WBCs using miRNeasy Tissue/Cells Advanced mini-Kit (QIAGEN, Hilden, Germany) [11]. Extracted RNA was converted to cDNA using TaqMan Advanced miRNA cDNA Synthesis Kit (Thermo Fisher, Waltham, MA, USA) according to manufacturer's instructions.

2.5. Real-Time Quantitative PCR

PCR reaction was done using TaqMan Fast Advanced Master Mix (Applied Biosystems, Vilnius, Lithuania) and TaqMan Advanced miRNA Assay (Applied Biosystems, Vilnius, Lithuania) for miRNAs hsa-26a-5p, hsa-21-5p, hsa-155-5p, hsa-miR-let-7d-5p, hsa-29a-5p, hsa-145-5p, hsa-150-5p and hsa-23a-5p according to the manufacturer's instructions (Applied Biosystems, Vilnius, Lithuania). Hsa-16-5p served for normalization of miRNA RT-qPCR expression analysis [11,12].

2.6. Isolation of Dermal Fibroblasts

Cells were isolated from the skin of healthy subjects who underwent abdominoplasty in the Plastic Surgery Unit of Meir Medical Center and who provided signed informed consent. The skin was cleaned with 70% ethanol, layered in PBS-penicillin streptomycin 1% and the fat was removed using scissors. Biopsy was cut into 25–30 small pieces (~4 mm), using a Uni Punch (Premier Company, Mumbai, Maharashtra, India), placed in a tube with 7 mL collagenase mix (0.02 g collagenase type 2 (Gibco, New York, NY, USA), 0.001 g DNase (Roche, Mannheim, Germany), 2 mL Trypsin X10 (Biological Industries, Beit Haemek, Israel), 0.1 gr bovine serum albumin (Sigma-Aldrich, St. Louis, MO, USA), in 20 mL Dulbecco's modified eagle medium (DMEM; Biological Industries, Beit Haemek, Israel) vortexed for 30 s, cultured in an incubator containing a rotating device for 45 min, at 37 °C, shaken and incubated for an additional 45 min. Following incubation, 7 mL of full DMEM (DMEM supplemented with 10% fetal bovine serum (FBS), 2 mM l-glutamine and antibiotics (100 U/mL penicillin, 100 µg/mL streptomycin, nystatin), all from Biological Industries, Beit Haemek, Israel) was added to each test tube and centrifuged (1700 rpm for 8 min). The top fluid was removed and the remaining fluid was cultured in plates with full medium containing 20% FBS in a CO2 incubator at 37 °C. Cells were observed every 2 days and new 20% FBS medium was added. After 4 days, the pieces were discarded and when cells reached 50% confluence, media were replaced with 10% full DMEM.

2.7. Cell Culture

Dermal fibroblasts were cultured in full DMEM. Cells were incubated at 37 °C in 5% CO2 and split twice a week using trypsin, upon reaching 80–90% confluence. Cells were used in the experiments in passages 4–9.

2.8. Exposure of Dermal Fibroblasts to TGF-β and to Macitentan

Cells were cultured in 96- or 24-well plates with full DMEM that contained 1% FBS. Twenty-four hours later, new medium (control) or TGF-β (10 ng/mL, Peprotech, East Windsor, NJ, USA), with or without macitentan (1) (1 μM) Selleck chemicals, Houston, TX, USA) or DMSO (Sigma-Aldrich, St. Louis, MO, USA) in the same concentration as control (1:50,000), were added to the cells for 48 h or 72 h. Then, the following procedures were performed: (1) Cells were harvested and counted using an automatic cell counter (NanoEntek, Waltham, MA, USA) and trypan-blue (Gibco, New York, NY, USA (72 h); (2) proteins were isolated from the cells for evaluation of αSMA and collagen I levels (western blot, 48 h) and stored at −80 °C; (3) RNA was isolated from the cells for future miRNA analysis (48 h).

2.9. Cell Count

Trypan blue (Biological Industries, Beit Haemek, Israel) was mixed in a 1:1 ratio with the cells. Then the cells were counted using automatic cell counter (NanoEntek, Waltham, MA, USA). Live cells remained unstained, while dead cells assimilated the dye. The cell counter counted the number of dyed and undyed cells in each sample. Using these values, it calculated the percentages of dead cells in the culture.

2.10. Protein Extraction

Dermal fibroblasts were lysed in a lysis buffer (50 mM Hepes, 150 mM NaCl, 1% Triton X-100, 0.1% SDS, 50 mM NaF, 10 mM NaPPi, 2 mM $NaVO_3$, 10 mM EDTA, 2 mM EGTA, 1 mM PMSF and 10 μg/mL Leu-peptin) for 10 min on ice, and centrifuged (15 min, 12,000 RPM, 4 °C). Protein levels were determined by using Pierce BCA protein assay kit (Pierce, Rockford, IL, USA), according to manufacturer's instructions.

2.11. Western Blotting

To assess the levels of αSMA and collagen I in the fibroblasts, protein lysates were mixed (1:5) with sample buffer (250 mM Tris-HCl, pH 6.8, 400 mM DTT, 140 mM SDS, 60% glycerin, 0.02% bromophenol blue) and denatured for 10 min at 65 °C. Proteins (25 μg) from each sample were separated by electrophoresis on SDS-PAGE and wet transferred to a PVDF membrane. Transfer efficiency was validated with Ponceau staining (Sigma-Aldrich, St. Louis, MO, USA). After blocking the non-specific binding sites with 5% milk powder (Difco, Saint-Ferreol, France) in Tris Buffer Saline containing 0.1% Tween (TBS-T) (Milipore Sigma, Molsheim, France), the membranes were incubated with the primary antibodies at 4 °C overnight. We used the following antibodies: Mouse anti-human alpha smooth muscle actin (αSMA, IgG2a, clone: 1A4, ab7817; Abcam, Cambridge, England), rabbit anti-human collagen type I, (IgG, ab34710, Abcam) and rabbit mouse anti-tubulin (clone: B-5-1-2, T516, Sigma Aldrich, St. Louis, MO, USA). Primary antibody was rinsed with TBS-T and TBS (Milipore Sigma)). Bound antibodies were visualized using peroxidase-conjugated secondary antibody goat anti-mouse IgG and goat anti-rabbit IgG (Jackson ImmunoResearch Laboratories, West Grove, PA, USA), followed by enhanced chemiluminescence HRP substrate detection (Milipore Sigma, Waltham, MA, USA). Optical densities were visualized and measured as arbitrary units using an LAS-3000 Imager (Fujifilm, Tokyo, Japan). Results were normalized to tubulin using a multi-gauge V3.0 program (Fujifilm).

2.12. Bioinformatics

Targets of the miRNA were found by using 2 software packages (targetscan, http://www.targetscan.org, accessed on 30 October 2021 and mirbd, http://mirdb.org, accessed on 30 October 2021).

Specific sites for miRNA let7d:

http://www.targetscan.org/cgi-bin/targetscan/vert_71/targetscan.cgi?mirg=hsa-miR-let-7d-5p (accessed on 30 October 2021) http://mirdb.org/cgi-bin/search.cgi?searchType=miRNA&full=mirbase&searchBox=MIMAT0000065 (accessed on 30 October 2021)

Specific sites for miR-26:

http://mirdb.org/cgi-bin/search.cgi?searchType=miRNA&full=mirbase&searchBox=MIMAT0000082 (accessed on 30 October 2021)

http://www.targetscan.org/cgi-bin/targetscan/vert_71/targetscan.cgi?mirg=hsa-miR-26a-5p (accessed on 30 October 2021)

A list of genes involved in (1) endothelial cell biology, (2) myofibroblast cell biology and (3) immune system biology was made by using these websites:

Specific sites for endothelial cell biology:

http://ezbiosystems.com/servicescontent.asp?d_id=174 (accessed on 30 October 2021)

https://www.sciencellonline.com/PS/GK019.pdf (accessed on 30 October 2021)

Specific sites for myofibroblast cell biology:

https://maayanlab.cloud/Harmonizome/gene_set/myofibroblast/TISSUES+Text-mining+Tissue+Protein+Expression+Evidence+Scores (accessed on 30 October 2021)

https://maayanlab.cloud/Harmonizome/gene_set/myofibroblast/GeneRIF+Biological+Term+Annotations (accessed on 30 October 2021)

Specific sites for Immune system biology

https://www.sciencellonline.com/PS/GK072.pdf (accessed on 30 October 2021)

https://www.sciencellonline.com/PS/GK039.pdf (accessed on 30 October 2021)

https://www.thermofisher.com/order/catalog/product/4370573#/4370573 (accessed on 30 October 2021)

Then, Venny software: https://bioinfogp.cnb.csic.es/tools/venny/, (accessed on 30 October 2021) was used to create 6 lists of genes which are common to the miRNA targets and the 3 types of cells. Toppgene software https://toppgene.cchmc.org/navigation/database.jsp, (accessed on 30 October 2021) was used to find annotations for the gene lists.

2.13. Statistical Analysis

Demographic, clinical and laboratory characteristics of the patients were presented as medians and percentages, each when appropriate. Due to the small number of patients, the median and 25–75 percentile was used to demonstrate the distribution of non-categorical variables. Percentages were used for categorical variables. Mann–Whitney U tests were used to compare differences between the two independent groups (SSc-noPAH and SSC-PAH) for the continuous, dependent variables. Chi-square tests were used to test relationships between categorical variables and the two independent groups. Spearman's Rho correlations were used to evaluate the association between the research variables. A partial Spearman's Rho correlation was used to evaluate the association between the study variables, adjusted for confounding variables. Cell culture experiments with dermal fibroblasts were conducted 4 times and miRNAs in WBCs were evaluated 7 to 8 times. Paired student t-tests were applied to analyze differences between paired data. All statistical analyses were performed using SPSS-PC statistical software, version 27.0 (IBM Corp., Armonk, NY, USA). Two-sided tests of significance $p < 0.05$ were used in all analyses.

3. Results

3.1. Patient Characteristics

To find miRNAs that characterize SSc-PAH, we extracted RNA from SSc patients with and without PAH. The detailed clinical features of these patients are presented in Table 1. There was no significant difference in the duration of illness between the groups However, SSc-PAH patients were diagnosed with SSc at an older age than SSc-noPAH patients (65 years vs. 46 years, $p < 0.05$). Furthermore, as expected, median echo pulmonary artery pressure results (mPAP) were higher in the SSc-PAH than in the SSc-noPAH group (65 mmHg vs. 28 mmHg, respectively, $p < 0.05$; Table 1). Median diffusing capacity of lung for carbon monoxide (DLCO) and six-minute walk (6 MW) test results were lower in the SSc-PAH group (DLCO: 43% vs. 59% and 6 MW: 259 m vs. 513 m ($p < 0.05$ for both), further detailed in Table 1). Among the laboratory tests, only the median level of anti-centromere antibodies differed between groups, with higher levels in the SSc-PAH than in the SSc-noPAH group (positive in 83% vs. 12%, respectively, $p < 0.05$).

3.2. Literature Search: Circulating miRNAs That May Differentiate between SSc and SSc-PAH Patients

Previous studies have demonstrated that miRNAs are involved in the development of SSc. Therefore, we examined the literature and generated a list of seven miRNAs. The seven miRNAs were chosen following a literature search for miRNAs that were (a) found in the circulation of SSc-PAH or SSc-noPAH patients (b) associated with fibrosis (c) correlated with disease prognosis. Table 2 indicates the miRNAs, biological material (circulating blood/cells) and what was reported in these studies. We hypothesized that some of these miRNAs may distinguish between SSc patients with and without PAH.

Table 2. miRNAs that characterize pulmonary fibrosis, PAH or are involved in the development of SSc.

miRNA	Previous Knowledge	Biological Material	References
miR-23a	Biomarkers in idiopathic PAH	Circulating blood	[13]
miR-29a	Increased in SSc Induce ventricular hypertrophy and fibrosis	Circulating blood Skin	[14–16]
miR-26a	Reduced in patients with PAH	Circulating blood	[17]
miR-150	Reduced level is associated with poor survival in PAH	Circulating blood Dermal fibroblasts	[18,19]
miR-21	Enhances vascular cell proliferation	Circulating blood	[20–22]
miR-let-7d	Negatively correlated with severity of PAH in patients with SSc	Skin Circulating blood	[23,24]
miR-155	Drives fibrosis	Fibroblasts	[25]

miR: miRNA; PAH: Pulmonary arterial hypertension; SSc: Systemic sclerosis.

3.3. miRNA Pattern in SSc-noPAH Versus SSc-PAH Patients

To evaluate the levels of the miRNAs shown in Table 2 in the circulating blood of SSc-noPAH and SSc-PAH patients, we isolated RNA from patients' plasma. First, we measured the level of the control miR-16 in both groups using qRT-PCR. As previously described [11,12], we found similar levels in both groups (SSc-noPAH CT = 25.8, SSc-PAH CT = 26.1) and concluded that miR-16 can serve as a normalizing miRNA. Then, we measured the levels of the seven miRNAs relative to the normalizing miR-16 using qRT-PCR (Figure 1). miR-23 was not detected in the patients' plasma. miR-29 levels were very low in the plasma and PCR results were found in only about 60% of the samples; therefore, it was excluded from the analysis. The most significant difference in miRNA expression between the two groups was that of miR-let-7d, which was decreased in the PAH group ($p = 0.15$, Figure 1). miR-let-7d levels were homogeneous in the PAH group. Nevertheless, high heterogeneity was found between the levels of all other miRNAs, both in the SSc-PAH group and especially in the SSc-noPAH group. Since the miRNA levels varied greatly, particularly in the SSc-noPAH group, as described in Figure 2, we suspected that the physiological conditions of the patients in this group were not consistent and that among the SSc-noPAH patients who were not diagnosed with SSC-PAH, were some with early onset PAH. Therefore, we assumed that this subset of patients expressed miRNAs at different levels than those who had not yet developed symptoms of PAH.

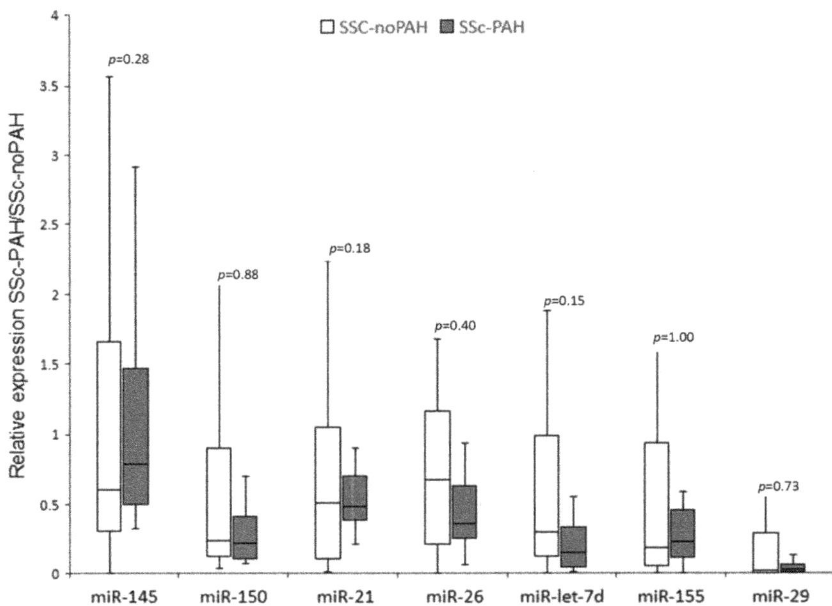

Figure 1. Median miRNA levels in the plasma of SSc patients with and without PAH. RNA was isolated from plasma of 25 SSc-noPAH and 12 SSc-PAH patients. The levels of seven miRNAs were evaluated by qRT-PCR. miR-16 served for normalization. The boxplot presents five values (minimum: (Q1–1.5*IQR); first quartile (Q1/25th percentile); median; third quartile (Q3/75th percentile); and maximum (Q3 + 1.5*IQR)) of the miRNAs. miR: miRNA; PAH: Pulmonary arterial hypertension; SSc-noPAH: Patients with SSc without PAH; SSc-PAH: Patients with SSc and PAH.

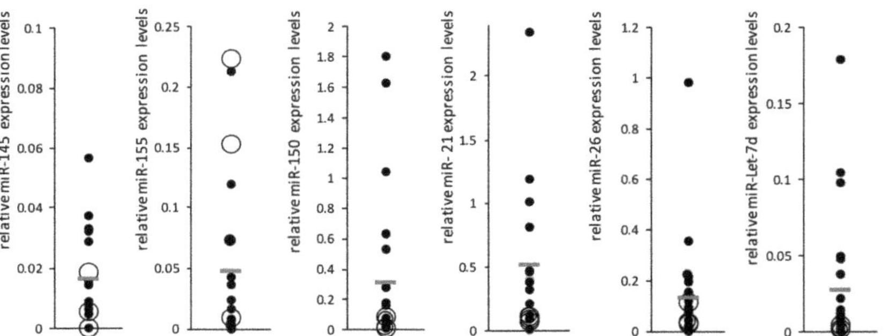

Figure 2. Mean miRNA levels in the SSc-noPAH group. miRNA levels in the plasma of 25 SSc-noPAH patients were evaluated using qRT-PCR. miR-16 served for normalization. Patients underwent echocardiography before blood sampling and at the end of the study. miRNA values of the patients with stable echo results during this time are marked in full black dots (n = 21). The miRNA levels of patients whose echo levels increased during this period are marked by empty circles (n = 4). All y-axis values are presented relative to miR-16. The red line indicates the mean value of each group.

SSc patients undergo routine monitoring to evaluate their condition every few months. One of the tests used is echocardiography (echo), which measures mean pulmonary artery pressure; elevated levels imply disease development. Therefore, data from patients in the SSc-noPAH group (from the time of blood sample collection for miRNA analysis until the end of the experiment, ~1–2 years) were analyzed to assess whether changes in their echocardiograms might indicate the development of PAH. Among the SSc-noPAH patients, we found three whose mean pulmonary arterial pressure had increased during the study by >5 mmHg and one whose pulmonary pressure could not be assessed using echo due to technical issues (lack of tricuspid regurgitation) but had developed PAH. miRNA levels of all patients are shown in Figure 2 as full black dots and the four patients with increased pulmonary artery pressure or who developed PAH (described above) are marked as an empty larger circle. Interestingly, it was found that the levels of four miRNAs (miR-let-7d, miR-26, miR-150 and miR-21) in these four patients (empty circles) were relatively low and similar to those of PAH patients. Among these four patients, two developed PAH.

As the four patients whose medical condition deteriorated during the study had similar miRNA levels as the SSc-PAH patients, but were similar in other parameters (such as lung function) to those of SSc patients who did not have PAH, we termed them SSc-susPAH (suspected PAH). We decided to exclude this subset of patients from both groups and from further analysis.

Additional statistical tests (Mann–Whitney) that compared the miRNA levels (21, 7, 150, 155, 145 and 26) of the SSc-noPAH group (without SSc-susPAH) and the PAH group, demonstrated a significant decrease in the level of miR-let-7d ($p < 0.05$) and a trend toward decreased levels of miR-26 ($p = 0.14$) in the PAH group.

Moreover, Spearman correlation tests were used to examine whether there was a relation between development of PAH and the patients' demographic parameters, the presence or absence of autoantibodies (anti-centromere antibody, anti-DNA-topo-isomerase I and anti-RNA-polymerase III), CRP levels, the level of miRNAs in the plasma and the pulmonary function test results (DLCO, 6MW). We found positive correlations between PAH and age (rs = 0.56, $p < 0.05$), anti-centromere antibody positivity (rs = 0.67, $p < 0.05$) and, as expected, to pulmonary artery pressure (mPAP, rs = 0.82, $p < 0.05$). A negative correlation was found with miR-let-7d plasma levels (rs = -0.41, $p < 0.05$) and as expected, with lung function tests (DLCO: rs = -0.53, $p < 0.05$, 6MW: rs = -0.55, $p < 0.05$).

We also performed a partial Spearman's Rho correlation between the PAH complication and the miRNAs, placing age and sex as confounding factors. We chose only these confounders as limiting factors because other parameters that were correlated with PAH (previous section, lung function and mPAP) are part of the definition of the disease. Partial Spearman's Rho correlation, which measures the association between the PAH phenotype to miR-let-7d and miR-26 levels, while adjusting for age, showed that these correlations were significant and not age-dependent ($p = 0.011$ and $p = 0.045$, respectively).

3.4. Partial Spearman's Rho Correlation between miRNA Level and Laboratory Parameters

We hypothesized that antibodies and complement might contribute to the generation of the pro-inflammatory/profibrotic tissue of patients. We also assumed that these changes would be early events in the disease and might be affected by the miRNAs. Therefore, we examined whether there were correlations between miR-let-7d and miR-26 and the patients' laboratory parameters (C3, C4, CRP, IgG1, IgG2, IgG3, IgG4, IgG and autoantibodies (anti-centromere, anti-DNA-topo-isomerase I and anti-RNA-polymerase III)). A partial Spearman's Rho correlation that measured the relations between levels of miR-let-7d and miR-26, between miR-let-7d and IgG4, between miR-let-7d and CRP and between miR-26 and IgG2, while adjusting for age and sex, showed that all correlations were statistically significant ($p < 0.05$) and were not conditional on age or sex (Table 3). Furthermore, a significant partial correlation was found between miR-26 level and C3 ($p < 0.05$) and a trend between miR-let-7d and C3 ($p = 0.07$), while adjusting for age, sex and C4 (which is correlated with C3, as described in Table 3).

Positive correlations were found between miR-26 and miR-let-7d and the C3 level. Since the miRNA levels were lower in PAH patients, we examined whether the level of C3 in the blood of SSc-PAH and SSc-susPAH patients was also lower than in SSc-noPAH patients. We found that the mean C3 level was 109 mg/dL among the SSc-noPAH patients, 97 mg/dL among the SSc-PAH patients and 87 mg/dL in the SSc-susPAH group. While the directionality of these results is as expected, the differences did not reach statistical significance, perhaps, due to the small cohort size.

Table 3. Correlation tests between miRNA levels and IgG and C3/4 levels.

Variables		Confounders-Z	r(s)xy.z	*p*-Value	*n*
X	**Y**				
miR-let-7d	miR-26	Age, Sex	0.431	0.006	31
miR-let-7d	C3	Age, Sex, C4	0.437	0.070	16
miR-let-7d	IgG4	Age, Sex	−0.626	0.017	12
miR-let-7d	CRP	Age, Sex	0.342	0.032	28
miR-26	C3	Age, Sex, C4	0.468	0.050	16
miR-26	IgG2	Age, Sex	−0.626	0.017	12

C3: Complement component 3; C4: Complement component 4; CRP: C-reactive protein; miR: miRNA; *n*: Number of patients; r(s)xy.z = partial Spearman's Rho correlation coefficient.

In conclusion, these data suggest that there is an association between low expression of miR-26 and miR-let-7d and the PAH complication, and that there is a correlation between these miRNAs, and parameters associated with inflammation (CRP) and complement activation. These associations may suggest that these miRNAs are involved in mediating the activity of the complement system. These assumptions will need to be proven in future studies, using appropriate biological systems.

3.5. Finding Targets of miR-Let-7d and miR-26 Involved in Immune, Endothelial and Myofibroblast Cell Biology

We surmised that if we identify known targets of let-7 and miR-26 that are also relevant to the cell biology of the major players in SSc, we may learn how these miRNAs contribute to its development. Therefore, we conducted a bioinformatics analysis of the microRNAs' targets using the software packages targetscan (http://www.targetscan.org, accessed on 30 October 2021) and mirbd (http://mirdb.org, accessed on 30 October 2021).

The resulting lists for each miRNA were combined to include all implicated genes (461 for miR-let-7d and 1,112 for miR-26). We also collected established characteristic gene lists for immune cells (233 genes), endothelial cells (130 genes) and myofibroblasts (477 genes). Using Venny software, we found genes common to each cell type with each miRNA (yielding six lists of 7–30 genes, Tables 4 and 5). Then, we used Toppgene software to find GO biological annotations that match each of the six lists. We speculated that the annotations would suggest biological processes in which the miRNAs are involved. The results shown in Table 4, Table 5, Tables S1 and S2 suggest that both miRNAs are involved in cytokine signaling (miR-let-7d in IL-10 and CCR7 and miR-26 in IL18R1 and IRF4). Moreover, the analysis suggests that miR-26 is involved in blood vessel biology and in motility processes, while miR-let-7d appears to be significant in the processes of myofibroblast proliferation and endothelial cell proliferation and apoptosis (Tables S1 and S2). Our bioinformatics study also showed that both miRNAs regulate genes that belong to the TGF-β signaling pathway or to the SMAD binding gene sets (M39432 and GO:0046332, respectively; genes are highlighted in gray in Tables 4 and 5) but mostly in relation to the myofibroblast cells. Moreover, miR-let-7d regulates the expression of endothelin 1 gene (EDN1, bold in Table 4). Endothelin 1 (ET-1) is a potent vasoconstrictor that is elevated in the plasma of patients with PAH [26], is known to potentiate the induced effect of TGF-β on mesenchymal cells and on the expression of profibrotic genes and proteins.

Table 4. Genes controlled by miR-let-7d and involved in the biology of immune, myofibroblast and endothelial cells.

miR-let-7d (461 Genes)		
Cells (Number of Genes)	**Number of Genes**	**Genes in Venny**
Immune cells (233 genes)	16	*CCR7, FAS, EDN1, RAG1, ICOS, FASLG, IL10, RORC, MAPK8, MASP1, NAP1L1, TAB2, BCL2L1, COL4A5, IL13, IL6*
Myofibroblasts (447 genes)	19	*COL1A2, RANBP2, TGFBR1, SCN5A, IGF1, TP53, CASP3, ITGB3, COL3A1, NGF, HAS2, P4HA2, MAP8, CCND1, CDKN1A, SMUG1, CCL7, IL10, HMGA2, EDN1*
Endothelial cells (130 genes)	7	*ITGP3, TNFRSF1, CASP3, FASLG, MAPK8, EDN1, PIK3CA, CCND2, COL4A1, COL1A2, COL3A1*

Venny software was used to find genes common to miR-let-7d targets and to the lists of the three types of cells (immune, myofibroblast and endothelial cells) yielding three lists of genes. Genes marked in gray belong to the "TGF-β signaling pathway" or to the "SMAD binding" gene sets.

Table 5. Genes controlled by miR-26 and involved in the biology of immune, myofibroblast and endothelial cells.

miR-26 (1112 Genes)		
Cells (Number of Genes)	**Number of Genes**	**Genes in Venny**
Immune cells (233 genes)	8	*MAP3K7, TRAF3, IRF4, SELP, TAB3, ILI8R1, ICOS, PYGS2*
Myofibroblasts (447 genes)	30	*HAS2, GSK3B, COL1A2, PCNA, SCN5A, CREB1, LTBP1, COL12A1, PTEN, JAG1, ESR1, PALLD, KCNJ2, PAK2, MLANA, SNTG1, SCL25A16, HMGA2, HDAC9, MIB1, PGR, MYLK3, SFTPB, ROCK1, WNT5A, MAP3K2, SMAD4, PURA, PURB, TMTC3*
Endothelial cells (130 genes)	7	*ADAM17, SELP, CDH2, PTGS2, PECAM1, EREG, PDGRFRA*

Venny software was used to find genes common to miR-26 targets and to the lists of the three types of cells (immune, myofibroblast and endothelial cells), yielding three lists of genes. Genes marked in gray belong to the "TGF-β signaling pathway" or to the "SMAD binding" gene sets.

The complement pathway did not appear in the annotations because the gene lists of the immune system cells contained C3, while the gene lists of miR-26 and miR-let-7d contained the complement components C1 and C2.

3.6. WBCs of SSc-noPAH and SSc-PAH Patients Have Equal Levels of miR-26 and miR-let-7d

We aimed to identify which cells are involved in mediating the reduced levels of miR-26 and miR-let-7d expression found in the plasma of SSc-PAH patients. Our clinical correlations suggested involvement of miRNAs in activating the complement system and antibody levels (levels of both miRNAs correlated with C3 and miR-26 also with IgG2 levels) and the bioinformatic analysis suggested involvement of the miRNAs in cytokine signaling. These factors are mediators of the immune system. Therefore, we primarily opted to find whether WBCs of SSc-PAH patients contain lower miR-26 and miR-let-7d levels compared to those of SSc-noPAH patients who do not have PAH. Unexpectedly, we found that they express equal amounts of the miRNAs (Figure 3).

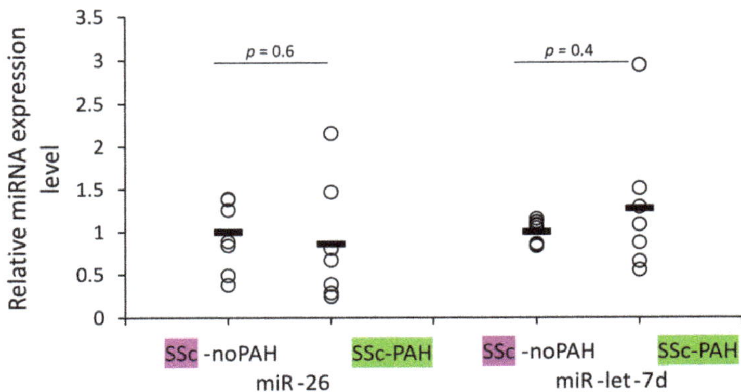

Figure 3. miR-26 and miR-let-7d levels in WBCs of SSc-noPAH and SSc-PAH patients. WBCs were isolated from 7 SSc-PAH and 8 SSc-noPAH patients. RNA was isolated from the cells and the levels of miR-26 and miR-let-7d were evaluated with qRT-PCR. miR-16 served as the normalizing miRNA. The graph represents miRNA expression levels relative to miR-16. Horizontal bars represent the mean values of the miRNAs.

3.7. Exposure of Dermal Fibroblasts to Tgf-β and Their Differentiation into Myofibroblasts Was Accompanied by Decreased miR-Let-7d and miR-26 Levels

Myofibroblasts induce a severe fibrotic process, affecting the fibroproliferative changes in the skin, internal organs and pulmonary arterioles in SSc [27,28]. They originate from resident fibroblasts, bone marrow-derived cells and the conversion of endothelial cells into activated fibroblasts [28]. A key growth factor for myofibroblast formation is TGF-β, which induces collagen production and alpha smooth muscle actin (αSMA) expression in the fibroblasts [3]. The bioinformatic analysis suggested that miR-let-7d and miR-26 regulate genes involved in the TGF-β pathway, especially in myofibroblast cells (Tables 4 and 5).

We suspected that exposing fibroblasts to TGF-β might reduce the levels of miR-26 and miR-let-7d in the cells and their secretion in the circulation. Therefore, we isolated dermal fibroblasts from normal skin biopsies, exposed them to TGF-β and induced their differentiation into myofibroblasts, as was evident by the increase in the number of cells (40%↑, 72 h, $p < 0.05$, data not shown) and increased expression of αSMA and collagen I (48 h, $p = 0.05$, Figure 4A–C). Total RNA was isolated from the cells and miR-26 and miR-let-7d levels were measured. As an additional control, we measured the level of miR-21, which is a pro-fibrotic miRNA [29]. We found small but consistent decreases in miR-let-7d levels (~10%, $p < 0.05$) and ~40% reduction in the level of miR-26 ($p < 0.05$) compared to cells that were not exposed to TGF-β, whereas miR-21 levels tended to increase in those cells (Figure 4D, black and white bars, respectively). These results support our hypothesis that the presence of low levels of miR-26 and miR-let-7d in the bloodstream is at least partly due to their low production by myofibroblast cells.

3.8. The Effect of Macitentan (ET-1 Receptor Antagonist) on Myofibroblast miR-26 and miR-Let-7d Levels

Dermal fibroblasts produce endothelin-1 (ET-1) [27], which plays a key role in the biology of PAH [30]. Previous publications showed that macitentan, an ET-1 receptor antagonist approved for PAH therapy, interferes with ET-1 and TGF-β-induced fibroblast activation [27]. We examined whether exposing fibroblasts cultured with TGF-β to macitentan affects the levels of miR-26 and miR-let-7d, and found that the addition of the drug together with TGF-β prevented the decrease in the levels of miR-26 and miR-let-7d observed when the cells were exposed to TGF-β only (Figure 4D).

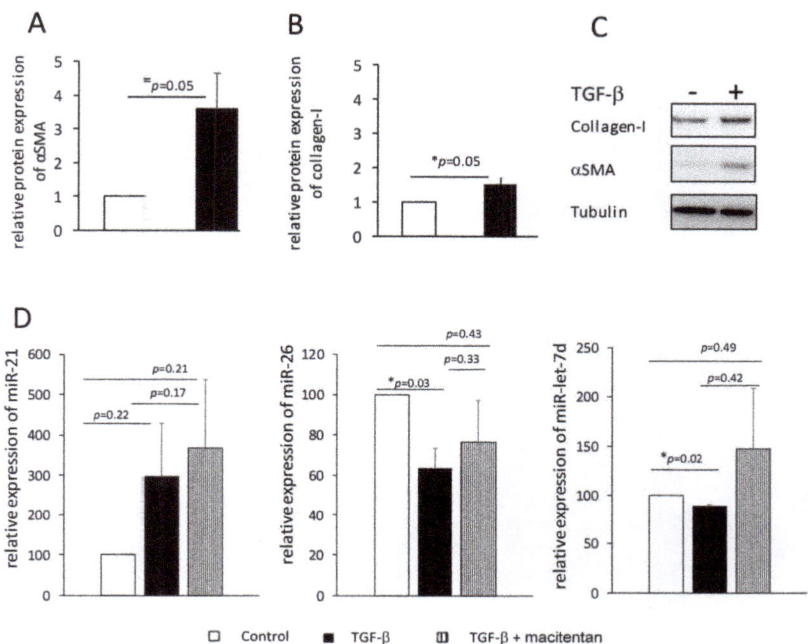

Figure 4. TGF-β induced myofibroblast differentiation is accompanied by reduced miR-26 and miR-let-7d levels. Dermal fibroblasts were isolated from healthy patients undergoing plastic surgery. The cells were exposed to TGF-β (10 ng/mL) for 48 h or 72 h. Then the cells were either harvested and counted (72 h) or harvested and their proteins and RNA isolated (48 h). The levels of the myofibroblast markers αSMA (**A**) and collagen I (**B**) were assessed by western blot, and the levels of miR-21, miR-26 and miR-let-7d were assessed with qRT-PCR (**D**). (**C**) is a representative blot of αSMA and collagen in cells that were treated or not treated with TGF-β ($n = 4$). Columns represent mean and whiskers represent standard errors. *—Statistically significant results, $p \leq 0.05$.

4. Discussion

The idea of using miRNAs as biological markers has been suggested in the past in the context of various pathologies [31]. In the current study, we searched for miRNAs that are expressed differently in SSc-PAH compared to SSc-noPAH patients. Our initial observation was that miRNA expression levels vary greatly among patients with SSc-noPAH. A more thorough examination of the SSc-noPAH group showed that while most SSc-noPAH patients presented with a stable mean pulmonary artery pressure, some had a relatively rapid rise in this variable between their first and last echo tests included in this study. These patients were termed SSc-susPAH (suspected to develop PAH). A comparison between SSc-noPAH and SSc-PAH patients showed differences in age at diagnosis, clinical parameters related to the disease (echo and lung function tests), anti-centromere levels and in the levels of miR-let-7d and miR-26. These clinical differences were similar to those described in the literature [4,32]. To the best of our knowledge, this is the first study to show that miR-26 levels in the plasma of SSc-PAH patients is lower than in the plasma of SSc-noPAH patients. Regarding miR-let-7d, our observations agree with those of a study by Wuttge et al. that showed that the level of miR-let-7d is lower in SSc-PAH compared to SSc-noPAH patients [33].

Echocardiography, which is a recommended modality for early detection of PAH, was used to differentiate between SSc-noPAH and SSc-susPAH patients [34]. Even a borderline elevation in mean pulmonary artery pressure from 21 to 24 mmHg may reflect an intermediate stage on the continuum between normal pressure and PAH [35]. Previous publications also showed that SSc patients with increasing systolic pulmonary artery pressure over a short period have a higher risk of mortality from PAH [7,36]. We found three SSc-susPAH patients with increased mean pulmonary artery pressures in the one- to two-year period between the initial blood collection and the end of the study. Interestingly, in these patients and in another SSc-noPAH patient whose mean pulmonary artery pressure at the end of the study could not be evaluated by echo due to technical problems, plasma levels of miR-let-7d and miR-26 were similar to those measured in SSc-PAH patients and were significantly lower than those measured in SSc-noPAH patients. At the end of the study period, two of these four patients developed PAH and one died from this complication. These results suggest that low levels of miR-let-7d and miR-26 miRNAs that characterize SSc-PAH patients, can be detected in SSc-noPAH patients even before pulmonary pressure rises. Despite this interesting observation, the limited number of patients (4) does not allow unequivocal conclusions to be drawn. A future, multicenter study with a larger patient population may reconfirm these findings.

To further assess the involvement of miR-let-7d and miR-26 in the pathogenesis of SSc and PAH, miRNA levels and the clinical and laboratory data of the patients were correlated. Interestingly, we found that the miR-let-7d and miR-26 levels correlated with C3. Using the http://mirdb.org database, we found that C2 is a target of miR-let-7d and C1s is a target of miR-26. Protease C1s cleaves C4 and C2 to generate the C3 convertase that cleaves C3, resulting in a decrease in C3. These processes may explain the positive correlation between miR-let-7d and miR-26 levels and C3. The relation between activation of the complement system to PAH was only recently suggested by demonstrating that vascular-specific, immunoglobulin-driven dysregulated complement signaling triggers and maintains pulmonary vascular remodeling and PAH [37]. It was also suggested that plasma complement signaling, including factors in the alternative pathway, is a prognostic factor for survival in patients with idiopathic PAH.

As a correlation was found between the miR-let-7d and miR-26 and C3 and between miR-26 and IgG2 that can activate the classic complement system, it was tempting to speculate that the lower levels of miR-26 and miR-let-7d are an early event in the development of SSc-PAH, at the stage of autoimmune attack. Our bioinformatic search also suggested involvement of these miRNAs in mediating the immune response in SSc. Chemokine (C-C Motif) Receptor 7 (CCR7), a miR-let-7d target expressed by WBCs, was previously found to mediate PAH in murine models. C-C Motif Chemokine Ligand 21 (CCL21) the ligand of CCR7, is a promising marker for predicting the risk of SSc-related PAH and PAH progression [38]. Furthermore, interleukin-18 (IL-18) Receptor 1 (IL-18R1) is a miR-26 target expressed by WBCs. Previous publications showed that IL-18 levels are altered in SSc patients [39,40] and that there is a positive correlation between serum IL-18 binding protein (IL-18BPa) levels and right ventricular systolic pressure [39].

The clinical correlations and bioinformatic analysis encouraged us to check whether miR-let-7d and miR-26 levels in WBCs of SSc-noPAH patients are higher than in SSc-PAH patients. Contrary to our expectations, similar levels of the miRNAs were found in the cells. This suggests that the low level of miRNAs in the circulating blood of PAH patients is not due to decreased production by WBCs. We performed the test on a mixed population of WBCs that was not separated into the different types of immune system cells. miR-26 or miR-let-7d levels might be reduced in one of the immune cell populations and the low levels might be masked by their expression in other WBCs; a possibility that should be explored in the future.

Our bioinformatic analyses also suggested that miR-26 might be involved in blood vessel morphogenesis and cell motility. While we did not continue our study in this direction, it is tempting to think that this miRNA is involved in the process of endothelial mesenchymal transition (EndMT). This process demands acquisition of mesenchymal characteristics by endothelial cells, such as cell migration [41] and ends with altered vessel morphology. It has been reported that EndMT may also have a pivotal role during PAH pathogenesis, promoting endothelial cell dysfunction [42]. Our bioinformatic analysis also suggested that miR-let-7d and miR-26 are involved in the biology of myofibroblast cells by regulating genes in the TGF-β and endothelin pathways. This is in accordance with literature that showed that collagen, a target protein of both miRNAs, is regulated by the TGF-β pathway [29]. This suggests that exposure of fibroblasts to TGF-β and/or endothelin 1, might alter miR-let-7d and miR-26 production. Indeed, we found that TGF-β significantly reduced the level of both miRNAs in dermal fibroblasts. These observations suggest that myofibroblasts are at least partially responsible for reduced miR-let-7d and miR-26 levels in the plasma of SSc-PAH patients. This agrees with reports that showed that miR-let-7d and miR-26 expression are downregulated in the skin of SSc patients [29,43] and that miR-let-7d is reduced in the skin of SSc-PAH patients even more than it is among SSc-noPAH patients [23].

Additionally, we found a correlation between the two microRNAs (miR-let-7d and miR-26). An interaction between these two miRNAs was previously described [44].in publication who showed that miR-26 disrupts the Lin28B/miR-let-7d circuit [44]. The Lin28B gene is negatively regulated by miRNAs (such as miR-26) and its overexpression is linked to the repression of let-7 miRNAs [44]. A previous publication also showed that Lin28b serves as a molecular switch for tuning the tolerance threshold during B cell receptor (BCR) repertoire selection early in life [45]. Therefore, it is enticing to speculate that this protein may contribute to the development of SSc, in which the appearance of autoantibodies is the first clinical phenomenon observed.

This study examined the level of miRNAs in patients treated in the rheumatology clinic at our medical center and its major limitation is the small cohort. An additional limitation is that only seven miRNAs were tested, while previous articles showed involvement of additional miRNAs in PAH and systemic sclerosis.

Despite the small group of patients, our results suggest that miR-let-7d and miR-26 are serological markers that differentiate SSc-noPAH from SSc-PAH patients. Moreover, decreased expression of the miRNAs among SSc-susPAH patients also suggests that they may be early predictors of the development of PAH, an assumption that warrants further testing with larger cohorts.

Supplementary Materials: The following supporting information can be downloaded at: https://www.mdpi.com/article/10.3390/biomedicines10030629/s1, Table S1: GO biological annotations for genes common to the miR-let-7d targets and to the lists of genes of immune, endothelial and myofibroblast cells; Table S2: GO biological annotations for genes common to the miR-26 targets and to the lists of genes of immune, endothelial and myofibroblast cells.

Author Contributions: Conceptualization and Supervision, S.T.M., N.R. and Y.L.; Project administration S.T.M. and N.R.; Investigation and Methodology, M.Z.L., M.Y.K., T.H.-L., A.S. and A.B.; Analysis and interpretation of data S.T.M., M.Z.L., T.H.-L. and L.D., Writing original draft, S.T.M. and M.Z.L.; Visualization and revision of the work, Y.L., N.R., M.Y.K., T.H.-L., L.D., A.S. and A.B.; Resources, Y.L. All authors have read and agreed to the published version of the manuscript.

Funding: This study was partly sponsored by Actelion Pharmaceuticals.

Institutional Review Board Statement: The study was approved by the Ethics Committee of Meir Medical Center (Approval number 0308-18-MMC).

Informed Consent Statement: Informed consent was obtained from all subjects involved in the study.

Data Availability Statement: Data are contained within the article except for the bioinformatics analysis data extracted from the computer sites described in the Methods and Materials (Bioinformatics) sections.

Conflicts of Interest: The authors declare no conflict of interest. The sponsors had no role in the design, execution, interpretation or writing of the study.

Abbreviations

αSMA	alpha smooth muscle actin
ACR	American College of Radiologists
BCR	B cell receptor
BP	Binding protein
CCL21	C-C Motif Chemokine Ligand 21
CCR7	C-C Motif chemokine Receptor 7
CT	Cycle threshold
DLCO	Diffusing capacity of lung for carbon monoxide
DMEM	Dulbecco's Modified Eagle Medium
DMSO	Dimethyl sulfoxide
DTT	Dithiothreitol
Echo	Echocardiography
EDN1	Gene symbol of endothelin 1
EDTA	Ethylenediaminetetraacetic acid
EndMT	Endothelial mesenchymal transition
ET-1	Endothelin 1
EULAR	European League Against Rheumatism
Fev1	Forced expiratory volume in one second
Fvc	Forced vital capacity
GO	Gene ontology
h	Hour
hsa	Homo sapiens
IgG	Immunoglobulin G
IL	Interleukin
IRF4	Interferon Regulatory Factor 4
miRNA, miR	microRNA
PAH	Pulmonary arterial hypertension
PBS	Phosphate Buffered Saline
PVDF	membrane: Poly(vinylidene fluoride)
SSc	Systemic sclerosis
SSc-noPAH	SSc patients without PAH
SSc-PAH	SSc patients with PAH
SSc-susPAH	SSc patients suspected to develop PAH
qRT-PCR	Real time quantitative polymerase chain reaction
rs	Spearman's Rank calculation
RT cDNA	Reverse transcription complementary DNA
SDS	Sodium dodecyl sulfate
TGF-β	Transforming growth factor beta
Tris	Tris(Hydroxymethyl)aminomethane
WBCs	White blood cells

References

1. Zhang, L.; Wu, H.; Zhao, M.; Lu, Q. Meta-Analysis of Differentially Expressed Micrornas in Systemic Sclerosis. *Int. J. Rheum. Dis.* **2020**, *23*, 1297–1304. [CrossRef] [PubMed]
2. Ramirez, A.; Varga, J. Pulmonary arterial hypertension in systemic sclerosis: Clinical manifestations, pathophysiology, evaluation, and management. *Treat. Respir. Med.* **2004**, *3*, 339–352. [CrossRef] [PubMed]
3. Asano, Y. Systemic sclerosis. *J. Dermatol.* **2018**, *45*, 128–138. [CrossRef] [PubMed]
4. Schachna, L.; Wigley, F.M.; Chang, B.; White, B.; Wise, R.A.; Gelber, A.C. Age and risk of pulmonary arterial hypertension in scleroderma. *Chest* **2003**, *124*, 2098–2104. [CrossRef]
5. Humbert, M.; Yaici, A.; Groote, P.; Montani, D.; Sitbon, O.; Launay, D.; Gressin, V.; Guillevin, L.; Clerson, P.; Simon-neau, G.; et al. Screening for pulmonary arterial hypertension in patients with systemic sclerosis: Clinical characteristics at diagnosis and long-term survival. *Arthritis Rheum.* **2011**, *63*, 3522–3530. [CrossRef]
6. Yaghi, S.; Novikov, A.; Trandafirescu, T. Clinical update on pulmonary hypertension. *J. Investig. Med.* **2020**, *68*, 821–827. [CrossRef]
7. Kampolis, C.; Plastiras, S.C.; Vlachoyiannopoulos, P.G.; Moyssakis, I.; Tzelepis, G.E. The presence of anti-centromere antibodies may predict progression of estimated pulmonary arterial systolic pressure in systemic sclerosis. *Scand. J. Rheumatol.* **2008**, *37*, 278–283. [CrossRef]

8. Rice, L.M.; Mantero, J.C.; Stratton, E.A.; Warburton, R.; Roberts, K.; Hill, N.; Simms, R.W.; Domsic, R.; Farber, H.W.; Layfatis, R. Serum biomarker for diagnostic evaluation of pulmonary arterial hypertension in systemic sclerosis. *Arthritis Res. Ther.* **2018**, *20*, 185. [CrossRef]
9. Odler, B.; Foris, V.; Gungl, A.; Muller, V.; Hassoun, P.M.; Kwapiszewska, G.; Olschewski, H.; Kovacs, G. Biomarkers for Pulmonary Vascular Remodeling in Systemic Sclerosis: A Pathophysiological Approach. *Front. Physiol.* **2018**, *9*, 587. [CrossRef]
10. Henry, T.W.; Mendoza, F.A.; Jimenez, A.S. Role of microRNA in the pathogenesis of systemic sclerosis tissue fibrosis and vasculopathy. *Autoimmun. Rev.* **2019**, *18*, 102396. [CrossRef]
11. Drobna, M.; Szarzyńska-Zawadzka, B. Identification of Endogenous Control miRNAs for RT-qPCR in T-Cell Acute Lymphoblastic Leukemia. *Int. J. Mol. Sci.* **2018**, *19*, 2858. [CrossRef] [PubMed]
12. McDermott, M.A.; Kerin, M.J.; Miller, N. Identification and validation of miRNAs as endogenous controls for RQ-PCR in blood specimens for breast cancer studies. *PLoS ONE* **2013**, *8*, e83718. [CrossRef] [PubMed]
13. Sarrion, I.; Milian, L.; Juan, G.; Ramon, M.; Furest, I.; Carda, C.; Backx, P.; Cortijo, G.J.; Mata, R.M. Role of circulating miRNAs as biomarkers in idiopathic pulmonary arterial hypertension: Possible relevance of miR-23a. *Oxid. Med. Cell Longev.* **2015**, *2015*, 792846. [CrossRef] [PubMed]
14. Han, W.; Han, Y.; Liu, X.; Shang, X. Effect of miR-29a inhibition on ventricular hypertrophy induced by pressure overload. *Cell Biochem. Biophys.* **2015**, *71*, 821–826. [CrossRef] [PubMed]
15. Gallant-Behm, C.L.; Piper, J.; Lynch, J.M.; Seto, A.G.; Hong, S.J.; Mustoe, T.A.; Maari, C.; Pestano, L.A.; Dalby, C.M.; Jackson, A.L.; et al. MicroRNA-29 Mimic (Remlarsen) Represses Extracellular Matrix Expression and Fibroplasia in the Skin. *J. Investig. Dermatol.* **2019**, *139*, 1073–1081. [CrossRef]
16. Maurer, B.; Stanczyk, J.; Jungel, A.; Akhmetshina, A.; Trenkmann, M.; Brock, M.; Kowal-Bielecka, O.; Gay, R.E.; Michel, B.A.; Distler, J.H.W.; et al. MicroRNA-29, a key regulator of collagen expression in systemic sclerosis. *Arthritis Rheum.* **2010**, *62*, 1733–1743. [CrossRef]
17. Schlosser, K.; White, R.J.; Stewart, D.J. miR-26a linked to pulmonary hypertension by global assessment of circulating extracellular microRNAs. *Am. J. Respir. Crit. Care Med.* **2013**, *188*, 1472–1475. [CrossRef]
18. Rhodes, C.J.; Wharton, J.; Boon, R.A.; Roexe, T.; Tsang, H.; Wojciak-Stothard, B.; Chakrabarti, A.; Howard, L.S.; Gibbs, J.S.; Lawrie, A.; et al. Reduced microRNA-150 is associated with poor survival in pulmonary arterial hypertension. *Am. J. Respir. Crit. Care Med.* **2013**, *187*, 294–302. [CrossRef]
19. Honda, N.; Jinnin, M.; Kira-Etoh, T.; Makino, K.; Kajihara, I.; Makino, T.; Fukushima, S.; Inoue, Y.; Okamoto, Y.; Ha-segawa, M.; et al. miR-150 down-regulation contributes to the constitutive type I collagen overexpression in scleroderma dermal fibroblasts via the induction of integrin β3. *Am. J. Pathol.* **2013**, *182*, 206–216. [CrossRef]
20. Green, D.E.; Murphy, T.C.; Kang, B.; Searles, C.D.; Hart, C.M. PPARγ Ligands Attenuate Hypoxia-Induced Proliferation in Human Pulmonary Artery Smooth Muscle Cells through Modulation of MicroRNA-21. *PLoS ONE* **2015**, *10*, e0133391. [CrossRef]
21. Zhu, H.; Luo, H.; Li, Y.; Zhou, Y.; Jiang, Y.; Chai, J.; Xiao, X.; You, Y.; Zuo, X. MicroRNA-21 in scleroderma fibrosis and its function in TGF-β-regulated fibrosis-related genes expression. *J. Clin. Immunol.* **2013**, *33*, 1100–1109. [CrossRef] [PubMed]
22. Drake, K.M.; Zygmunt, D.; Mavrakis, L.; Harbor, P.; Wang, L.; Comhair, S.A.; Erzurum, S.C.; Aldred, M.A. Altered MicroRNA processing in heritable pulmonary arterial hypertension: An important role for Smad-8. *Am. J. Respir. Crit. Care Med.* **2011**, *184*, 1400–1408. [CrossRef] [PubMed]
23. Izumiya, Y.; Jinnn, M.; Kimura, Y.; Wang, Z.; Onoure, Y.; Anatani, S.; Araki, S.; Ihn, H.; Ogawa, H. Expression of Let-7 family microRNAs in skin correlates negatively with severity of pulmonary hypertension in patients with systemic scleroderma. *Int. J. Cardiol. Heart Vasc.* **2015**, *8*, 98–102. [PubMed]
24. Ou, M.; Li, X.; Cui, S.; Zhao, S.; Tu, J. Emerging roles of let-7d in attenuating pulmonary arterial hypertension via suppression of pulmonary artery endothelial cell autophagy and endothelin synthesis through ATG16L1 downregulation. *Int. J. Mol. Med.* **2020**, *46*, 83–96. [CrossRef]
25. Artlett, C.M.; Sassi-Gaha, S.; Hope, J.L.; Feghali-Bostwick, C.A.; Katsikis, P.D. Mir-155 is overexpressed in systemic sclerosis fibroblasts and is required for NLRP3 inflammasome-mediated collagen synthesis during fibrosis. *Arthritis Res. Ther.* **2017**, *19*, 144. [CrossRef]
26. Khadka, A.; Brashier, D.B.; Tejus, A.; Sharma, A.K. Macitentan: An important addition to the treatment of pulmonary arterial hypertension. *J. Pharmacol. Pharm.* **2015**, *6*, 53–57. [CrossRef]
27. Cipriani, P.; Di Benedetto, P.; Ruscitti, P.; Verzella, D.; Fischietti, M.; Zazzeroni, F.; Liakouli, V.; Carubbi, F.; Berardicurti, O.; Alesse, E.; et al. Macitentan inhibits the transforming growth factor-β profibrotic action, blocking the signaling mediated by the ETR/TβRI complex in systemic sclerosis dermal fibroblasts. *Arthritis Res. Ther.* **2015**, *17*, 247. [CrossRef]
28. Hong, O.K.; Lee, S.; Yoo, S.J.; Lee, M.; Kim, M.; Baek, K.; Song, K.; Kwon, H. Gemigliptin Inhibits Interleukin-1β-Induced Endothelial-Mesenchymal Transition via Canonical-Bone Morphogenetic Protein Pathway. *Endocrinol. Metab.* **2020**, *35*, 384–395. [CrossRef]
29. Bagnato, G.; Roberts, W.N.; Roman, J.; Gangemi, S. A systematic review of overlapping microRNA patterns in systemic sclerosis and idiopathic pulmonary fibrosis. *Eur. Respir. Rev.* **2017**, *26*, 160125. [CrossRef]
30. Varga, J. Endothelin-I receptor antagonist for the treatment of pulmonary arterial hypertension in systemic sclerosis. *Curr. Rheumatol. Rep.* **2003**, *5*, 145–146. [CrossRef]

31. Popescu Driga, M.; Catalin, B.; Olaru, D.G.; Slowik, A.; Plesnila, N.; Hermann, D.M.; Popa-Wagner, A. The Need for New Biomarkers to Assist with Stroke Prevention and Prediction of Post-Stroke. Therapy Based on Plasma-Derived Extracellular Vesicles. *Biomedicines* **2021**, *9*, 1226. [CrossRef] [PubMed]
32. Nunes, J.P.L.; Cunha, A.C.; Meirinhos, T.; Nunes, A.; Araujo, P.M.; Godinho, A.R.; Vilela, E.M.; Vaz, C. Prevalence of auto-antibodies associated to pulmonary arterial hypertension in scleroderma—A review. *Autoimmun. Rev.* **2018**, *17*, 1186–1201. [CrossRef] [PubMed]
33. Wuttge, D.M.; Carlsen, A.L.; Teku, G.; Wildt, M.; Radegran, G.; Vihinen, M.; Heegaard, N.H.; Hesselstrand, R. Circulating Plasma microRNAs In Systemic Sclerosis-Associated Pulmonary Arterial Hypertension. *Rheumatology* **2021**, *61*, 309–318. [CrossRef] [PubMed]
34. Weatherald, J.; Montani, D.; Jevnikar, M.; Jais, X.; Savale, L.; Humbert, M. Screening for pulmonary arterial hypertension in systemic sclerosis. *Eur. Respir. Rev.* **2019**, *28*, 190023. [CrossRef]
35. Visovatti, S.H.; Distler, O.; Coghlan, J.G.; Denton, C.P.; Grunig, E.; Bonderman, D.; Muller-Ladner, U.; Pope, J.E.; Vonk, M.C.; Seibold, J.R.; et al. Borderline pulmonary arterial pressure in systemic sclerosis patients: A post-hoc analysis of the DETECT study. *Arthritis Res. Ther.* **2014**, *16*, 493. [CrossRef]
36. MacGregor, A.J.; Canavan, R.; Knight, C.; Denton, C.P.; Davar, J.; Black, C.M. Pulmonary hypertension in systemic sclerosis: Risk factors for progression and consequences for survival. *Rheumatology* **2001**, *40*, 453–459. [CrossRef]
37. Frid, M.G.; Thurman, J.M.; Hansen, K.C.; Maron, B.A.; Stenmark, K.R. Inflammation, immunity, and vascular remodeling in pulmonary hypertension; Evidence for complement involvement? *Glob. Cardiol. Sci. Pract.* **2020**, *2020*, e202001. [CrossRef]
38. Hoffmann-Vold, A.M.; Hesselstrand, R.; Fretheim, H.; Ueland, T.; Andreassen, A.K.; Brunborg, K.; Palchevsky, V.; Midtvedt, Ø.; Garen, T.; Aukrust, P.; et al. CCL21 as a Potential Serum Biomarker for Pulmonary Arterial Hypertension in Systemic Sclerosis. *Arthritis Rheumatol.* **2018**, *70*, 1644–1653. [CrossRef]
39. Lin, E.; Vincent, F.B.; Sahhar, J.; Ngian, G.; Kandane-Rathnayake, R.; Mende, R.; Morand, E.F.; Lang, T.; Harris, J. Analysis of serum interleukin (IL)-1α, IL-1β and IL-18 in patients with systemic sclerosis. *Clin. Transl. Immunol.* **2019**, *8*, e1045. [CrossRef]
40. Nakamura, K.; Asano, Y.; Taniguchi, T.; Minatsuki, S.; Inaba, T.; Maki, H.; Hatano, M.; Yamashita, T.; Saigusa, R.; Ichimura, Y.; et al. Serum levels of interleukin-18-binding protein isoform a: Clinical association with inflammation and pulmonary hypertension in systemic sclerosis. *J. Dermatol.* **2016**, *43*, 912–918. [CrossRef]
41. Alvandi, Z.; Bischoff, J. Endothelial-Mesenchymal Transition in Cardiovascular Disease. *Arterioscler. Thromb. Vasc. Biol.* **2021**, *41*, 2357–2369. [CrossRef] [PubMed]
42. Good, R.B.; Gilbane, A.J.; Trinder, S.L.; Denton, C.P.; Coghlan, G.; Abraham, D.J.; Holmes, A.M. Endothelial to Mesenchymal Transition Contributes to Endothelial Dysfunction in Pulmonary Arterial Hypertension. *Am. J. Pathol.* **2015**, *185*, 1850–1858. [CrossRef] [PubMed]
43. Zhu, Y.; Zhang, J.; Hu, X.; Wang, Z.; Wu, S.; Yi, Y. Extracellular vesicles derived from human adipose-derived stem cells promote the exogenous angiogenesis of fat grafts via the let-7/AGO1/VEGF signalling pathway. *Sci. Rep.* **2020**, *10*, 5313. [CrossRef] [PubMed]
44. Liang, H.; Liu, S.; Chen, Y.; Bai, X.; Liu, L.; Dong, Y.; Hu, M.; Su, X.; Chen, Y.; Huangfu, L.; et al. miR-26a suppresses EMT by disrupting the Lin28B/let-7d axis: Potential cross-talks among miRNAs in IPF. *J. Mol. Med.* **2016**, *94*, 655–665. [CrossRef]
45. Vergani, S.; Yuan, J. Developmental changes in the rules for B cell selection. *Immunol. Rev.* **2021**, *300*, 194–202. [CrossRef]

 biomedicines

Article

Inhibition of Hsp90 Counteracts the Established Experimental Dermal Fibrosis Induced by Bleomycin

Hana Štorkánová [1,2], Lenka Štorkánová [1], Adéla Navrátilová [1,2], Viktor Bečvář [1], Hana Hulejová [1], Sabína Oreská [1,2], Barbora Heřmánková [3], Maja Špiritović [1,3], Radim Bečvář [1,2], Karel Pavelka [1,2], Jiří Vencovský [1,2], Jörg H. W. Distler [4], Ladislav Šenolt [1,2] and Michal Tomčík [1,2,*]

1 Institute of Rheumatology, 12800 Prague, Czech Republic; storkanova@revma.cz (H.Š.); storkanoval@revma.cz (L.Š.); navratilova@revma.cz (A.N.); becvarv@revma.cz (V.B.); hulejova@revma.cz (H.H.); oreska@revma.cz (S.O.); spiritovic@revma.cz (M.Š.); becvar@revma.cz (R.B.); pavelka@revma.cz (K.P.); vencovsky@revma.cz (J.V.); senolt@revma.cz (L.Š.)
2 Department of Rheumatology, First Faculty of Medicine, Charles University, 12800 Prague, Czech Republic
3 Department of Physiotherapy, Faculty of Physical Education and Sport, Charles University, 16252 Prague, Czech Republic; hermankova@revma.cz
4 Department of Internal Medicine III and Institute for Clinical Immunology, University of Erlangen-Nuremberg, 91054 Erlangen, Germany; joerg.distler@uk-erlangen.de
* Correspondence: tomcik@revma.cz; Tel.: +420-234-075-101

Abstract: Our previous study demonstrated that heat shock protein 90 (Hsp90) is overexpressed in the involved skin of patients with systemic sclerosis (SSc) and in experimental dermal fibrosis. Pharmacological inhibition of Hsp90 prevented the stimulatory effects of transforming growth factor-beta on collagen synthesis and the development of dermal fibrosis in three preclinical models of SSc. In the next step of the preclinical analysis, herein, we aimed to evaluate the efficacy of an Hsp90 inhibitor, 17-dimethylaminoethylamino-17-demethoxygeldanamycin (17-DMAG), in the treatment of established experimental dermal fibrosis induced by bleomycin. Treatment with 17-DMAG demonstrated potent antifibrotic and anti-inflammatory properties: it decreased dermal thickening, collagen content, myofibroblast count, expression of transforming growth factor beta receptors, and pSmad3-positive cell counts, as well as leukocyte infiltration and systemic levels of crucial cytokines/chemokines involved in the pathogenesis of SSc, compared to vehicle-treated mice. 17-DMAG effectively prevented further progression and may induce regression of established bleomycin-induced dermal fibrosis to an extent comparable to nintedanib. These findings provide further evidence of the vital role of Hsp90 in the pathophysiology of SSc and characterize it as a potential target for the treatment of fibrosis with translational implications due to the availability of several Hsp90 inhibitors in clinical trials for other indications.

Keywords: heat shock protein 90; systemic sclerosis; established dermal fibrosis; treatment

Citation: Štorkánová, H.; Štorkánová, L.; Navrátilová, A.; Bečvář, V.; Hulejová, H.; Oreská, S.; Heřmánková, B.; Špiritović, M.; Bečvář, R.; Pavelka, K.; et al. Inhibition of Hsp90 Counteracts the Established Experimental Dermal Fibrosis Induced by Bleomycin. *Biomedicines* **2021**, *9*, 650. https://doi.org/10.3390/biomedicines9060650

Academic Editor: Stefano Bellosta

Received: 11 May 2021
Accepted: 2 June 2021
Published: 7 June 2021

1. Introduction

Systemic sclerosis (SSc, scleroderma) is a rare chronic autoimmune connective tissue disease of a complex etiopathogenesis characterized by vasculopathy, dysregulation of the immune system, and tissue fibrosis [1]. Fibrosis of the skin and internal organs, such as the lungs, gastrointestinal tract, heart, and kidneys, is the most characteristic feature of SSc [1]. During the course of the disease, the abnormally activated innate and acquired immune system affects resident fibroblasts, which are the critical cellular contributors to tissue fibrosis in SSc [1,2]. Chronically activated fibroblasts release an excessive amount of collagen and extracellular matrix (ECM) proteins, which leads to severe dysfunction of the affected tissues [2,3]. Multiple lines of evidence suggest that transforming growth factor-beta (TGF-β) plays a crucial role in the activation of fibroblasts and the development of tissue fibrosis in SSc [3]. TGF-β is able to induce transdifferentiation of fibroblasts into contractile cells called myofibroblasts, the microfilaments of which consist of alpha-smooth muscle actin (aSMA) and non-muscle myosin type II [3,4]. Myofibroblasts increase the production of collagen, fibronectin, proteoglycans, and other components of the ECM [3,5]. Moreover, TGF-β induces heterotetramerization of TGF-β-receptor type I (TβRI) and II (TβRII) and leads to activation of several intracellular pathways, particularly the ones mediated by small mothers against decapentaplegic homolog 3 (Smad3) and other kinases [3,5]. However, myofibroblasts have also been attributed an immunomodulatory role since they express interleukin (IL)-1, IL-6, IL-8, and monocyte chemoattractant protein 1 (MCP-1, CCL2) [6,7]. Despite recent substantial advancements shedding light on the pathophysiology of tissue fibrosis in SSc [8], an effective and available clinical treatment with disease-modifying and survival-improving effects has yet to be determined [9].

Heat shock protein 90 (Hsp90) belongs to the ubiquitous molecular chaperone family of heat shock proteins (Hsp), which are produced by a wide range of cells under conditions of cellular stress [10]. Hsp play an important role in protein–protein interactions, such as recognizing and binding to non-natively folded proteins, assisting in achieving their proper conformation, and preventing against their irreversible degradation and activation [11,12]. Hsp90 interacts with a wide range of substrate proteins, including kinases, transcription factors, steroid hormone receptors, and E3 ubiquitin ligases [13]. These client proteins are essential for many biological processes such as cell cycle and growth, apoptosis, cytoskeletal rearrangement, and many others [13–15]. Furthermore, Hsp90 mediates the activation and maturation of antigen-presenting cells and the induction of proinflammatory cytokines [16,17]. Hsp90 has been demonstrated to participate in autoimmune response, oncogenesis, viral infections, and neurodegenerative diseases [13,16,18–22]. In addition, Hsp90 contributes to stabilization and activation of TGF-β receptors (TβRI and TβRII) and Src kinases, which are intracellular mediators for profibrotic TGF-β signaling in SSc [23–25]. In our recent study, we described increased expression of Hsp90 in the involved skin of patients with SSc, in SSc dermal fibroblasts and in experimental dermal fibrosis in a TGF-β-dependent manner [26]. Furthermore, the inhibition of Hsp90 with a semi-synthetic derivative of geldanamycin, 17-dimethylaminoethylamino-17-demethoxy-geldanamycin (17-DMAG), demonstrated antifibrotic effects in vitro and prevented the development of dermal fibrosis in three preclinical models mimicking various stages of SSc [26].

In this study, as the next step of the preclinical analysis, we aimed to evaluate the efficacy of 17-DMAG in the established experimental dermal fibrosis, which better reflects the routine clinical practice, where most SSc patients seen by a rheumatologist have already developed tissue fibrosis. Therefore, we used a modified murine model of bleomycin-induced dermal fibrosis, in which the administration of bleomycin was prolonged to six weeks, and the treatment onset with 17-DMAG was delayed to the last three weeks of the ongoing bleomycin challenge [27,28]. To analyze both antifibrotic and anti-inflammatory properties exerted by 17-DMAG, we used the traditional outcome measures assessing dermal thickness, collagen content, activation of fibroblasts and of TGF-β pathway, including expression of pSmad3, TβRI, and TβRII, as well as markers of local and systemic inflammation, including selected cytokines and chemokines with established proinflammatory roles in the pathogenesis of SSc [1,2,7,29]. Given that several inhibitors of Hsp90 have already been tested in numerous clinical trials for other indications, this study could provide a novel therapy for the treatment of fibrosis in SSc with high translational potential [30–33].

2. Methods

2.1. Treating Established Bleomycin-Induced Dermal Fibrosis

A modified bleomycin model was used to analyze the efficacy of the Hsp90 inhibitor 17-DMAG in the regression of preestablished fibrosis [28,34–36]. Robust dermal fibrosis was first induced by injecting bleomycin for the first three weeks solely without treatment. A three-week course of treatment was then initiated at the start of the fourth week of the experiment, while bleomycin injections were continued for the remaining three weeks of the whole six-week experiment. The outcomes were analyzed six weeks after the first injection of bleomycin [28,34–36]. The following five groups of six-week-old male C57BL/6 mice (Velaz, s.r.o., Prague, Czech Republic) were included (Figure 1). The modified bleomycin model was performed as described previously [28,36,37], and can be briefly summarized as follows:

(1) The first control group was administered subcutaneous injections of 100 µL 0.9% NaCl every other day for six weeks, and served as a control for treatment with bleomycin (n = 8).
(2) The second control group of mice (n = 8) was subcutaneously injected with bleomycin for the first three weeks and with 0.9% NaCl for the last three weeks. The level of achieved dermal fibrosis in this group after the first three weeks of subcutaneous bleomycin injections represents the pretreatment level of established dermal fibrosis.
(3) Dermal fibrosis was induced by subcutaneous injections of bleomycin (Bleomedac, Medac GmbH, Wedel, Germany) dissolved in 0.9% sodium chloride (NaCl, B. Braun Medical s.r.o., Prague, Czech Republic) at a concentration of 0.5 mg/mL [38,39]. One hundred microliters of bleomycin was administered into the defined area of 1 cm^2 at the upper back every other day for six weeks (n = 8 mice) [38]. Vehicle treatment in groups 1–3 was performed with Dulbecco's Phosphate Buffered Saline (PBS, Lonza, Walkersville, MD, USA), 100 µL intraperitoneally, every third day in the last three weeks of the six-week experiment.
(4) The main treatment group was challenged with bleomycin for six weeks as described above. In the last three weeks of this six-week period, mice (n = 8) were treated intraperitoneally every third day with 100 µL of 17-DMAG (InvivoGen, San Diego, CA, USA) at a concentration of 25 mg/kg (5 mg/mL in PBS, Lonza) [26]. The dose of 17-DMAG used in this study had previously been shown to effectively inhibit Hsp90 in vivo [40], and to be well-tolerated in a chronic dose regimen of up to 180 days in vivo [41].
(5) For the control treatment, we chose a small-molecule competitive inhibitor of non-receptor tyrosine kinases (nRTKs), nintedanib, as an established antifibrotic agent (kindly provided by Boehringer Ingelheim Pharma GmbH & Co.KG, Ingelheim am Rhein, Germany). These mice (n = 8) were challenged with bleomycin for six weeks as described above, and in the last three weeks of this six-week period, nintedanib 50 mg/kg (100 µL diluted in deionized water) was administered twice daily perorally [28,42].

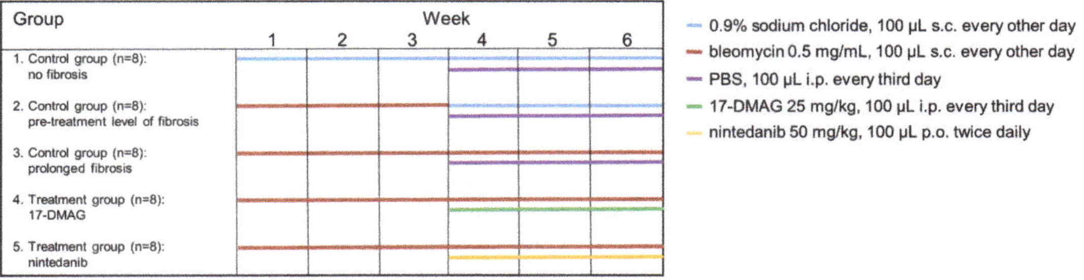

Figure 1. Design of the modified model of bleomycin-induced experimental dermal fibrosis. PBS, phosphate buffered saline; 17-DMAG, 17-dimethylaminoethylamino-17-demethoxy-geldanamycin (inhibitor of heat shock protein 90); s.c., subcutaneously administered; i.p., intraperitoneally administered; p.o., perorally administered.

This project (reference number AZV 16-33542A) and all animal experiments included were approved by the Ethics Committee of the Institute of Rheumatology in Prague (reference number 5689/2015, approved on 6 June 2015) and the Ministry of Education, Youth and Sports of the Czech Republic (reference number MSMT-9445/2018-7, approved on 5 May 2018). Animal experiments were conducted in accordance with relevant national legislation on the use of animals for research and complied with the commonly accepted 3Rs.

2.2. Histological Analysis of Dermal Thickness

Dermal thickness was assessed as described previously [43]. Injected skin areas were excised, then fixed in 4% formalin for 8 h and embedded in paraffin. Five-micrometer-thick sections were stained with hematoxylin–eosin and visualized by a BX53 microscope with a DP80 Digital Microscope Camera and CellSens Standard Software 3.1-Build 21199 (Olympus, Philadelphia, PA, USA) at 100-fold magnification. Dermal thickness was analyzed by two experienced examiners blinded to the treatment by measuring the maximum distance between the epidermal–dermal junction and the dermal–subcutaneous fat junction at four sites in four consecutive skin sections of the lesional skin from each mouse. The mean value for each mouse was used [43].

2.3. Assessment of the Number of Infiltrating Leukocytes

Infiltrating leukocytes in the lesional murine skin were quantified in hematoxylin and eosin-stained sections as described previously [43]. The assessment of the number of mononuclear/inflammatory cells (at 400-fold magnification) was performed by two experienced examiners blinded to the treatment in the full thickness of the dermis at eight high-power fields from different tissue sites in four consecutive skin sections of the lesional skin from each mouse. The mean value for each mouse was used [43].

2.4. Hydroxyproline Assay

The collagen content in lesional skin samples was determined by hydroxyproline assay as described previously [44]. Briefly, each punch biopsy specimen was digested in 6 M HCl for three hours at 120 °C, and the pH of the samples was adjusted to 7 with 6 M NaOH. Afterward, samples were mixed with 0.06 M chloramine T and incubated for 20 min at room temperature. Subsequently, 20% p-dimethylaminobenzaldehyde and 3.15 M perchloric acid were added, and samples were incubated for an additional 20 min at 60 °C. The absorbance was measured at 557 nm with a microplate spectrophotometer (SUNRISE; Tecan, Grödig, Austria) [43]. Two punch biopsies from the central third of the lesional skin of each mouse were analyzed, and the hydroxyproline content was normalized to the dry weight of each biopsy. The mean value was used. For direct visualization of collagen fibers, trichrome staining was performed using the Blue Masson's Trichrome Stain Kit (Sigma-Aldrich, St. Louis, MO, USA). Stained skin sections were visualized with a BX53 microscope with a DP80 Digital Microscope Camera and CellSens Standard Software 3.1-Build 21199 (Olympus, Philadelphia, PA, USA) at 100-fold magnification [43].

2.5. Immunohistochemistry Staining for Aplha-Smooth Muscle Actin (aSMA)

Assessment of myofibroblast counts was performed as described previously [43]. In brief, for the detection of alpha-smooth muscle actin (aSMA)-positive myofibroblasts, skin sections were deparaffinized, followed by incubation with 5% horse serum in PBS for 1 h to block nonspecific binding and incubation with 3% H_2O_2 for 10 min to block endogenous peroxidase activity. The cells positive for aSMA in the lesional skin sections were detected by incubation with mouse monoclonal anti-aSMA antibody (1:1000, clone 1A4, Sigma-Aldrich, St. Louis, MO, USA) for three hours at room temperature. Irrelevant isotype-matched antibodies were used as controls. Horseradish peroxidase-labeled polyclonal rabbit anti-mouse antibodies (1:200, Dako, Glostrup, Denmark) were then used as secondary antibodies for incubation for one hour at room temperature. The skin sections were then counterstained with Mayer's hematoxylin solution and visualized with a BX53 microscope with a DP80 Digital Microscope Camera and CellSens Standard Software v.3.1-Build 21199 (Olympus, Philadelphia, PA, USA) at 100-, 200-, and 400-fold magnification. The assessment of the number of myofibroblasts (at 200-fold magnification) was performed by two experienced examiners blinded to the treatment in the full thickness of the dermis at four sites in four consecutive skin sections of the lesional skin from each mouse. The mean value for each mouse was used [43].

2.6. Immunofluorescence Staining

Immunofluorescence staining was performed as described previously [45,46]. For the identification of cells expressing the phosphorylated Smad3 (pSmad3) [45], or TGF-β-receptor type I (TβRI) and II (TβRII) [46] in the lesional skin, deparaffinized sections were incubated with antigen retrieval solution (Dako. Glostrup, Denmark) for 15 min at 95 °C, followed by incubation with 5% horse serum in PBS for 1 h to block nonspecific binding. Phosphorylated Smad3-positive cells were detected by incubation with rabbit monoclonal anti-pSmad3 antibodies (phospho S423 + S425; 1:50, Abcam, Cambridge, UK) overnight at 4 °C [45]. TβRI- or TβRII-positive cells were detected by incubation with rabbit polyclonal anti-TβRI antibodies (PA5-32631; 1:50, Invitrogen, Carlsbad, CA, USA) or rabbit polyclonal anti-TβRII antibodies (AB186838; 1:100, Abcam, Cambridge, UK) overnight at 4 °C [46]. Irrelevant isotype-matched antibodies were used as controls. The skin samples were then incubated with polyclonal goat anti-rabbit Alexa Fluor 488 secondary antibodies (1:200, Abcam, Cambridge, UK) at room temperature for 1 h. The cell nuclei were stained using 4′,6-diamidino-2-phenylindole (DAPI, 1:1000, Invitrogen, Carlsbad, CA, USA) for 10 min at room temperature. Finally, the skin sections were mounted with Fluoromount aqueous mounting medium (Sigma-Aldrich, St. Louis, MO, USA) [45]. Images were captured at 400-fold magnification using appropriate fluorescence filters on a BX53 microscope with a DP80 Color lens camera using CellSens Standard software v.3.1-Build 21199 (Olympus, Philadelphia, PA, USA). The assessment of the number of TβRI-, TβRII-, and pSmad3-positive cells was performed by two experienced examiners blinded to the treatment at four sites in four consecutive skin sections of the lesional skin from each mouse. The count of positive cells was normalized to the number of cells and is presented as TβRI, TβRII-, and pSmad3-positive cell percentage. The mean value for each mouse was used [45].

2.7. Measurement of Inflammatory Cytokines/Chemokines in the Serum

The concentration of selected inflammatory cytokines and chemokines in the serum of mice was analyzed as described previously [47] by a commercially available Bio-Plex Pro™ Mouse Cytokine 23-plex Assay (BIO-RAD, Irvine, CA, USA) according to the manufacturer's instructions. The Bio-Plex Pro™ Mouse Cytokine 23-plex Assay measures the concentration of 23 cytokines and chemokines: IL-1α, IL-1β, IL-2, IL-3, IL-4, IL-5, IL-6, IL-9, IL-10, IL12p40, IL-12p70, IL-13, IL-17A, eotaxin, granulocyte colony-stimulating factor (G-CSF), granulocyte-macrophage colony-stimulating factor (GM-CSF), interferon-γ (IFN-γ), keratinocytes-derived chemokine (KC, also known as chemokine (C-X-C) motif ligand 1 (CXCL1)), monocyte chemoattractant protein (MCP)-1 (CCL2), macrophage inflammatory proteins (MIP)-1α (CCL3) and MIP-1β (CCL4), regulated on activation/normal T cell expressed and secreted (RANTES, CCL5), and tumor necrosis factor (TNF). The absorbance of the Bio-Plex Pro™ Mouse Cytokine 23-plex Assay was evaluated at Luminex BIO-PLEX 200 System (Bio-Rad, Hercules, CA, USA) [47]. Samples were measured as duplicates, and the mean value was used.

2.8. Safety of 17-DMAG in Mice

The safety of 17-DMAG in mice was assessed as described in previous studies of novel potential antifibrotic agents in a modified bleomycin model [27,28,36,37]. All studied mice were examined daily for any kind of distress, e.g., changes in physical activity, behavior, food and water consumption, quality of stools, and the quality and texture of their fur. Furthermore, the weight of the mice was assessed weekly using calibrated scales from the first bleomycin injection until sacrifice. During necropsy, internal organs were inspected for any macroscopically visible changes, e.g., formation of tumors, hemorrhages, presence of pus, or other significant pathologies [27,28,36,37].

2.9. Statistical Analysis

All analyses and graphs were conducted using GraphPad Prism 5 (v.5.02; GraphPad Software, La Jolla, CA, USA). Basic descriptive statistics (mean, standard error of the mean (SEM), skewness, and kurtosis) were computed for all variables, which were subsequently tested for normality using the Kolmogorov–Smirnov and Shapiro–Wilk tests. Differences between groups were analyzed by unpaired t-test with Welch's correction or the Mann–Whitney U test. Statistical significance was set at $p < 0.05$. Data are presented as mean ± SEM.

3. Results

3.1. Treatment with 17-DMAG Prevents Progression and May Induce Regression of Preestablished Bleomycin-Induced Skin Fibrosis

Using the modified bleomycin model of experimental dermal fibrosis, we investigated the effect of Hsp90 inhibitor 17-DMAG on the progression of dermal fibrosis, and of particular interest, the treatment of established skin fibrosis, since the 17-DMAG treatment was initiated after the development of dermal fibrosis. The efficacy of 17-DMAG was then compared to that of the established antifibrotic agent nintedanib.

Compared to mice treated with NaCl for six weeks (group 1), the challenge with bleomycin in the first three weeks (group 2) resulted in an expected development of skin fibrosis, manifested by an increase in dermal thickening by 51.5 ± 3.9% ($p < 0.0001$), hydroxyproline content by 52.4 ± 13.2% ($p = 0.0014$), and myofibroblast count by 150.3 ± 12.5% ($p < 0.0001$) (Figures 2–4). Extended bleomycin treatment during the subsequent three weeks (group 3) led to further progression of skin fibrosis, with an additional increase in dermal thickening, hydroxyproline content, and myofibroblast count ($p < 0.05$ for all, compared to group 2; $p < 0.001$ for all, compared to group 1) (Figures 2–4).

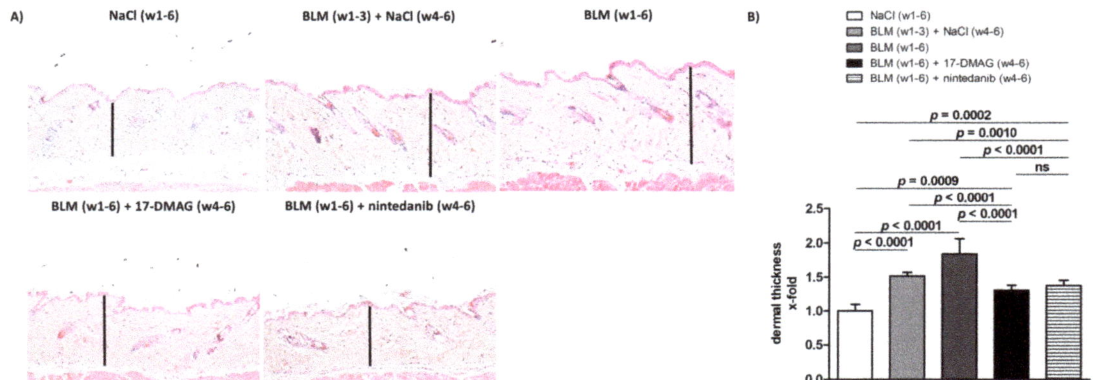

Figure 2. Treatment with 17-DMAG prevents further progression and may induce regression of dermal thickening induced by bleomycin. (**A**) Representative images of hematoxylin and eosin-stained skin sections are shown. Original magnification ×100. Vertical bars represent the dermal thickness. (**B**) Treatment with 17-DMAG prevents further progression and may induce regression of dermal thickening induced by bleomycin. The extent of the protective effects of 17-DMAG is comparable to the effect of the treatment with nintedanib. Columns represent the mean, and whiskers represent the standard error of the mean. w, week; NaCl, sodium chloride; BLM, bleomycin; 17-DMAG, 17-dimethylaminoethylamino-17-demethoxygeldanamycin (inhibitor of heat shock protein 90); ns, not significant ($p \geq 0.05$); $n = 8$ mice in each group.

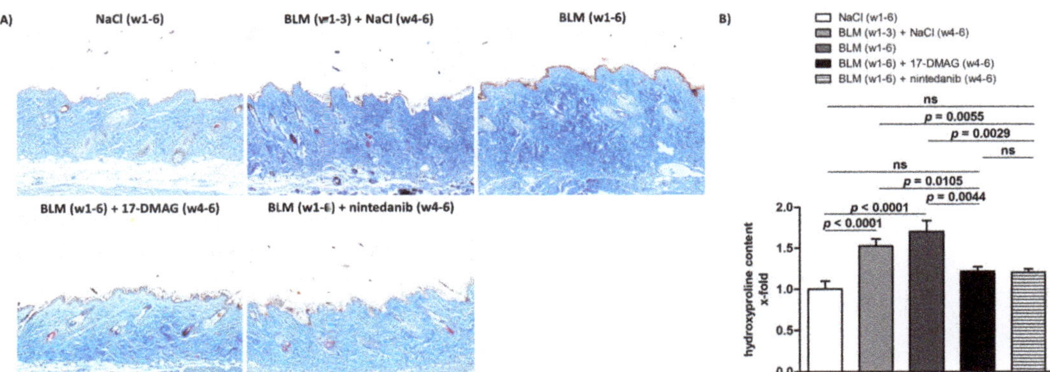

Figure 3. Treatment with 17-DMAG prevents further progression and may induce regression of collagen accumulation induced by bleomycin. (**A**) Representative images of Blue Masson's Trichrome-stained skin sections are shown. Collagen bundles are stained blue. Original magnification ×100. (**B**) Treatment with 17-DMAG prevents further progression and may induce regression of collagen accumulation (analyzed by hydroxyproline content) induced by bleomycin. The extent of the protective effects of 17-DMAG is comparable to the effect of the treatment with nintedanib. Columns represent the mean and whiskers represent the standard error of the mean. w, week; NaCl, sodium chloride; BLM, bleomycin; 17-DMAG, 17-dimethylaminoethylamino-17-demethoxygeldanamycin (inhibitor of heat shock protein 90); ns, not significant ($p \geq 0.05$); $n = 8$ mice in each group.

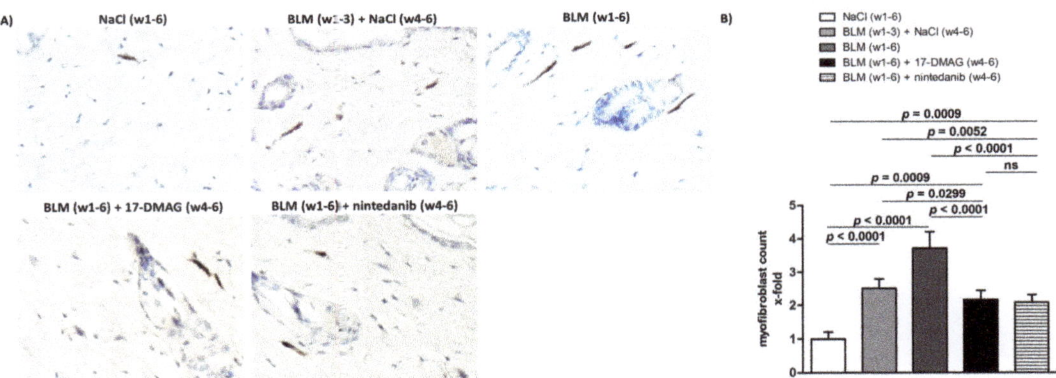

Figure 4. Treatment with 17-DMAG prevents further progression and may induce regression of proliferation of myofibroblasts induced by bleomycin. (**A**) Representative images of α-smooth muscle actin (aSMA)-stained skin sections are shown. aSMA-positive cells are stained brown, nuclei are counterstained blue by hematoxylin. Original magnification ×400. (**B**) Treatment with 17-DMAG prevents further progression and may induce regression of the proliferation of myofibroblasts induced by bleomycin. The extent of the protective effects of 17-DMAG is comparable to the effect of the treatment with nintedanib. Columns represent the mean and whiskers represent the standard error of the mean. w, week; NaCl, sodium chloride; BLM, bleomycin; 17-DMAG, 17-dimethylaminoethylamino-17-demethoxy-geldanamycin (inhibitor of heat shock protein 90); ns, not significant ($p \geq 0.05$); $n = 8$ mice in each group.

Treatment with 17-DMAG (group 4) during the last three weeks of bleomycin treatment inhibited this progression despite concurrent bleomycin injections (Figures 2–4). Compared with vehicle-treated mice challenged with bleomycin for six weeks (group 3), 17-DMAG treatment (group 4) prevented the increase in skin thickness by 53.5 ± 8.2% ($p < 0.0001$) (Figure 2A,B), hydroxyproline content by 48.8 ± 14.4% ($p = 0.0044$) (Figure 3B), and myofibroblast count by 154.5 ± 19.8% ($p < 0.0001$) (Figure 4A,B). Thus, treatment with

17-DMAG effectively prevented the progression of established dermal fibrosis induced by bleomycin.

Furthermore, the extent of dermal fibrosis upon 17-DMAG treatment (group 4) decreased even below the pretreatment levels represented by mice treated with bleomycin for three weeks followed by NaCl for three weeks (group 2): skin thickness decreased by 21.1 ± 3.3% ($p < 0.0001$) (Figure 2A,B), hydroxyproline content by 30.8 ± 10.4% ($p = 0.0105$) (Figure 3B), and myofibroblast count by 33.5 ± 13.9% ($p = 0.0299$) (Figure 4A,B). Thus, treatment with 17-DMAG may induce regression of established dermal fibrosis induced by bleomycin.

Interestingly, the decreased extent of fibrosis in the 17-DMAG-treated mice (group 4), as demonstrated by the abovementioned skin thickness, hydroxyproline content, and myofibroblast count, was comparable to that observed in mice treated with the established antifibrotic agent nintedanib (group 5) ($p = 0.0919$, $p = 0.8551$, $p = 0.4776$, respectively) (Figures 2–4).

3.2. 17-DMAG Reduces the Activation of TGF-β Smad Signaling in Bleomycin-Induced Dermal Fibrosis

In addition, given the Hsp90-mediated stabilization of TβRI and TβRII [25], the abrogation of profibrotic effects of TGF-β upon 17-DMAG treatment demonstrated in our previous study [26], and the suppression of TGF-β Smad signaling upon Hsp90 inhibition demonstrated by Noh et al. [48], we aimed to investigate the impact of 17-DMAG on the activation of TGF-β signaling by assessing the expression of TGF-β type I and II receptors and the intracellular pathway mediated by Smad3. Compared to mice treated with NaCl for six weeks (group 1), we observed an expected activation of TGF-β signaling with an increased percentage of TβRI-, TβRII-, and pSmad3-positive cells in the lesional skin upon challenge with bleomycin for three weeks (group 2, by 108.3 ± 16.5%, $p < 0.0001$; 122.2 ± 29.9%, $p = 0.0020$; 90.6 ± 13.2%, $p < 0.0001$, respectively) or six weeks (group 3, by 239.9 ± 30.7%, $p < 0.0001$; 206.4 ± 32.0%, $p < 0.0001$; 181.9 ± 26.1%, $p < 0.0001$, respectively) (Figure 5A–F). Indeed, treatment with 17-DMAG (group 4) significantly decreased the percentage of TβRI-, TβRII-, and pSmad3-positive cells (by 199.5 ± 31.8%, $p < 0.0001$; 163.0 ± 33.7%, $p = 0.0003$; 130.3 ± 28.4%, $p = 0.0005$ compared to group 3, and by 67.9 ± 18.6%, $p = 0.0026$; 78.8 ± 33.9%, $p = 0.0357$; 39.0 ± 16.9%, $p = 0.0363$ compared to group 2, respectively), as well as the intensity of staining (Figure 5A–F). Thus, treatment with 17-DMAG strongly downregulated TGF-β/pSmad3 signaling in the lesional skin of mice challenged with bleomycin to below pretreatment levels. Nintedanib treatment (group 5) demonstrated mild reduction in TGF-β/pSmad3 signaling upon bleomycin challenge ($p < 0.05$ for all three outcomes compared to group 3).

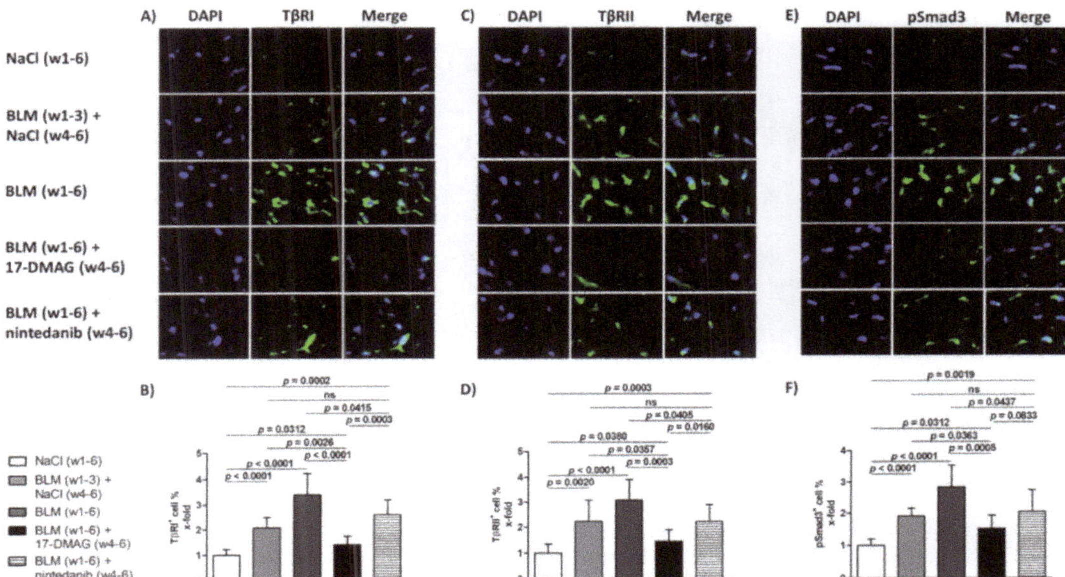

Figure 5. Treatment with 17-DMAG inhibits the TGF-β/Smad signaling induced by bleomycin. Representative images of TβRI- (**A**), TβRII- (**C**), and phosphorylated Smad3 (pSmad3)-stained skin sections are shown (**E**). TβRI-, TβRII-, and pSmad3-positive cells are stained green, and nuclei are stained blue by DAPI. Original magnification ×400. Treatment with 17-DMAG decreased the accumulation of TβRI- (**B**), TβRII- (**D**), and pSmad3 (**F**) induced by bleomycin. Columns represent the mean and whiskers represent the standard error of the mean. w, week; NaCl, sodium chloride; BLM, bleomycin; 17-DMAG, 17-dimethylaminoethylamino-17-demethoxygeldanamycin (inhibitor of heat shock protein 90); TβRI, TGF-β receptor type I; TβRII, TGF-β receptor type II; ns, not significant ($p \geq 0.05$); $n = 8$ mice in each group.

3.3. 17-DMAG Treatment Reduces Local and Systemic Inflammation in Bleomycin-Induced Dermal Fibrosis

The mouse model of bleomycin-induced experimental dermal fibrosis mimics the early stages of SSc, characterized by perivascular inflammatory infiltrates in the dermis containing leukocytes, including T and B lymphocytes, macrophages, eosinophils, and mast cells, which stimulate fibroblast activation and collagen synthesis by releasing profibrotic cytokines and growth factors [2,7,34,49]. Given the crucial role of Hsp90 in the innate and adaptive immune system [16,17], and to investigate whether 17-DMAG affects the outcome of bleomycin-induced dermal fibrosis partially by regulating inflammatory infiltration, we further quantified the number of leukocytes in the lesional skin. Compared to mice treated with NaCl for six weeks (group 1), we observed expectedly elevated numbers of leukocytes infiltrating the lesional skin upon challenge with bleomycin for three weeks (group 2, by 164.4 ± 18.7%, $p < 0.0001$) or six weeks (group 3, by 253.9 ± 22.6%, $p < 0.0001$) (Figure 6A). Treatment with 17-DMAG (group 4) significantly decreased the leukocyte count (by 182.9 ± 25.5%, $p < 0.0001$ compared to group 3, and by 93.4 ± 22.2%, $p = 0.0009$ compared to group 2) (Figure 6A). Thus, treatment with 17-DMAG strongly reduced the inflammatory infiltration in the lesional skin of mice challenged with bleomycin to below pretreatment levels. Nintedanib treatment (group 5) demonstrated only a mild reduction in the inflammatory infiltration in the lesional skin upon bleomycin challenge (by 62.6 ± 33.3%, $p = 0.0803$ compared to group 3) (Figure 6A).

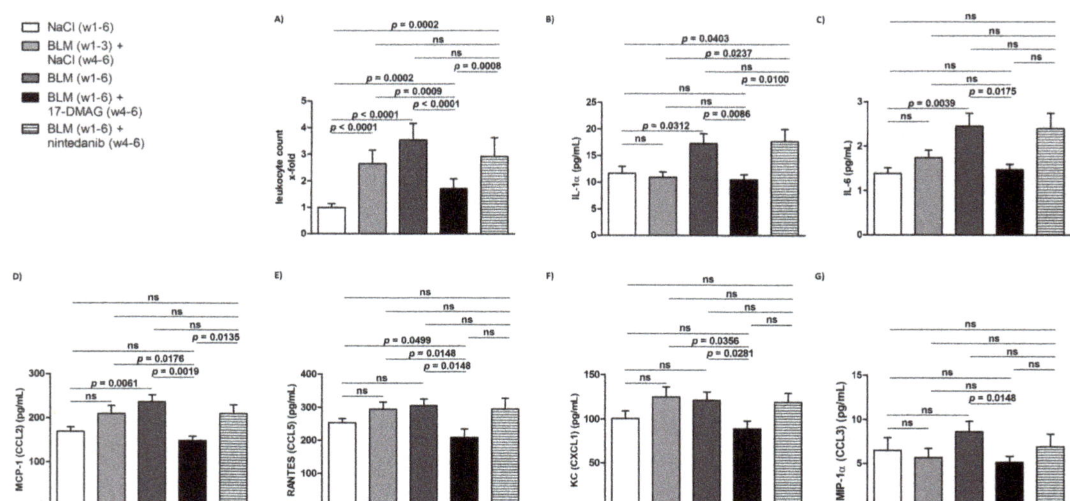

Figure 6. Treatment with 17-DMAG strongly downregulates the local and systemic proinflammatory response induced by bleomycin. Treatment with 17-DMAG significantly reduced the number of leukocytes infiltrating the lesional skin (**A**), and the serum levels of interleukin (IL)-1α (**B**), IL-6 (**C**), monocyte chemoattractant protein-1 (MCP-1, CCL2) (**D**), regulated on activation/normal T cell expressed and secreted (RANTES, CCL5) (**E**), keratinocytes-derived chemokine (KC, CXCL1) (**F**), and macrophage inflammatory proteins-1α (MIP1α, CCL3) (**G**). Columns represent the mean and whiskers represent the standard error of the mean. w, week; NaCl, sodium chloride; BLM, bleomycin; 17-DMAG, 17-dimethylaminoethylamino-17-demethoxygeldanamycin (inhibitor of heat shock protein 90); ns, not significant ($p \geq 0.05$); $n = 8$ mice in each group.

Given the documented systemic manifestations of prolonged subcutaneous administration of bleomycin in mice [34,39,50], we determined the systemic levels of selected proinflammatory cytokines and chemokines which have been implicated in the pathophysiology of SSc [1,2,7,29], in order to further assess the potential regulation of inflammatory infiltration and the associated fibroblast activation in the lesional skin mediated by Hsp90. Compared to NaCl-treated mice (group 1), in mice challenged with bleomycin for six weeks (group 3), we found an increase in serum levels of IL-1α, IL-6, and MCP-1 (CCL2), which was statistically significant ($p < 0.05$ for all), and a trend to increased serum levels of CCL5, CXCL1, and CCL3, which did not reach the level of statistical significance ($p = 0.074$, $p = 0.128$, $p = 0.130$, respectively) (Figure 6B–G). Nevertheless, 17-DMAG treatment (group 4) resulted in a significant decrease ($p < 0.05$ for all, compared to group 3) in serum levels of all these proinflammatory cytokines/chemokines to levels comparable to those observed in the NaCl-treated mice (group 1) (Figure 6B–G). Thus, treatment with 17-DMAG strongly downregulated the systemic proinflammatory response induced by extended subcutaneous administration of bleomycin in experimental dermal fibrosis. No differences were observed in the remaining cytokines or chemokines measured by the Bio-Plex Pro™ Mouse Cytokine 23-plex Assay ($p > 0.05$ for all). No differences in serum levels of any of these cytokines/chemokines were demonstrated in mice treated with nintedanib (group 5) upon bleomycin challenge (Figure 6B–G).

4. Discussion

In this study, we demonstrate for the first time that treatment with 17-DMAG, an Hsp90 inhibitor, effectively prevents progression and may induce regression of established experimental dermal fibrosis induced by bleomycin. The extent of reduction of dermal fibrosis mediated by 17-DMAG is comparable to the effects of the treatment with nintedanib, an established antifibrotic agent. The antifibrotic effects of 17-DMAG in this modified bleomycin model are not only mediated by direct inhibition of fibroblast activation via the downregulation of TGF-β/Smad signaling, but also indirectly via a potent suppression of the local and systemic inflammatory response. This dual mode of action translates into potent antifibrotic effects of 17-DMAG.

A large body of evidence has demonstrated the crucial role of Hsp90 in tumorigenesis and metastasis, particularly by maintaining homeostasis of several oncogenic proteins, including a number of transcription factors and kinases [18,19,22]. Thus, over the last two decades, several Hsp90 inhibitors have been tested in clinical trials, predominantly in solid tumors and hematological malignancies [18,30,33]. 17-DMAG is a second-generation inhibitor of Hsp90 characterized by higher water solubility, improved bioavailability, reduced toxicity, and higher therapeutic efficacy than its predecessors [32]. It suppresses the ATPase activity of Hsp90 and consequently leads to misfolding, ubiquitylation, and degradation of the client proteins of Hsp90 by the proteasome [31,32].

Compared to the well-documented role of Hsp90 in cancer, the understanding of the role of Hsp90 in tissue fibrosis is very limited. Hsp90 was upregulated in the lung biopsies of patients with idiopathic pulmonary fibrosis and in the isolated fibroblasts from the fibrotic lung lesions. Furthermore, inhibition of Hsp90 by 17-N-allylamino-17-demethoxy-geldanamycin (17-AAG) efficiently reduced TGF-β-driven activation of fibroblasts and production of ECM in vitro and attenuated progression of established fibrosis in a mouse model of pulmonary fibrosis [51]. More recently, five other studies have confirmed potent antifibrotic effects of various Hsp90 inhibitors, including 17-DMAG, in several preclinical models of pulmonary fibrosis [52–56]. Similarly, the treatment with 17-AAG blocked the TGF-β-induced production of ECM in renal fibroblasts in vitro and suppressed renal fibrosis in a murine model of unilateral ureteral obstruction [48]. Inhibition of Hsp90 with an engineered protein inhibitor reduced the TGF-β-mediated profibrotic events in cardiac fibroblasts in vitro and ameliorated experimental murine myocardial fibrosis induced by angiotensin-II [57]. Our recent study on the role of Hsp90 in SSc demonstrated an increased expression of Hsp90 in the SSc skin and dermal fibroblasts and in experimental dermal fibrosis in a TGF-β-dependent manner [26]. Treatment with 17-DMAG effectively abrogated the profibrotic effects of TGF-β in cultured dermal fibroblasts and prevented the development of experimental dermal fibrosis in three different murine models of SSc [26].

In line with these findings, in this study, we provide evidence that treatment with 17-DMAG not only effectively prevents further progression, but may also promote regression of established experimental dermal fibrosis induced by bleomycin. However, given the limited capacity of group 2 to precisely represent the pretreatment level of fibrosis, regression of the established experimental dermal fibrosis induced by 17-DMAG treatment should be interpreted with caution. Compared to vehicle-treated mice, treatment with 17-DMAG decreased dermal thickening, hydroxyproline content, and myofibroblast count to below pretreatment levels. Given the proven efficacy of several potential antifibrotic therapies in the preclinical model of established dermal fibrosis [8], we selected nintedanib as a positive treatment control. Nintedanib is a small-molecule competitive inhibitor of nonreceptor tyrosine kinases (nRTKs), such as lymphocyte-specific protein tyrosine kinase (Lck), tyrosine-protein kinase Lyn (Lyn), and Src. It also inhibits receptor tyrosine kinases (RTKs), such as platelet derived growth factor (PDGF) receptor α and β, fibroblast growth factor (FGF) receptor 1–3, vascular endothelial growth factor (VEGF) receptor 1–3, and fms-like tyrosine kinase 3 (FLT-3) [58]. Moreover, nintedanib has previously demonstrated efficacy in the regression of preestablished experimental dermal fibrosis induced by bleomycin [28], as well as in other preclinical models of tissue fibrosis [28,42,58], and was recently approved by Food and Drug Administration (FDA) for slowing the rate of decline in lung function in adults with SSc-associated interstitial lung disease (ILD) [59]. Interestingly, in this study, the extent of the antifibrotic effects of 17-DMAG treatment was comparable to that of nintedanib treatment. Previous studies on nintedanib in preclinical models of tissue fibrosis have attributed its antifibrotic effects to several possible mechanisms. In vitro, direct antifibrotic effects of nintedanib were mediated by the inhibition of PDGF- and TGF-β-induced proliferation, migration and activation of fibroblasts, as well as myofibroblast differentiation and collagen release [28]. In vivo, possible indirect mechanisms include the anti-inflammatory effects mediated by the inhibition of Lck and Lyn kinases [28], or by inhibiting the alternative activation of macrophages, along with decreased serum levels of M-CSF and VEGF in an Fra-2 mouse model [42]. In our present study, in contrast to 17-DMAG, the antifibrotic effects of nintedanib were not associated with major alterations in serum levels of proinflammatory cytokines or chemokines.

In addition to the findings from our previous study [26], the antifibrotic effects of 17-DMAG observed in this study were mediated by the inhibition of TGF-β/Smad signaling, as was evidenced by the decreased percentage of TβRI-, TβRII-, and pSmad3-positive cells to below pretreatment levels. Our findings are in line with previously published observations in experimental renal and myocardial fibrosis [48,57]. Noh et al. demonstrated that 17-AAG treatment suppressed TGF-β-induced Smad signaling in renal fibroblasts via a mechanism dependent on proteasome-mediated degradation of TβRII [48]. Caceres et al. showed that inhibition of Hsp90 in myocardial fibroblasts resulted in the disruption of the TGFβRI-Hsp90 complex and a reduction in TGF-β/Smad signaling [57]. Nevertheless, the detailed mechanisms of the TGF-β signaling suppression by 17-DMAG in bleomycin-induced dermal fibrosis have yet to be elucidated by further studies.

Given the documented role of Hsp90 in the innate and adaptive immune system [16,17], we aimed to examine whether the effect of 17-DMAG on bleomycin-induced dermal fibrosis is partially mediated by the regulation of the immune response. The early stages of SSc are characterized by increased number and activation of inflammatory cells infiltrating the involved skin, which release proinflammatory and profibrotic cytokines that further stimulate the production of ECM in resident fibroblasts [2,7,49]. Similar to SSc, repeated subcutaneous administration of bleomycin in the experimental model of dermal fibrosis increases the number and activation of leukocytes infiltrating the lesional skin [34,60]. Our results demonstrate that 17-DMAG treatment significantly reduced the number of leukocytes infiltrating the lesional skin challenged with bleomycin, even to pretreatment levels. Our findings are in agreement with previous studies demonstrating anti-inflammatory effects of Hsp90 inhibitors, which significantly improved the clinical course in animal models of several autoimmune inflammatory diseases, such as autoimmune encephalomyopathy, rheumatoid arthritis, systemic lupus erythematosus-like autoimmune disease, and epidermolysis bullosa acquisita [61–64]. In addition, our recently published study demonstrated elevated systemic levels of Hsp90 in SSc patients, particularly in those with elevated C-reactive protein levels [65].

To further evaluate the Hsp90-mediated regulation of inflammatory infiltration and the associated fibroblast activation in the lesional skin challenged with bleomycin, we assessed the systemic levels of selected proinflammatory cytokines and chemokines, which represent the links between the immune response and the development of fibrosis in SSc [7,29]. Prolonged subcutaneous administration of bleomycin in the murine model of experimental dermal fibrosis also results in systemic manifestations, in particular, lung fibrosis exhibiting thickened alveolar walls with cellular infiltrates [60]. However, intratracheal (or intranasal) administration (single-dose or modified repeated-dose protocol) of bleomycin in rodents represents the golden standard in preclinical models of experimental lung fibrosis [66], and the antifibrotic effects of 17-DMAG in such models have been previously demonstrated [55]. Therefore, the impact of 17-DMAG on lung fibrosis was not assessed in this study. Nevertheless, in bleomycin-challenged vehicle-treated mice, we demonstrate a significant increase in serum levels of IL-1α, IL-6, and MCP-1 (CCL2), and a trend to increased serum levels of RANTES (CCL5), KC (CXCL1), and MIP-1α (CCL3). Treatment with 17-DMAG significantly reduced the serum levels of all these proinflammatory cytokines and chemokines to an extent comparable to the levels of NaCl-challenged mice. The role of these cytokines and chemokines in SSc has already been well established [1,2,7,29]. Expression of IL-1α is spontaneously increased in SSc fibroblasts and additionally induces expression of IL-6, PDGF, and the fibrogenic phenotype of SSc fibroblasts [67,68]. IL-6 has a crucial role in the development of fibrosis and inflammation in SSc, substantiated by its increased levels in serum, skin, SSc fibroblasts, and peripheral blood mononuclear cells (PBMCs). Moreover, inhibition of IL-6 by tocilizumab demonstrated promising effects in stabilizing SSc-ILD in a randomized controlled trial [7,69,70]. MCP-1 (CCL2) is one of the most robustly described chemokines with an established role in mediating fibrotic immune responses in SSc, and its levels are increased in SSc serum, skin, and bronchoalveolar lavage (BAL) fluid [29]. Elevated levels of RANTES (CCL5) and/or MIP-1α (CCL3) have been found in SSc serum, BAL fluid, skin, and PBMCs [29]. Increased serum levels of CXCL1 were associated with deteriorated lung function in SSc patients [71]. However, the potential impact of 17-DMAG in the preclinical models of SSc on serum levels of other cytokines and chemokines, which play vital roles in the pathogenesis of SSc and the progression of fibrosis, such as TGF-β [3] and IL-8 [71], needs to be elucidated by further investigation. Several studies involving different Hsp90 inhibitors have already demonstrated antifibrotic effects mediated by the inhibition of the TGF-β pathway in experimental models of lung fibrosis [51–56]. Interestingly, in the study by Sibinska et al. [55], only the lower dose of 17-DMAG (i.e., 10 mg/kg vs. 25 mg/kg) significantly decreased both the bronchoalveolar lavage fluid and serum levels of TGF-β in the bleomycin-induced pulmonary fibrosis. Similarly, inhibition of Hsp90 reduced local or systemic levels of IL-8 in experimental models of various inflammatory and tumorous conditions [72–76]. However, further studies are required to elucidate the mechanisms underlying the suppression of local and systemic inflammatory response mediated by 17-DMAG, as well as the potential link between the humoral or cellular immune response and the progression of dermal fibrosis induced by bleomycin.

The treatment with 17-DMAG was well tolerated without obvious signs of toxicity on clinical examination, or on gross necropsy. No reduction in weight was detected in 17-DMAG-treated mice compared to mice injected subcutaneously with bleomycin only. Nevertheless, data on the most common dose-limiting toxicities of 17-DMAG are already available from seven clinical trials in hematological malignancies and solid tumors, and symptoms include fatigue, nausea, vomiting, diarrhea, anorexia, and liver enzyme disturbances [32]. Therefore, our findings on the efficacy of 17-DMAG in the treatment of established experimental dermal fibrosis need to be validated by further studies, preferably using the more recently developed Hsp90 inhibitors with improved safety profile [33].

5. Conclusions

We demonstrate that treatment with 17-DMAG, an Hsp90 inhibitor, effectively prevented further progression and may induce regression of established bleomycin-induced dermal fibrosis with a comparable outcome to that of nintedanib, a well-established antifibrotic agent. Antifibrotic effects of 17-DMAG were mediated by direct inhibition of fibroblast activation via the suppression of TGF-β/Smad signaling and indirectly by reduction of the local and systemic inflammatory response induced by bleomycin. The treatment with 17-DMAG was well tolerated without obvious clinical signs of toxicity. Our findings further support the vital role of Hsp90 in the pathophysiology of SSc, thus providing a potential target for the treatment of fibrosis with translational implications due to the availability of several Hsp90 inhibitors in clinical trials for other indications. In addition, given our recently published results [65], which demonstrated increased systemic levels of Hsp90 in SSc patients, their association with systemic inflammation, skin and lung involvement, and their ability to predict the treatment response in SSc-ILD, Hsp90 could be a potential biomarker for stratification of SSc patients suitable for targeted antifibrotic treatment.

Author Contributions: Conceptualization, H.Š., J.H.W.D., L.Š. (Ladislav Šenolt), and M.T.; data curation, H.Š., A.N., V.B., S.O., R.B., and M.T.; formal analysis, A.N., S.O., B.H., M.Š., R.B., K.P., J.V., J.H.W.D., and L.Š. (Ladislav Šenolt); funding acquisition, K.P., J.V., and M.T.; investigation, L.Š. (Lenka Štorkanova), A.N., V.B., H.H., and B.H.; methodology, L.Š. (Lenka Štorkanova), A.N., V.B., H.H., M.Š., and M.T.; project administration, M.T.; resources, M.T.; supervision, M.T.; writing—original draft, H.Š. and M.T.; writing—review and editing, H.Š., L.Š. (Lenka Štorkanova), A.N., V.B., H.H., S.O., B.H., M.Š., R.B., K.P., J.V., J.H.W.D., L.Š. (Ladislav Šenolt), and M.T. All authors have read and agreed to the published version of the manuscript.

Funding: This research was funded by the Ministry of Health of the Czech Republic 00023728, grant nr. 16-33542A, and SVV 260523. This research received no external funding from Boehringer Ingelheim Pharma GmbH & Co.KG, Ingelheim am Rhein, Germany.

Institutional Review Board Statement: The study was conducted according to the guidelines of the Declaration of Helsinki, and approved by the Ethics Committee of the Institute of Rheumatology in Prague (reference number 5689/2015, approved on 6 June 2015) and the Ministry of Education, Youth and Sports of the Czech Republic (reference number MSMT-9445/2018-7, approved on 5 May 2018). Animal experiments were conducted in accordance with relevant national legislation on the use of animals for research and complied with the commonly accepted 3Rs.

Informed Consent Statement: Not applicable.

Data Availability Statement: Individual anonymized data will not be shared. Pooled study data, protocol, or statistical analysis plans can be shared upon request at tomcik@revma.cz.

Acknowledgments: The authors would like to thank Boehringer Ingelheim Pharma GmbH & Co.KG, Ingelheim am Rhein, Germany for kindly providing nintedanib for the experiments, Růžena Paroubková for technical assistance, and Zainab Mansoor and Xiao Švec for language editing.

Conflicts of Interest: The authors declare no conflict of interest.

Abbreviations

Hsp90	heat shock protein 90
SSc	systemic sclerosis
17-DMAG	17-dimethylaminoethylamino-17-demethoxy-geldanamycin
pSmad3	phosphorylated small mothers against decapentaplegic homolog 3
ECM	extracellular matrix
TGF-β	transforming growth factor-beta
aSMA	alpha-smooth muscle actin
TβRI	transforming growth factor-beta type I
TβRII	transforming growth factor-beta type II
PDGF	platelet-derived growth factor
IL	interleukin
MCP-1	monocyte chemoattractant protein 1
CCL	CC chemokine
Hsp	heat shock proteins
TNF	tumor necrosis factor
Src	proto-oncogene tyrosine-protein kinase Src
C57BL/6	laboratory strain of mice referred to as black 6
NaCl	sodium chloride
PBS	phosphate buffered saline
nRTKs	nonreceptor tyrosine kinases
Lck	lymphocyte-specific protein tyrosine kinase
Lyn	tyrosine-protein kinase Lyn

RTKs	receptor tyrosine kinases
FGF	fibroblast growth factor
VEGF	vascular endothelial growth factor
FLT-3	fms-like tyrosine kinase 3
FDA	Food and Drug Administration
ILD	interstitial lung disease
AZV	Czech Health Research Council
MSMT	Ministry of Education, Youth and Sports of the Czech Republic
3Rs	Replacement of animals by alternatives wherever possible, Reduction in number of animals used, and Refinement of experimental conditions and procedures to minimize the harm to animals
HCl	hydrogen chloride
pH	potential of hydrogen
NaOH	sodium hydroxide
H2O2	hydrogen peroxide
HRP	horseradish peroxidase
DAPI	4′,6-diamidino-2-phenylindole
G-CSF	granulocyte colony-stimulating factor
GM-CSF	granulocyte-macrophage colony-stimulating factor
IFN-γ	interferon-γ
KC	keratinocytes-derived chemokine
CXCL1	chemokine C-X-C motif ligand 1
MIP-1α	macrophage inflammatory protein-1α, also referred to as CCL3
MIP-1β	macrophage inflammatory protein-1β, also referred to as CCL4
RANTES	regulated on activation/normal T cell expressed and secreted, also referred to as CCL5
SEM	standard error of the mean
P	*p*-value
17-AAG	17-N-allylamino-17-demethoxy-geldanamycin
PBMCs	peripheral blood mononuclear cells
BAL	bronchoalveolar lavage, w: week
BLM	bleomycin, NINT: nintedanib
N	number of mice
s.c.	subcutaneous
i.p.	intraperitoneal
p.o.	peroral

References

1. Denton, C.P.; Khanna, D. Systemic sclerosis. *Lancet* **2017**, *390*, 1685–1699. [CrossRef]
2. Stern, E.P.; Denton, C.P. The pathogenesis of systemic sclerosis. *Rheum. Dis. Clin. N. Am.* **2015**, *41*, 367–382. [CrossRef] [PubMed]
3. Distler, J.H.W.; Gyorfi, A.H.; Ramanujam, M.; Whitfield, M.L.; Konigshoff, M.; Lafyatis, R. Shared and distinct mechanisms of fibrosis. *Nat. Rev. Rheumatol.* **2019**, *15*, 705–730. [CrossRef]
4. Bond, J.E.; Ho, T.Q.; Selim, M.A.; Hunter, C.L.; Bowers, E.V.; Levinson, H. Temporal spatial expression and function of non-muscle myosin ii isoforms iia and iib in scar remodeling. *Lab. Investig.* **2011**, *91*, 499–508. [CrossRef]
5. Van Caam, A.; Vonk, M.; van den Hoogen, F.; van Lent, P.; van der Kraan, P. Unraveling ssc pathophysiology; the myofibroblast. *Front. Immunol.* **2018**, *9*, 2452. [CrossRef]
6. Kendall, R.T.; Feghali-Bostwick, C.A. Fibroblasts in fibrosis: Novel roles and mediators. *Front. Pharmacol.* **2014**, *5*, 123. [CrossRef]
7. Raja, J.; Denton, C.P. Cytokines in the immunopathology of systemic sclerosis. *Semin. Immunopathol.* **2015**, *37*, 543–557. [CrossRef] [PubMed]
8. Distler, J.H.; Feghali-Bostwick, C.; Soare, A.; Asano, Y.; Distler, O.; Abraham, D.J. Review: Frontiers of antifibrotic therapy in systemic sclerosis. *Arthritis Rheumatol.* **2017**, *69*, 257–267. [CrossRef] [PubMed]
9. Poudel, D.R.; Derk, C.T. Mortality and survival in systemic sclerosis: A review of recent literature. *Curr. Opin. Rheumatol.* **2018**, *30*, 588–593. [CrossRef]
10. Schlesinger, M.J. Heat shock proteins. *J. Biol. Chem.* **1990**, *265*, 12111–12114. [CrossRef]
11. Santoro, M.G. Heat shock factors and the control of the stress response. *Biochem. Pharmacol.* **2000**, *59*, 55–63. [CrossRef]
12. Lindquist, S.; Craig, E.A. The heat-shock proteins. *Annu. Rev. Genet.* **1988**, *22*, 631–677. [CrossRef]

13. Biebl, M.M.; Buchner, J. Structure, function, and regulation of the hsp90 machinery. *Cold Spring Harb. Perspect. Biol.* **2019**, *11*, a034017. [CrossRef]
14. Burrows, F.; Zhang, H.; Kamal, A. Hsp90 activation and cell cycle regulation. *Cell Cycle* **2004**, *3*, 1530–1536. [CrossRef] [PubMed]
15. Echeverria, P.C.; Picard, D. Molecular chaperones, essential partners of steroid hormone receptors for activity and mobility. *Biochim. Biophys. Acta* **2010**, *1803*, 641–649. [CrossRef]
16. Jackson, S.E. Hsp90: Structure and function. *Top. Curr. Chem.* **2013**, *328*, 155–240.
17. Li, J.; Buchner, J. Structure, function and regulation of the hsp90 machinery. *Biomed. J.* **2013**, *36*, 106–117. [PubMed]
18. Mahalingam, D.; Swords, R.; Carew, J.S.; Nawrocki, S.T.; Bhalla, K.; Giles, F.J. Targeting hsp90 for cancer therapy. *Br. J. Cancer* **2009**, *100*, 1523–1529. [CrossRef]
19. Wong, D.S.; Jay, D.G. Emerging roles of extracellular hsp90 in cancer. *Adv. Cancer Res.* **2016**, *129*, 141–163. [PubMed]
20. Geller, R.; Taguwa, S.; Frydman, J. Broad action of hsp90 as a host chaperone required for viral replication. *Biochim. Biophys. Acta* **2012**, *1823*, 698–706. [CrossRef] [PubMed]
21. Kalia, S.K.; Kalia, L.V.; McLean, P.J. Molecular chaperones as rational drug targets for parkinson's disease therapeutics. *CNS Neurol. Disord. Drug Targets* **2010**, *9*, 741–753. [CrossRef] [PubMed]
22. Zuehlke, A.D.; Moses, M.A.; Neckers, L. Heat shock protein 90: Its inhibition and function. *Philos. Trans. R Soc. Lond. B Biol. Sci.* **2018**, *373*, 20160527. [CrossRef] [PubMed]
23. Koga, F.; Xu, W.; Karpova, T.S.; McNally, J.G.; Baron, R.; Neckers, L. Hsp90 inhibition transiently activates src kinase and promotes src-dependent akt and erk activation. *Proc. Natl. Acad. Sci. USA* **2006**, *103*, 11318–11322. [CrossRef] [PubMed]
24. Skhirtladze, C.; Distler, O.; Dees, C.; Akhmetshina, A.; Busch, N.; Venalis, P.; Zwerina, J.; Spriewald, B.; Pileckyte, M.; Schett, G.; et al. Src kinases in systemic sclerosis: Central roles in fibroblast activation and in skin fibrosis. *Arthritis Rheum.* **2008**, *58*, 1475–1484. [CrossRef]
25. Wrighton, K.H.; Lin, X.; Feng, X.H. Critical regulation of tgfbeta signaling by hsp90. *Proc. Natl. Acad. Sci. USA* **2008**, *105*, 9244–9249. [CrossRef]
26. Tomcik, M.; Zerr, P.; Pitkowski, J.; Palumbo-Zerr, K.; Avouac, J.; Distler, O.; Becvar, R.; Senolt, L.; Schett, G.; Distler, J.H. Heat shock protein 90 (hsp90) inhibition targets canonical tgf-beta signalling to prevent fibrosis. *Ann. Rheum. Dis.* **2014**, *73*, 1215–1222. [CrossRef]
27. Akhmetshina, A.; Venalis, P.; Dees, C.; Busch, N.; Zwerina, J.; Schett, G.; Distler, O.; Distler, J.H. Treatment with imatinib prevents fibrosis in different preclinical models of systemic sclerosis and induces regression of established fibrosis. *Arthritis Rheum.* **2009**, *60*, 219–224. [CrossRef]
28. Huang, J.; Beyer, C.; Palumbo-Zerr, K.; Zhang, Y.; Ramming, A.; Distler, A.; Gelse, K.; Distler, O.; Schett, G.; Wollin, L.; et al. Nintedanib inhibits fibroblast activation and ameliorates fibrosis in preclinical models of systemic sclerosis. *Ann. Rheum. Dis.* **2016**, *75*, 883–890. [CrossRef]
29. King, J.; Abraham, D.; Stratton, R. Chemokines in systemic sclerosis. *Immunol. Lett.* **2018**, *195*, 68–75. [CrossRef] [PubMed]
30. Gao, C.; Peng, Y.N.; Wang, H.Z.; Fang, S.L.; Zhang, M.; Zhao, Q.; Liu, J. Inhibition of heat shock protein 90 as a novel platform for the treatment of cancer. *Curr. Pharm. Des.* **2019**, *25*, 849–855. [CrossRef]
31. Li, L.; Wang, L.; You, Q.D.; Xu, X.L. Heat shock protein 90 inhibitors: An update on achievements, challenges, and future directions. *J. Med. Chem.* **2020**, *63*, 1798–1822. [CrossRef]
32. Mellatyar, H.; Talaei, S.; Pilehvar-Soltanahmadi, Y.; Barzegar, A.; Akbarzadeh, A.; Shahabi, A.; Barekati-Mowahed, M.; Zarghami, N. Targeted cancer therapy through 17-dmag as an hsp90 inhibitor: Overview and current state of the art. *Biomed. Pharmacother.* **2018**, *102*, 608–617. [CrossRef] [PubMed]
33. Sanchez, J.; Carter, T.R.; Cohen, M.S.; Blagg, B.S.J. Old and new approaches to target the hsp90 chaperone. *Curr. Cancer Drug Targets* **2020**, *20*, 253–270. [CrossRef] [PubMed]
34. Beyer, C.; Schett, G.; Distler, O.; Distler, J.H. Animal models of systemic sclerosis: Prospects and limitations. *Arthritis Rheum.* **2010**, *62*, 2831–2844. [CrossRef]
35. Beyer, C.; Reich, N.; Schindler, S.C.; Akhmetshina, A.; Dees, C.; Tomcik, M.; Hirth-Dietrich, C.; von Degenfeld, G.; Sandner, P.; Distler, O.; et al. Stimulation of soluble guanylate cyclase reduces experimental dermal fibrosis. *Ann. Rheum. Dis.* **2012**, *71*, 1019–1026. [CrossRef]
36. Dees, C.; Zerr, P.; Tomcik, M.; Beyer, C.; Horn, A.; Akhmetshina, A.; Palumbo, K.; Reich, N.; Zwerina, J.; Sticherling, M.; et al. Inhibition of notch signaling prevents experimental fibrosis and induces regression of established fibrosis. *Arthritis Rheum.* **2011**, *63*, 1396–1404. [CrossRef]
37. Dees, C.; Akhmetshina, A.; Zerr, P.; Reich, N.; Palumbo, K.; Horn, A.; Jungel, A.; Beyer, C.; Kronke, G.; Zwerina, J.; et al. Platelet-derived serotonin links vascular disease and tissue fibrosis. *J. Exp. Med.* **2011**, *208*, 961–972. [CrossRef]
38. Distler, J.H.; Jungel, A.; Huber, L.C.; Schulze-Horsel, U.; Zwerina, J.; Gay, R.E.; Michel, B.A.; Hauser, T.; Schett, G.; Gay, S.; et al. Imatinib mesylate reduces production of extracellular matrix and prevents development of experimental dermal fibrosis. *Arthritis Rheum.* **2007**, *56*, 311–322. [CrossRef] [PubMed]
39. Yamamoto, T.; Takagawa, S.; Katayama, I.; Yamazaki, K.; Hamazaki, Y.; Shinkai, H.; Nishioka, K. Animal model of sclerotic skin. I: Local injections of bleomycin induce sclerotic skin mimicking scleroderma. *J. Investig. Dermatol.* **1999**, *112*, 456–462. [CrossRef]

40. Lang, S.A.; Klein, D.; Moser, C.; Gaumann, A.; Glockzin, G.; Dahlke, M.H.; Dietmaier, W.; Bolder, U.; Schlitt, H.J.; Geissler, E.K.; et al. Inhibition of heat shock protein 90 impairs epidermal growth factor-mediated signaling in gastric cancer cells and reduces tumor growth and vascularization in vivo. *Mol. Cancer Ther.* **2007**, *6*, 1123–1132. [CrossRef]
41. Hertlein, E.; Wagner, A.J.; Jones, J.; Lin, T.S.; Maddocks, K.J.; Towns, W.H., 3rd; Goettl, V.M.; Zhang, X.; Jarjoura, D.; Raymond, C.A.; et al. 17-dmag targets the nuclear factor-kappab family of proteins to induce apoptosis in chronic lymphocytic leukemia: Clinical implications of hsp90 inhibition. *Blood* **2010**, *116*, 45–53. [CrossRef] [PubMed]
42. Huang, J.; Maier, C.; Zhang, Y.; Soare, A.; Dees, C.; Beyer, C.; Harre, U.; Chen, C.W.; Distler, O.; Schett, G.; et al. Nintedanib inhibits macrophage activation and ameliorates vascular and fibrotic manifestations in the fra2 mouse model of systemic sclerosis. *Ann. Rheum. Dis.* **2017**, *76*, 1941–1948. [CrossRef]
43. Avouac, J.; Furnrohr, B.G.; Tomcik, M.; Palumbo, K.; Zerr, P.; Horn, A.; Dees, C.; Akhmetshina, A.; Beyer, C.; Distler, O.; et al. Inactivation of the transcription factor stat-4 prevents inflammation-driven fibrosis in animal models of systemic sclerosis. *Arthritis Rheum.* **2011**, *63*, 800–809. [CrossRef]
44. Woessner, J.F., Jr. The determination of hydroxyproline in tissue and protein samples containing small proportions of this imino acid. *Arch. Biochem. Biophys.* **1961**, *93*, 440–447. [CrossRef]
45. Tomcik, M.; Palumbo-Zerr, K.; Zerr, P.; Sumova, B.; Avouac, J.; Dees, C.; Distler, A.; Becvar, R.; Distler, O.; Schett, G.; et al. Tribbles homologue 3 stimulates canonical tgf-beta signalling to regulate fibroblast activation and tissue fibrosis. *Ann. Rheum. Dis* **2016**, *75*, 609–616. [CrossRef]
46. Zheng, Z.; Nguyen, C.; Zhang, X.; Khorasani, H.; Wang, J.Z.; Zara, J.N.; Chu, F.; Yin, W.; Pang, S.; Le, A.; et al. Delayed wound closure in fibromodulin-deficient mice is associated with increased tgf-beta3 signaling. *J. Investig. Dermatol.* **2011**, *131*, 769–778. [CrossRef] [PubMed]
47. Kropackova, T.; Vernerova, L.; Storkanova, H.; Horvathova, V.; Vokurkova, M.; Klein, M.; Oreska, S.; Spiritovic, M.; Hermankova, B.; Kubinova, K.; et al. Clusterin is upregulated in serum and muscle tissue in idiopathic inflammatory myopathies and associates with clinical disease activity and cytokine profile. *Clin. Exp. Rheumatol.* **2020**. Epub ahead of print.
48. Noh, H.; Kim, H.J.; Yu, M.R.; Kim, W.Y.; Kim, J.; Ryu, J.H.; Kwon, S.H.; Jeon, J.S.; Han, D.C.; Ziyadeh, F. Heat shock protein 90 inhibitor attenuates renal fibrosis through degradation of transforming growth factor-beta type ii receptor. *Lab. Investig.* **2012**, *92*, 1583–1596. [CrossRef]
49. Prescott, R.J.; Freemont, A.J.; Jones, C.J.; Hoyland, J.; Fielding, P. Sequential dermal microvascular and perivascular changes in the development of scleroderma. *J. Pathol.* **1992**, *166*, 255–263. [CrossRef] [PubMed]
50. Yoshizaki, A.; Iwata, Y.; Komura, K.; Ogawa, F.; Hara, T.; Muroi, E.; Takenaka, M.; Shimizu, K.; Hasegawa, M.; Fujimoto, M.; et al. Cd19 regulates skin and lung fibrosis via toll-like receptor signaling in a model of bleomycin-induced scleroderma. *Am. J. Pathol.* **2008**, *172*, 1650–1663. [CrossRef] [PubMed]
51. Sontake, V.; Wang, Y.; Kasam, R.K.; Sinner, D.; Reddy, G.B.; Naren, A.P.; McCormack, F.X.; White, E.S.; Jegga, A.G.; Madala, S.K. Hsp90 regulation of fibroblast activation in pulmonary fibrosis. *JCI Insight* **2017**, *2*, e91454. [CrossRef]
52. Dong, H.; Luo, L.; Zou, M.; Huang, C.; Wan, X.; Hu, Y.; Le, Y.; Zhao, H.; Li, W.; Zou, F.; et al. Blockade of extracellular heat shock protein 90alpha by 1g6-d7 attenuates pulmonary fibrosis through inhibiting erk signaling. *Am. J. Physiol. Lung Cell Mol. Physiol.* **2017**, *313*, L1006–L1015. [CrossRef] [PubMed]
53. Li, X.; Yu, H.; Liang, L.; Bi, Z.; Wang, Y.; Gao, S.; Wang, M.; Li, H.; Miao, Y.; Deng, R.; et al. Myricetin ameliorates bleomycin-induced pulmonary fibrosis in mice by inhibiting tgf-beta signaling via targeting hsp90beta. *Biochem. Pharmacol.* **2020**, *178*, 114097. [CrossRef]
54. Marinova, M.; Solopov, P.; Dimitropoulou, C.; Colunga Biancatelli, R.M.L.; Catravas, J.D. Post-treatment with a heat shock protein 90 inhibitor prevents chronic lung injury and pulmonary fibrosis, following acute exposure of mice to hcl. *Exp. Lung Res.* **2020**, *46*, 203–216. [CrossRef]
55. Sibinska, Z.; Tian, X.; Korfei, M.; Kojonazarov, B.; Kolb, J.S.; Klepetko, W.; Kosanovic, D.; Wygrecka, M.; Ghofrani, H.A.; Weissmann, N.; et al. Amplified canonical transforming growth factor-beta signalling via heat shock protein 90 in pulmonary fibrosis. *Eur. Respir J.* **2017**, *49*, 1501941. [CrossRef]
56. Solopov, P.; Biancatelli, R.; Marinova, M.; Dimitropoulou, C.; Catravas, J.D. The hsp90 inhibitor, auy-922, ameliorates the development of nitrogen mustard-induced pulmonary fibrosis and lung dysfunction in mice. *Int. J. Mol. Sci.* **2020**, *21*, 4740. [CrossRef]
57. Caceres, R.A.; Chavez, T.; Maestro, D.; Palanca, A.R.; Bolado, P.; Madrazo, F.; Aires, A.; Cortajarena, A.L.; Villar, A.V. Reduction of cardiac tgfbeta-mediated profibrotic events by inhibition of hsp90 with engineered protein. *J. Mol. Cell Cardiol.* **2018**, *123*, 75–87. [CrossRef] [PubMed]
58. Wollin, L.; Wex, E.; Pautsch, A.; Schnapp, G.; Hostettler, K.E.; Stowasser, S.; Kolb, M. Mode of action of nintedanib in the treatment of idiopathic pulmonary fibrosis. *Eur. Respir J.* **2015**, *45*, 1434–1445. [CrossRef] [PubMed]
59. Distler, O.; Highland, K.B.; Gahlemann, M.; Azuma, A.; Fischer, A.; Mayes, M.D.; Raghu, G.; Sauter, W.; Girard, M.; Alves, M.; et al. Nintedanib for systemic sclerosis-associated interstitial lung disease. *N. Engl. J. Med.* **2019**, *380*, 2518–2528. [CrossRef]
60. Yamamoto, T. The bleomycin-induced scleroderma model: What have we learned for scleroderma pathogenesis? *Arch. Dermatol. Res.* **2006**, *297*, 333–344. [CrossRef]

61. Dello Russo, C.; Polak, P.E.; Mercado, P.R.; Spagnolo, A.; Sharp, A.; Murphy, P.; Kamal, A.; Burrows, F.J.; Fritz, L.C.; Feinstein, D.L. The heat-shock protein 90 inhibitor 17-allylamino-17-demethoxygeldanamycin suppresses glial inflammatory responses and ameliorates experimental autoimmune encephalomyelitis. *J. Neurochem.* **2006**, *99*, 1351–1362. [CrossRef] [PubMed]
62. Rice, J.W.; Veal, J.M.; Fadden, R.P.; Barabasz, A.F.; Partridge, J.M.; Barta, T.E.; Dubois, L.G.; Huang, K.H.; Mabbett, S.R.; Silinski, M.A.; et al. Small molecule inhibitors of hsp90 potently affect inflammatory disease pathways and exhibit activity in models of rheumatoid arthritis. *Arthritis Rheum.* **2008**, *58*, 3765–3775. [CrossRef]
63. Han, J.M.; Kwon, N.H.; Lee, J.Y.; Jeong, S.J.; Jung, H.J.; Kim, H.R.; Li, Z.; Kim, S. Identification of gp96 as a novel target for treatment of autoimmune disease in mice. *PLoS ONE* **2010**, *5*, e9792. [CrossRef]
64. Kasperkiewicz, M.; Muller, R.; Manz, R.; Magens, M.; Hammers, C.M.; Somlai, C.; Westermann, J.; Schmidt, E.; Zillikens, D.; Ludwig, R.J.; et al. Heat-shock protein 90 inhibition in autoimmunity to type vii collagen: Evidence that nonmalignant plasma cells are not therapeutic targets. *Blood* **2011**, *117*, 6135–6142. [CrossRef] [PubMed]
65. Storkanova, H.; Oreska, S.; Spiritovic, M.; Hermankova, B.; Bubova, K.; Komarc, M.; Pavelka, K.; Vencovsky, J.; Dislter, J.; Senolt, L.; et al. Plasma hsp90 levels in patients with systemic sclerosis and relation to lung and skin involvement: A cross-sectional and longitudinal study. *Sci. Rep.* **2020**, *11*, 1. [CrossRef]
66. Degryse, A.L.; Lawson, W.E. Progress toward improving animal models for idiopathic pulmonary fibrosis. *Am. J. Med. Sci.* **2011**, *341*, 444–449. [CrossRef] [PubMed]
67. Kawaguchi, Y.; Hara, M.; Wright, T.M. Endogenous il-1alpha from systemic sclerosis fibroblasts induces il-6 and pdgf-a. *J. Clin. Investig.* **1999**, *103*, 1253–1260. [CrossRef]
68. Kawaguchi, Y.; McCarthy, S.A.; Watkins, S.C.; Wright, T.M. Autocrine activation by interleukin 1alpha induces the fibrogenic phenotype of systemic sclerosis fibroblasts. *J. Rheumatol.* **2004**, *31*, 1946–1954.
69. Khanna, D.; Denton, C.P.; Jahreis, A.; van Laar, J.M.; Frech, T.M.; Anderson, M.E.; Baron, M.; Chung, L.; Fierlbeck, G.; Lakshminarayanan, S.; et al. Safety and efficacy of subcutaneous tocilizumab in adults with systemic sclerosis (fasscinate): A phase 2, randomised, controlled trial. *Lancet* **2016**, *387*, 2630–2640. [CrossRef]
70. Khanna, D.; Denton, C.P.; Lin, C.J.F.; van Laar, J.M.; Frech, T.M.; Anderson, M.E.; Baron, M.; Chung, L.; Fierlbeck, G.; Lakshminarayanan, S.; et al. Safety and efficacy of subcutaneous tocilizumab in systemic sclerosis: Results from the open-label period of a phase ii randomised controlled trial (fasscinate). *Ann. Rheum. Dis.* **2018**, *77*, 212–220. [CrossRef]
71. Furuse, S.; Fujii, H.; Kaburagi, Y.; Fujimoto, M.; Hasegawa, M.; Takehara, K.; Sato, S. Serum concentrations of the cxc chemokines interleukin 8 and growth-regulated oncogene-alpha are elevated in patients with systemic sclerosis. *J. Rheumatol.* **2003**, *30*, 1524–1528. [PubMed]
72. Tukaj, S.; Gruner, D.; Zillikens, D.; Kasperkiewicz, M. Hsp90 blockade modulates bullous pemphigoid igg-induced il-8 production by keratinocytes. *Cell Stress Chaperones* **2014**, *19*, 887–894. [CrossRef]
73. Chung, S.W.; Lee, J.H.; Choi, K.H.; Park, Y.C.; Eo, S.K.; Rhim, B.Y.; Kim, K. Extracellular heat shock protein 90 induces interleukin-8 in vascular smooth muscle cells. *Biochem. Biophys. Res. Commun.* **2009**, *378*, 444–449. [CrossRef] [PubMed]
74. Yeo, M.; Park, H.K.; Lee, K.M.; Lee, K.J.; Kim, J.H.; Cho, S.W.; Hahm, K.B. Blockage of hsp 90 modulates helicobacter pylori-induced il-8 productions through the inactivation of transcriptional factors of ap-1 and nf-kappab. *Biochem. Biophys. Res. Commun.* **2004**, *320*, 816–824. [CrossRef] [PubMed]
75. Di Martino, S.; Amoreo, C.A.; Nuvoli, B.; Galati, R.; Strano, S.; Facciolo, F.; Alessandrini, G.; Pass, H.I.; Ciliberto, G.; Blandino, G.; et al. Hsp90 inhibition alters the chemotherapy-driven rearrangement of the oncogenic secretome. *Oncogene* **2018**, *37*, 1369–1385. [CrossRef]
76. Hartman, M.L.; Rogut, M.; Mielczarek-Lewandowska, A.; Wozniak, M.; Czyz, M. 17-aminogeldanamycin inhibits constitutive nuclear factor-kappa b (nf-kappab) activity in patient-derived melanoma cell lines. *Int. J. Mol. Sci.* **2020**, *21*, 3749. [CrossRef]

Article

Plasma Metabolomic Profiling Reveals Four Possibly Disrupted Mechanisms in Systemic Sclerosis

Thomas Bögl [1], Franz Mlynek [1], Markus Himmelsbach [1], Norbert Sepp [2], Wolfgang Buchberger [1] and Marija Geroldinger-Simić [2,3,*]

[1] Institute of Analytical and General Chemistry, Johannes Kepler University Linz, 4040 Linz, Austria; thomas.boegl@jku.at (T.B.); franz.mlynek@jku.at (F.M.); markus.himmelsbach@jku.at (M.H.); wolfgang.buchberger@jku.at (W.B.)

[2] Department of Dermatology, Ordensklinikum Linz Elisabethinen, 4020 Linz, Austria; norbert.sepp@ordensklinikum.at

[3] Faculty of Medicine, Johannes Kepler University Linz, 4020 Linz, Austria

* Correspondence: marija.geroldinger-simic@jku.at

Abstract: Systemic sclerosis (SSc) is a rare systemic autoimmune disorder marked by high morbidity and increased risk of mortality. Our study aimed to analyze metabolomic profiles of plasma from SSc patients by using targeted and untargeted metabolomics approaches. Furthermore, we aimed to detect biochemical mechanisms relevant to the pathophysiology of SSc. Experiments were performed using high-performance liquid chromatography coupled to mass spectrometry technology. The investigation of plasma samples from SSc patients (n = 52) compared to a control group (n = 48) allowed us to identify four different dysfunctional metabolic mechanisms, which can be assigned to the kynurenine pathway, the urea cycle, lipid metabolism, and the gut microbiome. These significantly altered metabolic pathways are associated with inflammation, vascular damage, fibrosis, and gut dysbiosis and might be relevant for the pathophysiology of SSc. Further studies are needed to explore the role of these metabolomic networks as possible therapeutic targets of SSc.

Keywords: systemic sclerosis; metabolomics; LC-MS/MS; ion mobility; kynurenine pathway; urea cycle; lipids; gut dysbiosis

Citation: Bögl, T.; Mlynek, F.; Himmelsbach, M.; Sepp, N.; Buchberger, W.; Geroldinger-Simić, M. Plasma Metabolomic Profiling Reveals Four Possibly Disrupted Mechanisms in Systemic Sclerosis. *Biomedicines* **2022**, *10*, 607. https://doi.org/10.3390/biomedicines10030607

Academic Editor: Michal Tomcik

Received: 31 December 2021
Accepted: 26 February 2022
Published: 4 March 2022

1. Introduction

Systemic sclerosis (SSc) is a rare autoimmune disease, which is characterized by the production of autoantibodies, vasculopathy, and fibrosis [1,2]. The high morbidity and increased mortality make it a disease of great concern. Patients most commonly suffer skin thickening, digital ulcers, lung fibrosis, pulmonary arterial hypertension, Raynaud's phenomena, esophageal dysmotility or gut dysbiosis, and produce SSc-related autoantibodies [3–5]. The primary cause of death in SSc patients is lung fibrosis followed by pulmonary arterial hypertension or sepsis. The most prominent clinical feature is the fibrosis of the skin and/or internal organs, and skin involvement is a crucial sign for the early diagnosis [4,5]. The modified Rodnan Skin Score (mRSS) was introduced to evaluate skin involvement in SSc patients in clinical trials [6]. Depending on the skin and organ involvement, one can distinguish between limited cutaneous SSc (lcSSc), which manifests in only partial skin (sclerosis of face and distal extremities) and minor systemic involvement, diffuse cutaneous SSc (dcSSc), which includes extensive skin and systemic involvement, and non-cutaneous SSc (ncSSc), with no evident skin involvement [4].

No curative therapy for SSc exists, and in recent years, the vast majority of studies have been published on investigations of new therapeutic strategies [7–10]. Most of the therapeutic strategies are directed towards inflammatory and vascular pathways, and recently the antifibrotic drug nintedanib has been proven efficient for the therapy of lung fibrosis in SSc [11]. Due to numerous challenges with therapies (efficacy, side effects, multimorbidity, long-term survival) there is an urgent need to improve existing treatments and detect new possible therapeutic targets for SSc in the future.

Exploring the pathophysiology of SSc is crucial for the improvement of therapy and long-term survival of SSc patients. Metabolites are the final products of the pathophysiologic processes and may play a key role in establishing personalized treatment or biomarkers of drug response in SSc. Metabolomics is a method for describing metabolism by different technologies as published for autoimmune diseases like systemic lupus erythematosus, rheumatoid arthritis, and multiple sclerosis [12]. Mass spectrometry technology holds great potential for a metabolic investigation in a variety of clinical applications as well as different bio-samples [13,14].

Very little is known about the role of metabolites in the pathophysiology of SSc. Over the last five years, several studies were published discovering single metabolites possibly relevant for SSc, and rarely a study would define whole metabolomic pathways involved in the pathophysiology of SSc. In addition, metabolomic studies can sometimes be limited due to the applicability of statistics, which might be problematic due to the rare occurrence of SSc and small study cohorts. Recent data indicated significantly changed metabolites in serum of SSc patients involved in glycolysis, gluconeogenesis, glutamate, and pyruvate metabolism [15]. Analyses of plasma metabolites and fecal microbiota showed glycerophospholipids and benzene derivates to interact with certain fecal bacteria (Desulfovibrio), which may influence gut dysbiosis and inflammation in SSc [16]. Moreover, metabolic profiling of urine from SSc patients revealed deregulated fatty acid oxidation, which might be relevant for inflammation in SSc [17]. A recent study reported dysregulated carnitine in plasma and dendritic cells of SSc patients, and carnitines were suggested to increase inflammation in SSc [18]. Furthermore, altered amino acid metabolism (e.g., betaine, tryptophan, proline, glutamine) was detected in the plasma of SSc patients and has been attributed to changes in vascular endothelial dysfunction and inflammation during SSc [19]. Finally, data have been recently published on characteristic metabolomic changes for certain organ involvement during SSc, like pulmonary arterial hypertension [20] and lung fibrosis [21].

The aim of our study was to analyze metabolomic profiles of plasma from SSc patients by using targeted, and untargeted metabolomics approaches enabled by high-performance liquid chromatography coupled to mass spectrometry technology. Furthermore, we aimed to detect metabolomic networks relevant to the pathophysiology of SSc and generate hypotheses about new therapeutic targets for SSc.

2. Materials and Methods

2.1. Study Group and Sample Collection

The patients with SSc ($n = 52$) and control group ($n = 48$) were recruited consecutively at the Department of Dermatology, Ordensklinikum Linz in Austria. This study was approved by the Ethics Committee of the Johannes Kepler University Linz in Austria (study protocol number 1265/2019). Diagnosis of the SSc was made according to the criteria of the American College of Rheumatology (ACR) and the European League Against Rheumatism (EULAR) [4]. The inclusion criteria for patients were diagnosis of SSc according to ACR/EULAR criteria and age 18–90. Exclusion criteria for control group were acute infections, liver and/or kidney diseases and diabetes mellitus. Peripheral blood was collected in compliance with the Declaration of Helsinki (1975/83) by using BD-K2-Edta tubes (Fischer Scientific, Schwerte, Germany). After centrifugation of the blood, the resulting plasma was deep-frozen and stored at −80 °C until further treatment.

2.2. Sample Preparation

For protein precipitation 50 µL plasma sample is mixed with 150 µL of cooled 5% sulfosalicylic acid followed by 20 min equilibration on a thermo-shaker (4 °C, VWR, Vienna, Austria). Afterwards, samples are centrifuged (8 min, 4 °C, 4200 g, VWR, Vienna, Austria), and 30 μL of the supernatant is transferred to a HPLC vial with a 200 μL glass inlet that also contains 150 μL of acetonitrile as well as 20 μL of an internal standard (Sigma Aldrich, Vienna, Austria, Cell Free Amino Acid Mixture—13C,15N) and is rigorously mixed. The prepared sample is stored at −80 °C until analysis.

2.3. HPLC-MS Analysis

High-performance liquid chromatography (HPLC) was performed in a hydrophilic interaction chromatography (HILIC) mode using an XBridge BEH Amide column (2.1 mm × 150 mm, 2.5 μm, Waters, Vienna, Austria) connected to an XBridge Glycan BEH Amide pre-column (130 A. 2.5 μm, 2.1 mm × 5 mm, Waters, Vienna, Austria). HPLC separation was performed by using a gradient of 10 mM ammonium formate and 0.2% formic acid in 18 MΩ-water (solvent A), 0.2% formic acid in acetonitrile (solvent B) and 100 mM ammonium formate with 0.2% formic acid in 18 MΩ-water (solvent C). After injecting 5 μL sample the composition of the gradient was kept constant for 4 min at 4% A and 96% B. Subsequently, solvent A was increased to 18% within 12 min. Then, the gradient was changed within 6 min to a final composition of 40 % A, 40% B and 20% C, which was kept constant for 5 min, followed by switching back to the starting conditions and reconditioning the column for 10 min. These settings were found suitable for measuring polar metabolites such as amino acids and other small molecules.

The targeted method utilized a 1260 Infinity HPLC coupled to a 6460 triple quadrupole mass spectrometer (QQQ-MS) from Agilent Technologies (Waldbronn, Germany). Source parameters for this approach were as follows: gas temperature was set to 325 °C with a flow of 12 L min^{-1}, the nebulizer was set to 40 psig, sheath gas was set to 350 °C with a flow of 11 L min^{-1}, and the capillary voltage was set to 3500 V.

For untargeted metabolomics a 1290 Infinity HPLC coupled to a 6560 ion mobility quadrupole time-of-flight mass spectrometer (IMS-QTOF-MS) from Agilent Technologies (Waldbronn, Germany) was used. MS/MS experiments for untargeted metabolomics were performed on IMS-QTOF-MS. The IMS was performed by using 4 bit multiplexing with a trap fill time of 3900 μs and a trap release time of 250 μs, and used N_2 as drift gas. The measurement was done in positive mode in a range of 50 to 1000 m/z. The source settings have been as follows: gas temperature was set to 320 °C with a flow of 10 L min^{-1}, nebulizer was set to 45 psig, sheath gas was set to 320 °C with a flow of 10 L min^{-1}, and the capillary voltage was set to 4000 V.

2.4. Data Pre-Processing

Result files from targeted metabolomics were integrated and processed including an intensity correction with internal standards within the Mass Hunter Quantitative Analysis (Version B.09.00) for QQQ Software (Agilent Technologies, Waldbronn, Germany). Subsequently, results were exported for statistical analysis.

Data obtained by the untargeted approach were initially processed by PNNL Pre-Processor (v2019.08.17, Pacific Northwest National Laboratory, Richland, WA, USA). IM Reprocessor (Version 10.00) and IM-MS Browser (Version 10.00) (both Agilent Technologies, Waldbronn, Germany) were used for determining collision cross section $^{DT}CCS_{N2}$ values and applying mass correction. Feature extraction was based on the Mass Profiler Software algorithm (Version B.08.01, Agilent Technologies, Waldbronn, Germany). In the first approach no annotation was performed. After the first statistical analysis, the identification was completed by several steps. First, the exact mass was searched in the Human Metabolome Database (HMDB, accessed on 30 March 2021) [14]. Afterward, matches from HMDB were checked by a comparison of predicted CCS_{pred} values with experimental $^{DT}CCS_{N2}$ values [22]. In this manner, possible matches for each feature were reduced drastically. Subsequently, significant features were identified by matching MS^2 spectra with entries in the MassBank of North America (MoNA) (http://mona.fiehnlab.ucdavis.edu/, accessed on 30 March 2021) as well as matching calculated CCS_{pred} (http://allccs.zhulab.cn/ [22], accessed on 30 March 2021) values using AllCCS with experimental $^{DT}CCS_{N2}$ values.

2.5. Statistical Analysis

Merged results from targeted and untargeted analysis were finally analyzed by the online statistical analysis tool of MetaboAnalyst 4.0. (https/www.metaboanalyst.ca [23], accessed on 5 April 2021). Data were imported into the online platform using missing value estimation, sample normalization by median, data transformation by log transformation, and Pareto scaling. Retrieved p-values from univariate analysis were seen as significant if they were below the threshold of 0.05. Additionally, fold changes were seen as significant for up- or downregulation by surpassing an overall increment of 2, respectively. Outcome of these two tests is summarized in a volcano plot. For multi-group comparison, an ANOVA, with associated post hoc analysis, by using Fisher LSD test, was applied. Most noteworthy, dysregulated features are additionally visualized by boxplots. As a further visualization method, a heatmap with an agglomerative hierarchical cluster was chosen. This enables us to present samples or features with similarities close to each other and consequently, to visually recognize clusters. For visualization, Jupyter-Lab 3.2.4 and Python 3.9 were utilized. The following libraries were used: pandas 1.3.4, matplotlib 3.4.3, and seaborn 0.11.2.

3. Results

3.1. Study Group and Sample Analysis

The cohort of the collected plasma sample included SSc patients (n = 52) and the control group (n = 48). Females were more dominant in both groups, and the group ratio between females and males was consistent. Most of the SSc patients suffered from lcSSc (n = 39) followed by dcSSc (n = 11), while the number of ncSSc (n =2) was too small for intergroup comparisons. Further clinical information can be retrieved from Supplementary Table S1.

The data from the targeted and untargeted approaches were analyzed separately in the first evaluation. The employed method enabled a large spectrum of interesting pathophysiological molecules to be measured. Metabolites selected by the targeted method mainly consisted of physiologically important small polar molecules, which might need a higher sensitivity for a proper measurement. The targeted and the identified untargeted metabolites were combined for a final statistic. This was performed to generate a list with the most promising molecules for differentiation of the investigated groups.

3.2. Targeted Metabolomics

The results of the targeted approach are summarized in Table 1 and Supplementary Table S2. Listed metabolites are significantly up-/or down-regulated in terms of their p-values and fold change. Tryptophan was found to be significantly down-regulated, while kynurenine, which is also closely related to tryptophan through the kynurenine pathway (KP), was significantly up-regulated. Alanine, which also can be associated with the KP, was significantly reduced.

Table 1. Significantly changed metabolites identified by targeted metabolomics. Mass to charge ratio (m/z), retention time (RT), p-value (determined by a two-tailed Student's t-test), and fold change (ratio of SSc to control).

Exact Mass (m/z)	RT/ (min)	Name	p-Value	Fold Change (SSc/Control)
205.0971	14.7	Tryptophan	<0.0001	0.3117
90.0549	18.2	Alanine	<0.0001	0.3444
203.1503	23.2	Dimethylarginine	<0.0001	2.6918
209.0921	14.5	Kynurenine	0.0004	2.1049
176.1030	22.1	Citrulline	0.0016	2.3760
133.0972	23.9	Ornithine	0.0026	2.3604

In the case of dimethylarginine, which can influence the urea cycle, two forms are known, namely symmetric dimethylarginine (SDMA) and asymmetric dimethylarginine (ADMA). In our approach, both forms could not be separated chromatographically, and therefore, they are presented as the sum parameter "dimethylarginine", which was significantly up-regulated. Moreover, citrulline and ornithine, both main mediators in the urea cycle, were significantly up-regulated (see Table 1). Additional statistically significant metabolites are given in Supplementary Table S2.

3.3. Untargeted Metabolomics

Untargeted metabolomics analyses led to further relevant metabolites from the plasma of SSc patients. All the metabolites from Table 1 from the targeted approach were excluded, and only previously unrecognized features are shown in Table 2. Metabolites found to be regulated were lipids, especially lysophosphatidylcholines (LPCs), sphingomyelins (SMs), metabolites phenylacetylglutamine (PAG), and OH-tryptophan as well as acyl-carnitines (OH-butyrylcarnitine and OH-decanoylcarnitine) (see Table 2). Additionally, observed features with a p-value < 0.05 are given in Supplementary Table S3.

Table 2. Dysregulated metabolites were identified by the untargeted metabolomics approach. A feature consists of retention time (RT), measured mass to charge ratio (m/z), and the measured cross collision section ($^{DT}CCS_{N2}$). LPC = lysophosphatidylcholine, PAG = phenylacetylglutamine, Cer = ceramide, SM = sphingomyeline, CCS_{pred} was retrieved from AllCCS [22] predictor. p-value was determined by a two-tailed Student's t-test. Fold change was calculated from the ratio of SSc to control group.

Feature (RT_mz_$^{DT}CCS_{N2}$)	Metabolite ID	Name	CCS_{pred}	m/z_{calc}	p-Value	Fold Change (SSc/Control)
13.89_572.3679_236.76	HMDB0010401	LPC a 22:4 a (+H)	243.3	572.3711	<0.0001	0.4771
4.283_265.1168_160.02	HMDB06344	PAG (+H)	159.3	265.1183	<0.0001	2.1906
13.511_572.3676_238.65	HMDB0010401	LPC a 22:4 b (+H)	243.3	572.3711	0.0002	0.4958
13.619_673.5254_274.79	HMDB0240612	SM 32:2 (+H)	273.9	673.5279	0.0020	0.7053
15.977_221.0915_150.56	HMDB0000472	OH-tryptophan (+H)	149.7	221.0926	0.0055	0.6392
13.987_548.3692_234.01	HMDB0010392	LPC a 20:2 (+H)	240.7	548.3711	0.0079	0.5156
13.472_685.5613_286.32	HMDB0013464	SM 34:1 (−H2O +H)	283.5	685.5649	0.0101	0.9840
13.203_783.6341_296.53	HMDB0240670	SM 40:3 (+H)	291.2	783.6375	0.0152	0.7605
12.479_248.1485_156.00	HMDB0013127	OH-butyrylcarnitine (+H)	158.4	248.1498	0.0219	1.2434
7.364_332.2419_193.87	HMDB0061636	OH-decanoylcarnitine (+H)	189.3	332.2437	0.0396	1.6815

3.4. Metabolomic Patterns Altered in SSc Patients

After separate evaluations of targeted and untargeted approaches, the results were combined to gain an insight into the overall data. For a first overview, a heatmap for all identified features was generated based on selecting the 30 best molecules due to t-tests, where minor clustering of the SSc samples could be shown (Figure 1A).

p-values and fold change were used for creating a two-dimensional volcano plot (Figure 1B). With the applied thresholds, the significantly down-regulated features are depicted on the top-left, and the significantly up-regulated features are depicted on the top-right side. Closely related metabolites can be identified. First, a prominent imbalance of kynurenine and tryptophan was noticed, whereby down-regulation of tryptophan and up-regulation of kynurenine can be observed. Secondly, metabolites of the urea cycle were significantly up-regulated (ornithine, citrulline, and dimethylarginine). Additionally, dysregulation of the lipidome also seems to be present, along with significantly down-regulated LPC species. Another metabolite, phenylacetylglutamine (PAG), was significantly higher in plasma of patients with SSc than controls. Within Figure 1C, the statistically significant differences between metabolites in the plasma of SSc-patients and the control group were emphasized for tryptophan, ornithine, LPCs, and PAG molecules.

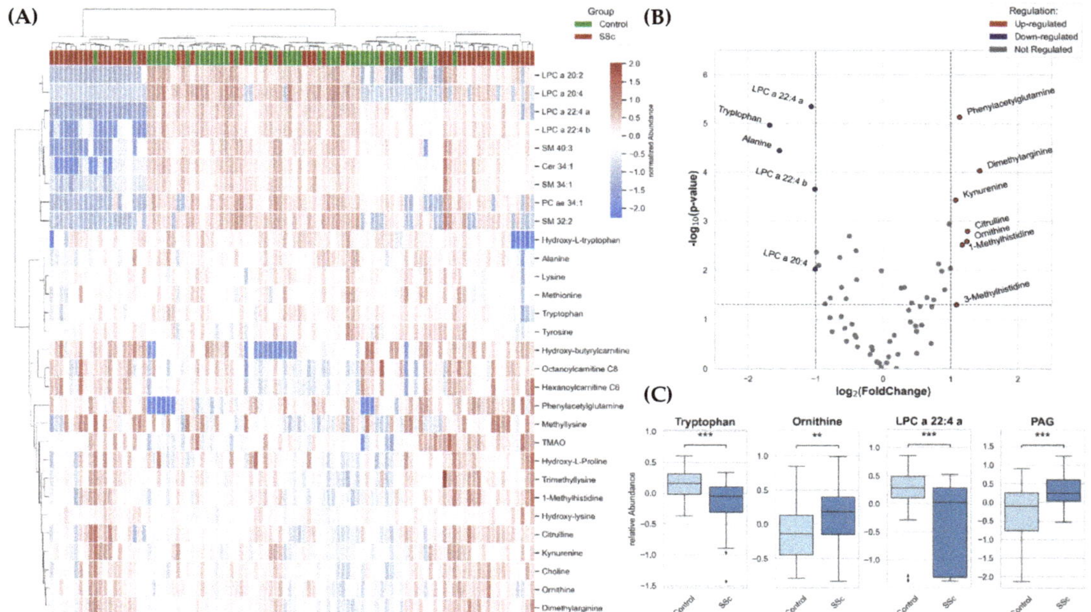

Figure 1. (**A**) Heatmap for all identified features from plasma of SSc patients ($n = 52$) and controls ($n = 48$) was generated based on a selection of the best 30 molecules due to t-tests using clustering method (Distance measure: Euclidean, Algorithm: Ward). (**B**) p-values and fold change were used for creating a two-dimensional volcano plot of summarized metabolites identified in metabolomics approaches. (**C**) Box Plots (Box = IQR, whiskers = 1.5 × IQR, horizontal bars = median) of significantly dysregulated metabolites. LPC = lysophosphatidylcholine, PAG = phenylacetylglutamine. p-values were determined using a two-tailed Student's t-test; $p < 0.05$ was considered significant (** $p < 0.01$, *** $p < 0.001$).

3.5. Altered Metabolism in lcSSc and dcSSc Patients

In a further approach, we compared the control group with lcSSc and dcSSc. We excluded ncSSc because a group with only two samples was not suitable for gaining a proper statistical distribution. A comparison of subgroups lcSSc and dcSSc with controls suggested that dysregulations in the observed metabolism were associated with worsening of the disease. This association becomes evident in Figure 2A,B: the metabolites kynurenine, citrulline, ornithine, and PAG are regulated the highest in dcSSc, while the lowest observed concentration is in the control samples.

3.6. Cross-Correlation between Lipids and Carnitines

Based on the earlier finding of highly influenced LPC species and carnitine species, a cross-correlation matrix was established to investigate the possibility of statistically insignificant enrichments. Lipid species marked on the x-axis in Figure 3 show a correlation to the acyl-carnitine species marked on the y-axis (marked with red squares within the matrix).

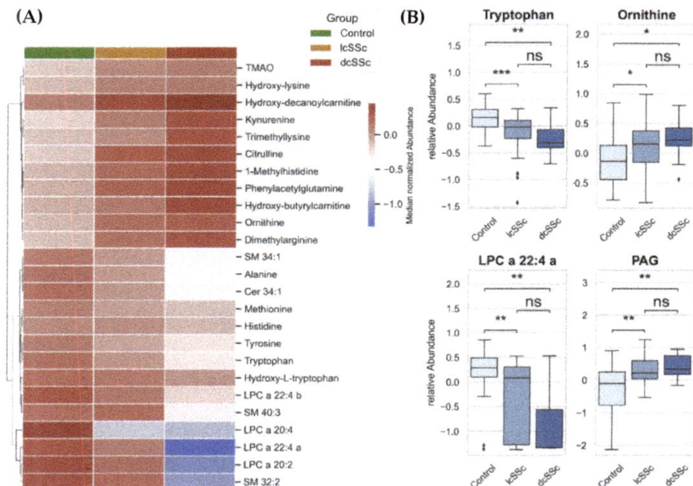

Figure 2. (**A**) Hierarchical heatmap showing a gradient of metabolites from plasma of control group, patients with limited cutaneous SSc and patients with diffuse cutaneous SSc. Distance measure: Euclidean, Clustering-Algorithm: Ward. (**B**) Box Plots (Box = IQR, whiskers = 1.5 × IQR, horizontal bars = median) of significantly changed metabolites: tryptophan, ornithine, LPC = lysophosphatidylcholine, and PAG = phenylacetylglutamine. p-values were determined using a two-tailed Student's t-test; $p < 0.05$ was considered significant (* $p < 0.05$, ** $p < 0.01$, *** $p < 0.001$, ns: not significant).

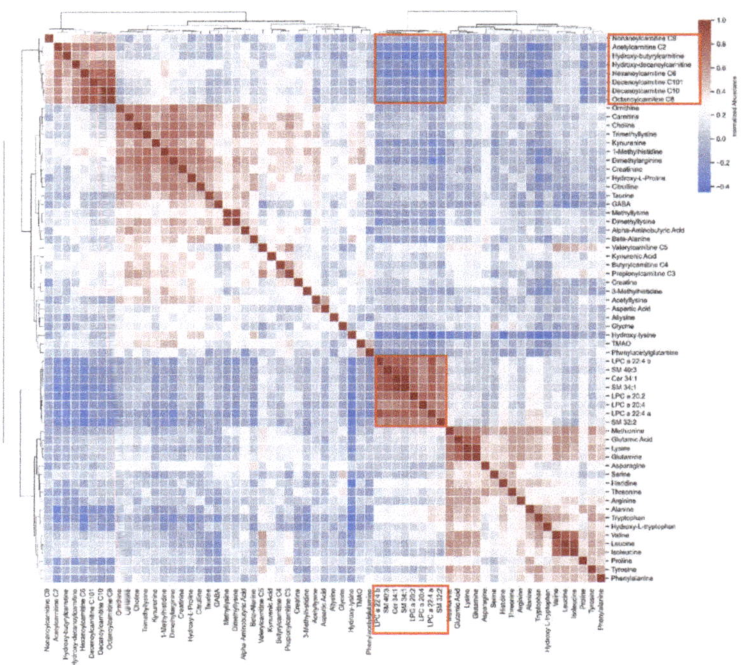

Figure 3. Correlation heatmap showing an interaction of lipids with acyl-carnitines (marked with red squares).

4. Discussion

In this study, we explored metabolites from the plasma of SSc patients and a control group by using targeted, and untargeted metabolomics approaches. The main results of our research indicate four possibly disrupted metabolic mechanisms in patients with SSc, namely an enhanced kynurenine pathway, a dysregulated urea cycle, a disrupted lipid metabolism, and a disturbed gut microbiome, which could be involved in the pathophysiology of SSc and might serve as potential targets of treatment for SSc in the future.

We found a statistically significant down-regulation of tryptophan and a statistically significant up-regulation of kynurenine, leading to the conclusion of a dysregulation of the kynurenine pathway in SSc patients compared to controls. Furthermore, OH-tryptophan was also found to be down-regulated within SSc patients. These results implied a depletion of tryptophan towards the kynurenine route, which could result in inflammatory processes (Figure 4).

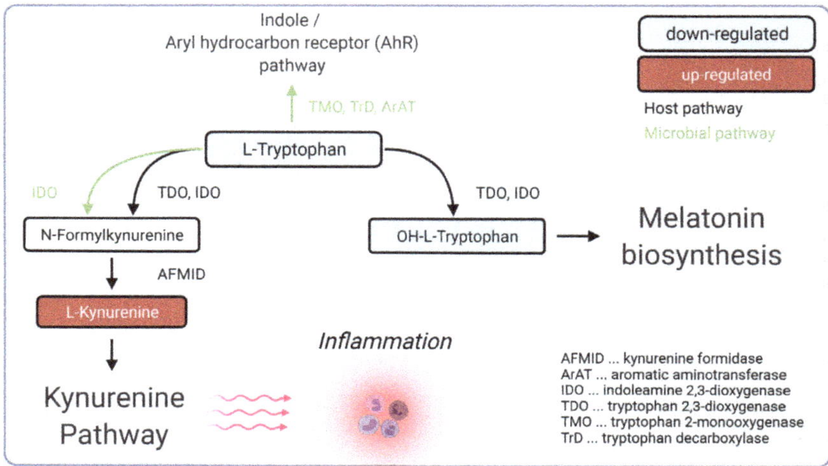

Figure 4. Hypotheses for the role of dysregulated tryptophan/kynurenine pathway during inflammation in SSc. Created with BioRender.com (accessed on 31 December 2021).

Previous studies have identified single metabolites of the kynurenine pathway as relevant for SSc [18,19,21,24]. Our data showed these findings combined and identified a dysregulation of the kynurenine pathway in SSc for the first time within one study.

The kynurenine pathway plays an essential role in various diseases such as allergies, autoimmune disorders, or neurodegeneration [25]. Tryptophan catabolism to kynurenine was shown to be regulated by the immune regulatory enzyme indolamine-2,3-dioxygenase (IDO1) [26]. Metabolites of the kynurenine pathway, especially kynurenine, were described to block T-cell proliferation and induce T-cell apoptosis [27–29]. It has been shown for several autoimmune diseases that overexpression of kynurenines leads to a dysregulation of regulatory T-cells mediated by pro-inflammatory cytokine cascades [25,30]. Furthermore, local depletion of tryptophan also leads to endothelial cell apoptosis [27,31]. Moreover, tryptophan metabolism is also closely connected to a dysfunctional microbiome in SSc [32–34].

As T-cell-mediated inflammation, endothelial cell dysfunction, and gut dysbiosis are essential mechanisms of SSc, the kynurenine pathway might play a crucial role in the pathogenesis of SSc and should further be explored as a possible therapeutic target for the therapy of SSc.

Our data showed up-regulated metabolites ornithine, citrulline in plasma of SSc patients, which are central protagonists in the urea cycle. Ornithine is responsible for the intra-mitochondrial binding of ammonia, whereby it is transformed to citrulline. Citrulline can then pass the mitochondrial membrane through the ornithine translocase (ORNT1) and will again be metabolized to ornithine, which can circulate back into the mitochondria. This metabolization aims to reduce ammonia levels in body fluids by producing urea and arginine. The responsible enzyme is arginase (ARG). ARG catabolizes arginine to ornithine and urea and, in a further stage, to polyamines. Furthermore, NO and citrulline can originate from arginine through NO-synthase (NOS) [35,36].

The overexpression of citrulline was described to indicate the availability of pro-inflammatory stimuli. ARG has been reported to be associated with inflammatory disorders. In psoriasis, an overexpression of ARG leads to decreasing nitric oxide (NO) levels, which modulate the immune responses in tissues [37]. NO dysregulation was identified as having a high impact on the modulation of T-cell responses, with local effects in a tissue due to its short lifetime [35]. Polyamines activity was increased in autoimmune disorders as a pro-inflammatory mediator due to its competition with the cellular methylation for S-adenosylmethionine [38,39].

Furthermore, we found up-regulation of dimethylarginine (DMA) in the plasma of SSc patients. Two different DMAs have been described: symmetric dimethylarginine (SDMA) and asymmetric dimethylarginine (ADMA). In our approach, no separation was achieved. Studies previously showed that ADMA is significantly increased in dcSSc [40,41]. Moreover, it has been shown that ADMA potentially inhibits the degradation of arginine by hindering the catalytic center of NOS [42]. Thus, the up-regulation of DMA might lead to the disruption of the urea cycle (based on an increase in ornithine and citrulline) and subsequently to the downstream production of pro-inflammatory molecules in SSc.

Fibrosis is a common signature of SSc. In this context, generated collagen is dominated by a high content of proline, which is in equilibrium with pyrroline-5-carboxylate (P5C). P5C can be further processed by ornithine aminotransferase (OAT) to ornithine. An association of collagen metabolism with the urea cycle was suggested in metastatic tumor disease [43].

We hypothesized that the urea cycle, particularly up-regulated ornithine and citrulline metabolites, could play an important role in inflammation during SSc. Furthermore, we suggest that in SSc, the urea cycle is fuelled in connection with collagen metabolism, whereby citrulline and ornithine concentrations are increased and further stimulate inflammation in SSc patients (Figure 5). Future studies are needed to test the possibility of influencing the urea cycle pathway to block fibrosis and inflammation in SSc.

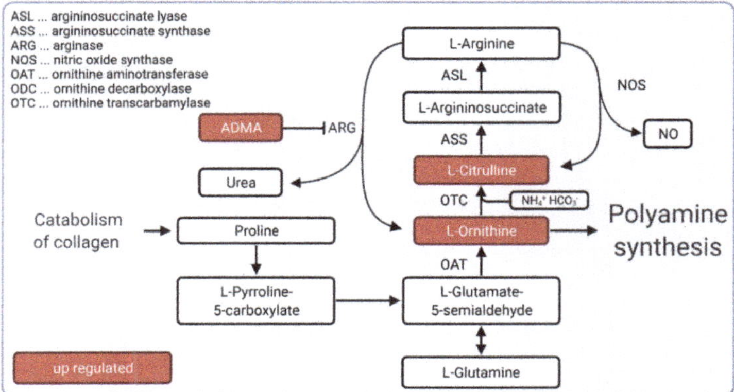

Figure 5. Hypotheses for the role of urea cycle during inflammation and fibrosis in SSc. Our hypothesis suggests that the urea cycle is fueled by collagen catabolism during SSc. Metabolites citrulline and ornithine and dimethylarginine (depicted as ADMA), which is known to inhibit arginase (ARG), were up-regulated in SSc patients. Created with BioRender.com (accessed on 31 December 2021).

We detected significant down-regulation of three lysophosphatidylcholines (such as LPC 22:4 a, LPC 22:4 b, and LPC 20:2) in patients with SSc compared to controls. LPCs are part of low-density lipoproteins (LDLs) and are involved in the physiological immune response due to their interaction with Toll-like receptors. It has been shown that LPCs affect oxidative stress, endothelial cells, and lymphocytes and can stimulate pro-inflammatory cytokines. [44] LPCs are chemo-attractants for macrophages and defective clearance through phagocytic cells has been reported to play an important role in systemic lupus erythematosus [45,46]. Our data on LPCs differ from the literature, by showing three down-regulated LPCs molecules indicating that the role of LPCs in SSs might be different for specific LPCs-species. Thus, LPCs might be involved in inflammation and endothelial damage during SSc disease, and further studies are needed to elucidate the role of different LPCs species in SSc and to investigate if LPCs might serve as potential therapeutic targets for SSc in the future.

Our metabolomics study further showed significant down-regulation of sphingomyelins (SM 34:1 and SM 40:3) in the plasma of SSc patients compared to the control group. Previous studies described SMs to be involved in controlling of fibrosis in the skin, lungs, and kidneys [47,48]. Furthermore, our previously published lipidomics study demonstrated significantly decreased SMs in the plasma of SSc patients with more intensive skin sclerosis (dcSSc or mRSS > 14; SM 30:1, SM 32:2, and SM 40:4) [49]. Therefore, it can be suggested that certain species of SMs might affect skin sclerosis of SSc patients. Further studies are needed to determine the possibility of targeting SMs as a new therapeutic strategy for organ fibrosis during SSc.

Our metabolomic analyses showed correlations of lipids such as SMs and LPCs with acyl-carnitines in the plasma of SSc patients. Even though acyl-carnitines are not regulated in a statistically significant manner, the correlation with lipids could indicate their involvement in the pathogenesis of SSc. Acyl-carnitines and lipid metabolism have been reported as dysregulated in autoimmune disorders, especially in combination with disrupted gut microbiome [50].

Acyl-carnitines function as a transporter of fatty acid chains from the cytosol to the inner mitochondria, where fatty acids are further processed by beta-oxidation. Beta-oxidation or fatty acid oxidation (FAO) is a well-known source of energy and, in combination with high glycolysis, FAO is directly linked to collagen production in fibrotic tissue [51,52]. Furthermore, perturbation of FAO has also been reported to be closely related to fibrosis [52–54]. Additionally, FAO can also enhance the release of pro-inflammatory cytokines from macrophages. Thus, in addition to LPCs and SMs, the correlation of lipid metabolism with acyl-carnitines, FAO, and the gut microbiome might be relevant for fibrosis and inflammation in patients with SSc and should further be explored in the future.

We identified phenylacetylglutamine (PAG) by the untargeted approach to be significantly up-regulated in patients with SSc compared with the control group. PAG is associated with changes in the gut microbiome, especially in combination with kynurenine pathway metabolites, as presented in our study [55,56]. Kynurenine pathway metabolites can increase due to the synthesis in bacteria, which might enhance cytokine production [30,50,56–58]. It has been proposed that fibrosis results in gastrointestinal tract dysmotility, and therefore can have a significant impact on the gut microbiome. Gut dysbiosis is a known feature of SSc [34] and PAG could be an indicator of gastrointestinal involvement [59]. Further studies are needed to explore the role of PAG as a therapeutic target of gastrointestinal symptoms in SSc.

Several clinical trials explored drugs that target metabolites that we found possibly disrupted in our SSc patients, underlining the importance of our data [12,60,61]. Regarding the kynurenine pathway, the quinoline-3-carboxamides like laquinimod and paquinimod, which are structurally similar to kynurenines, have been proven in clinical trials to have positive effects in patients with multiple sclerosis, systemic lupus erythematosus, or SSc [62,63]. Moreover, drugs that influence IDO1 have been shown to have a high potential in treating autoimmune diseases [64]. For example, tocilizumab, which indirectly affects IDO1, has shown positive effects on lung and skin fibrosis in SSc [65]. Regarding the urea cycle, pirfenidone, an ARG1 inhibitor, has shown positive results in patients with idiopathic pulmonary fibrosis, and therefore inhibition of ARG1 might be a valuable therapeutic aim in SSc [66]. On the other side, inhibition of ARG1 can also increase NO, which stimulates inflammation and fibrosis in the lung [67]. Thus, the equilibrium of metabolites is essential for biological effects of metabolic pathways. Further studies, including multi-centre -omics studies, animal models, and in vitro studies, are needed to explore relevant metabolites, their interactions, and their role as possible therapeutic targets in SSc.

5. Conclusions

In summary, our data showed four possibly dysregulated metabolic mechanisms in plasma from patients with SSc, namely a dysregulated kynurenine pathway, a dysregulated urea cycle, disrupted lipid metabolism, and a disturbed gut microbiome. An accelerated kynurenine metabolism could induce production of pro-inflammatory cytokines, T-cell-mediated inflammation, endothelial cell dysfunction, and gut dysbiosis in SSc. A dysregulated urea cycle could be involved in inflammation and fibrosis during SSc and might be stimulated due to excessive collagen metabolism, where proline, one of the main constituents of collagen, can increase the production of ornithine and citrulline. A disrupted lipid metabolism, with down-regulated LPCs and SMs in SSc patients, was correlated with acyl-carnitines and FAO, which might stimulate fibrosis and macrophage-mediated inflammation in SSc, especially in combination with a disrupted gut microbiome. Finally, up-regulated PAG could be involved in gut dysbiosis in patients with SSc-related gastrointestinal symptoms.

A limitation of our study is the fact that blood sampling was not done at the same time point during the day and we have no data regarding fasting before blood sampling. Furthermore, due to small numbers of patients in the clinical subgroups, we could not correlate metabolites to single organ involvement or medication in SSc patients. Moreover, our results showed possible changes in metabolic pathways in SSc patients from single cohort study. In the future, multi-centre studies including higher numbers of patients as well as animal-models and in vitro studies are needed to further explore these metabolic pathways and their role as possible therapeutic targets for SSc.

In conclusion, our study of plasma metabolites in patients with SSc identified four possibly disrupted metabolite mechanisms, which are associated with autoimmune inflammation, vascular damage, fibrosis, and gastrointestinal dysbiosis and might be relevant for the pathophysiology of SSc. Further studies are needed to evaluate the role of these metabolomic networks as potential treatment targets for SSc or as personalized biomarkers of drug response in SSc in the future.

Supplementary Materials: The following supporting information can be downloaded at: https://www.mdpi.com/article/10.3390/biomedicines10030607/s1, Table S1: Clinical information about sample cohort; Table S2: Significant regulated metabolites of the targeted approach due to p-value (<0.05)., Table S3: Significant regulated metabolites of the untargeted approach due to p-value (<0.05).

Author Contributions: Conceptualization, W.B., T.B. and M.G.-S.; formal analysis, T.B. and F.M.; investigation, T.B. and F.M.; recruitment of study group: M.G.-S., resources, W.B., N.S. and M.G.-S.; data curation, T.B.; writing—original draft preparation, T.B.; writing—review and editing, W.B., M.H., N.S. and M.G.-S.; visualization, T.B.; supervision, W.B. and M.H.; funding acquisition, W.B. and M.G.-S. All authors have read and agreed to the published version of the manuscript.

Funding: This research was supported in part by a Clinician scientist grant of the Austrian Society of Dermatology and Venereology and in part by Postdoctoral scholarship for medical doctors in Upper Austria. Open Access Funding by the Johannes Kepler University of Linz.

Institutional Review Board Statement: The study was conducted according to the guidelines of the Declaration of Helsinki and approved by the Ethics Committee of the Johannes Kepler University Linz in Austria (protocol 1265/2019 and amendments).

Informed Consent Statement: Informed consent was obtained from all subjects involved in the study.

Data Availability Statement: The data presented in this study are available on request from the corresponding authors M.G.S. and T.B.

Acknowledgments: Figures 3 and 4 were created with BioRender (https://biorender.com/, accessed on 31 December 2021). Open Access Funding by the University of Linz.

Conflicts of Interest: The authors declare no conflict of interest. The funders had no role in the design of the study; in the collection, analyses, or interpretation of data; in the writing of the manuscript, or in the decision to publish the results.

Abbreviations

ACR	American College of Rheumatology
ADMA	asymmetric dimethylarginine
AFMID	kynurenine formidase
AhR	Aryl hydrocarbon receptor
ANOVA	Analysis of variance
ArAT	aromatic aminotransferase
ARG	arginase
ASL	argininosuccinate lyase
ASS	argininosuccinate synthase
CCS	collision cross section
Cer	ceramide
dcSSc	diffuse cutaneous Systemic sclerosis
DMA	dimethylarginine
FAO	fatty acid oxidation
HILIC	hydrophilic interaction chromatography
HMDB	Human Metabolome Database
HPLC	high-performance liquid chromatography
IDO	indoleamine 2,3-dioxygenase
IMS-QTOF-MS	ion mobility quadrupole time-of-flight mass spectrometer
KP	kynurenine pathway
lcSSc	limited cutaneous Systemic sclerosis
LDL	low-density lipoproteins
LPC	lysophosphatidylcholine
m/z	mass to charge ratio
MoNA	MassBank of North America
mRSS	modified Rodnan Skin Score
ncSSc	non-cutaneous systemic sclerosis
NOS	nitric oxide synthase
NOS	NO-synthase
OAT	ornithine aminotransferase
ODC	ornithine decarboxylase
ORNT1	ornithine translocase
OTC	ornithine transcarbamylase
P5C	pyrroline-5-carboxylate
PAG	phenylacetylglutamine
QQQ-MS	triple quadrupole mass spectrometer
RT	retention time
SDMA	symmetric dimethylarginine
SM	sphingomyelin
SSc	systemic sclerosis
TDO	tryptophan 2,3-dioxygenase
TMO	tryptophan 2-monooxygenase
TrD	tryptophan decarboxylase

References

1. Varga, J.; Abraham, D. Systemic sclerosis: A prototypic multisystem fibrotic disorder. *J. Clin. Invest.* **2007**, *117*, 557–567. [CrossRef]
2. Allanore, Y.; Simms, R.; Distler, O.; Trojanowska, M.; Pope, J.; Denton, C.P.; Varga, J. Systemic sclerosis. *Nat. Rev. Dis. Primers* **2015**, *1*, 1–21. [CrossRef]
3. Denton, C.P.; Khanna, D. Systemic sclerosis. *Lancet* **2017**, *390*, 1685–1699. [CrossRef]
4. Van Den Hoogen, F.; Khanna, D.; Fransen, J.; Johnson, S.R.; Baron, M.; Tyndall, A.; Matucci-Cerinic, M.; Naden, R.P.; Medsger, T.A.; Carreira, P.E.; et al. 2013 classification criteria for systemic sclerosis: An American College of Rheumatology/European League against Rheumatism collaborative initiative. *Arthritis. Rheum.* **2013**, *65*, 2737–2747. [CrossRef]
5. Gabrielli, A.; Avvedimento, E.V.; Krieg, T. Scleroderma. *N. Engl. J. Med.* **2009**, *360*, 1989–2003. [CrossRef]
6. Khanna, D.; Furst, D.E.; Clements, P.J.; Allanore, Y.; Baron, M.; Czirjak, L.; Distler, O.; Foeldvari, I.; Kuwana, M.; Matucci-Cerinic, M.; et al. Standardization of the modified Rodnan skin score for use in clinical trials of systemic sclerosis. *Scleroderma Relat. Disord.* **2017**, *2*, 11–18. [CrossRef]
7. Hinchcliff, M.; O'Reilly, S. Current and Potential New Targets in Systemic Sclerosis Therapy: A New Hope. *Curr. Rheumatol. Rep.* **2020**, *22*, 1–7. [CrossRef]
8. Misra, D.P.; Ahmed, S.; Agarwal, V. Is biological therapy in systemic sclerosis the answer? *Rheumatol Int.* **2020**, *40*, 679–694. [CrossRef]
9. Sierra-Sepúlveda, A.; Esquinca-González, A.; Benavides-Suárez, S.A.; Sordo-Lima, D.E.; Caballero-Islas, A.E.; Cabral-Castañeda, A.R.; Rodríguez-Reyna, T.S. Systemic Sclerosis Pathogenesis and Emerging Therapies, beyond the Fibroblast. *Biomed. Res. Int.* **2019**, *2019*, 1–15. [CrossRef]
10. Zanatta, E.; Codullo, V.; Avouac, J.; Allanore, Y. Systemic sclerosis: Recent insight in clinical management. *Jt. Bone Spine* **2020**, *87*, 293–299. [CrossRef]
11. Distler, O.; Highland, K.B.; Gahlemann, M.; Azuma, A.; Fischer, A.; Mayes, M.D.; Raghu, G.; Sauter, W.; Girard, M.; Alves, M.; et al. Nintedanib for Systemic Sclerosis-Associated Interstitial Lung Disease. *N. Engl. J. Med.* **2019**, *380*, 2518–2528. [CrossRef] [PubMed]
12. Yoon, N.; Jang, A.-K.; Seo, Y.; Jung, B.H. Metabolomics in Autoimmune Diseases: Focus on Rheumatoid Arthritis, Systemic Lupus Erythematous, and Multiple Sclerosis. *Metabolites* **2021**, *11*, 812. [CrossRef] [PubMed]
13. Sandlers, Y. Amino Acids Profiling for the Diagnosis of Metabolic Disorders. *Biochem. Test.* **2019**, *12*, 1–21. [CrossRef]
14. Wishart, D.S. Metabolomics for Investigating Physiological and Pathophysiological Processes. *Physiol. Rev.* **2019**, *99*, 1819–1875. [CrossRef] [PubMed]
15. Murgia, F.; Svegliati, S.; Poddighe, S.; Lussu, M.; Manzin, A.; Spadoni, T.; Fischetti, C.; Gabrielli, A.; Atzori, L. Metabolomic profile of systemic sclerosis patients. *Sci. Rep.* **2018**, *8*, 1–11. [CrossRef]
16. Bellocchi, C.; Fernández-Ochoa, Á.; Montanelli, G.; Vigone, B.; Santaniello, A.; Milani, C.; Quirantes-Piné, R.; Borrás-Linares, I.; Ventura, M.; Segura-Carrettero, A.; et al. Microbial and metabolic multi-omic correlations in systemic sclerosis patients. *Ann. N. Y. Acad. Sci.* **2018**, *1421*, 97–109. [CrossRef]
17. Fernández-Ochoa, Á.; Quirantes-Piné, R.; Borrás-Linares, I.; Gemperline, D.; Alarcón Riquelme, M.E.; Beretta, L.; Segura-Carretero, A. Urinary and plasma metabolite differences detected by HPLC-ESI-QTOF-MS in systemic sclerosis patients. *J. Pharm. Biomed. Anal.* **2019**, *162*, 82–90. [CrossRef]
18. Ottria, A.; Hoekstra, A.T.; Zimmermann, M.; Van Der Kroef, M.; Vazirpanah, N.; Cossu, M.; Chouri, E.; Rossato, M.; Beretta, L.; Tieland, R.G.; et al. Fatty Acid and Carnitine Metabolism Are Dysregulated in Systemic Sclerosis Patients. *Front. Immunol.* **2020**, *11*, 1–12. [CrossRef]
19. Smolenska, Z.; Zabielska-Kaczorowska, M.; Wojteczek, A.; Kutryb-Zajac, B.; Zdrojewski, Z. Metabolic Pattern of Systemic Sclerosis: Association of Changes in Plasma Concentrations of Amino Acid-Related Compounds With Disease Presentation. *Front. Mol. Biosci.* **2020**, *7*, 1–7. [CrossRef]
20. Deidda, M.; Piras, C.; Cadeddu Dessalvi, C.; Locci, E.; Barberini, L.; Orofino, S.; Musu, M.; Mura, M.N.; Manconi, P.E.; Finco, G.; et al. Distinctive metabolomic fingerprint in scleroderma patients with pulmonary arterial hypertension. *Int. J. Cardiol.* **2017**, *241*, 401–406. [CrossRef]
21. Meier, C.; Freiburghaus, K.; Bovet, C.; Schniering, J.; Allanore, Y.; Distler, O.; Nakas, C.; Maurer, B. Serum metabolites as biomarkers in systemic sclerosis-associated interstitial lung disease. *Sci. Rep.* **2020**, *10*, 1–14. [CrossRef] [PubMed]
22. Zhou, Z.; Luo, M.; Chen, X.; Yin, Y.; Xiong, X.; Wang, R.; Zhu, Z.-J. Ion mobility collision cross-section atlas for known and unknown metabolite annotation in untargeted metabolomics. *Nat. Commun.* **2020**, *11*, 1–13. [CrossRef] [PubMed]
23. Chong, J.; Wishart, D.S.; Xia, J. Using MetaboAnalyst 4.0 for Comprehensive and Integrative Metabolomics Data Analysis. *Curr. Protoc. Bioinform.* **2019**, *68*, 1–128. [CrossRef]
24. Shi, B.; Wang, W.; Korman, B.; Kai, L.; Wang, Q.; Wei, J.; Bale, S.; Marangoni, R.G.; Bhattacharyya, S.; Miller, S.; et al. Targeting CD38-dependent NAD+ metabolism to mitigate multiple organ fibrosis. *iScience* **2021**, *24*, 1–13. [CrossRef] [PubMed]
25. Wang, Q.; Liu, D.; Song, P.; Zou, M.H. Tryptophan-kynurenine pathway is dysregulated in inflammation, and immune activation. *Front. Biosci.* **2015**, 1116–1143.
26. Mellor, A.L.; Munn, D.H. IDO expression by dendritic cells: Tolerance and tryptophan catabolism. *Nat. Rev. Immunol.* **2004**, *4*, 762–774. [CrossRef] [PubMed]
27. Mándi, Y.; Vécsei, L. The kynurenine system and immunoregulation. *J. Neural. Transm.* **2012**, *119*, 197–209. [CrossRef]

28. Fallarino, F.; Grohmann, U.; Vacca, C.; Bianchi, R.; Orabona, C.; Spreca, A.; Fioretti, M.C.; Puccetti, P. T cell apoptosis by tryptophan catabolism. *Cell. Death. Differ.* **2002**, *9*, 1069–1077. [CrossRef]
29. Puccetti, P.; Grohmann, U. IDO and regulatory T cells: A role for reverse signalling and non-canonical NF-κB activation. *Nat. Rev. Immunol.* **2007**, *7*, 817–823. [CrossRef]
30. Biernacki, T.; Sandi, D.; Bencsik, K.; Vécsei, L. Kynurenines in the Pathogenesis of Multiple Sclerosis: Therapeutic Perspectives. *Cells* **2020**, *9*, 1564. [CrossRef]
31. Sgonc, R.; Gruschwitz, M.S.; Boeck, G.; Sepp, N.; Gruber, J.; Wick, G. Endothelial cell apoptosis in systemic sclerosis is induced by antibody-dependent cell-mediated cytotoxicity via CD95. *Arthritis. Rheum.* **2000**, *43*, 2550–2562. [CrossRef]
32. Agus, A.; Planchais, J.; Sokol, H. Gut Microbiota Regulation of Tryptophan Metabolism in Health and Disease. *Cell. Host. Microbe.* **2018**, *23*, 716–724. [CrossRef] [PubMed]
33. Patrone, V.; Puglisi, E.; Cardinali, M.; Schnitzler, T.S.; Svegliati, S.; Festa, A.; Gabrielli, A.; Morelli, L. Gut microbiota profile in systemic sclerosis patients with and without clinical evidence of gastrointestinal involvement. *Sci. Rep.* **2017**, *7*, 1–11. [CrossRef]
34. Andréasson, K.; Alrawi, Z.; Persson, A.; Jönsson, G.; Marsal, J. Intestinal dysbiosis is common in systemic sclerosis and associated with gastrointestinal and extraintestinal features of disease. *Arthritis. Res. Ther.* **2016**, *18*, 1–8. [CrossRef]
35. Caldwell, R.W.; Rodriguez, P.C.; Toque, H.A.; Narayanan, S.P.; Caldwell, R.B. Arginase: A Multifaceted Enzyme Important in Health and Disease. *Physiol. Rev.* **2018**, *98*, 641–665. [CrossRef]
36. Watford, M. The Urea Cycle: Teaching intermediary metabolism in a physiological setting. *Biochem. Mol. Biol. Educ.* **2003**, *31*, 289–297. [CrossRef]
37. Bruch-Gerharz, D.; Schnorr, O.; Suschek, C.; Beck, K.-F.; Pfeilschifter, J.; Ruzicka, T.; Kolb-Bachofen, V. Arginase 1 Overexpression in Psoriasis. *Am. J. Clin. Pathol.* **2003**, *162*, 203–211. [CrossRef]
38. Brooks, W.H. Autoimmune diseases and polyamines. *Clin. Rev. Allergy. Immunol.* **2012**, *42*, 58–70. [CrossRef]
39. Bandyopadhyay, M.; Larregina, A.T. Keratinocyte-Polyamines and Dendritic Cells: A Bad Duet for Psoriasis. *Immunity* **2020**, *53*, 16–18. [CrossRef]
40. Vona, R.; Giovannetti, A.; Gambardella, L.; Malorni, W.; Pietraforte, D.; Straface, E. Oxidative stress in the pathogenesis of systemic scleroderma: An overview. *J. Cell. Mol. Med.* **2018**, *22*, 3308–3314. [CrossRef]
41. Zhu, H.; Chen, W.; Liu, D.; Luo, H. The role of metabolism in the pathogenesis of systemic sclerosis. *Metabolism* **2019**, *93*, 44–51. [CrossRef] [PubMed]
42. Napoli, C.; De Nigris, F.; Williams-Ignarro, S.; Pignalosa, O.; Sica, V.; Ignarro, L.J. Nitric oxide and atherosclerosis: An update. *Nitric Oxide* **2006**, *15*, 265–279. [CrossRef] [PubMed]
43. D'Aniello, C.; Patriarca, E.J.; Phang, J.M.; Minchiotti, G. Proline Metabolism in Tumor Growth and Metastatic Progression. *Front. Oncol.* **2020**, *10*, 1–14. [CrossRef] [PubMed]
44. Liu, P.; Zhu, W.; Chen, C.; Yan, B.; Zhu, L.; Chen, X.; Peng, C. The mechanisms of lysophosphatidylcholine in the development of diseases. *Life Sci.* **2020**, *247*, 1–12. [CrossRef] [PubMed]
45. Martinez, J. Prix Fixe: Efferocytosis as a Four-Course Meal. *Curr. Top. Microbiol. Immunol.* **2017**, *403*, 1–36. [CrossRef]
46. Grossmayer, G.E.; Keppeler, H.; Boeltz, S.; Janko, C.; Rech, J.; Herrmann, M.; Lauber, K.; Muñoz, L.E. Elevated Serum Lysophosphatidylcholine in Patients with Systemic Lupus Erythematosus Impairs Phagocytosis of Necrotic Cells In Vitro. *Front. Immunol.* **2017**, *8*, 1–11. [CrossRef]
47. Shea, B.S.; Tager, A.M. Sphingolipid regulation of tissue fibrosis. *Open Rheumatol. J.* **2012**, *6*, 123–129. [CrossRef]
48. Huwiler, A.; Pfeilschifter, J. Sphingolipid signaling in renal fibrosis. *Matrix Biol.* **2018**, *68–69*, 230–247. [CrossRef]
49. Geroldinger-Simić, M.; Bögl, T.; Himmelsbach, M.; Sepp, N.; Buchberger, W. Changes in Plasma Phospholipid Metabolism Are Associated with Clinical Manifestations of Systemic Sclerosis. *Diagnostics (Basel)* **2021**, *11*, 2116. [CrossRef]
50. He, J.; Chan, T.; Hong, X.; Zheng, F.; Zhu, C.; Yin, L.; Dai, W.; Tang, D.; Liu, D.; Dai, Y. Microbiome and Metabolome Analyses Reveal the Disruption of Lipid Metabolism in Systemic Lupus Erythematosus. *Front. Immunol.* **2020**, *11*, 1–13. [CrossRef]
51. McKleroy, W.; Lee, T.-H.; Atabai, K. Always cleave up your mess: Targeting collagen degradation to treat tissue fibrosis. *AJP* **2013**, *304*, L709–L721. [CrossRef] [PubMed]
52. Kang, H.M.; Ahn, S.H.; Choi, P.; Ko, Y.-A.; Han, S.H.; Chinga, F.; Park, A.S.D.; Tao, J.; Sharma, K.; Pullman, J.; et al. Defective fatty acid oxidation in renal tubular epithelial cells has a key role in kidney fibrosis development. *Nat. Med.* **2015**, *21*, 37–46. [CrossRef] [PubMed]
53. Mills, E.L.; O'Neill, L.A. Reprogramming mitochondrial metabolism in macrophages as an anti-inflammatory signal. *Eur. J. Immunol.* **2016**, *46*, 13–21. [CrossRef]
54. Zhao, X.; Kwan, J.Y.Y.; Yip, K.; Liu, P.P.; Liu, F.-F. Targeting metabolic dysregulation for fibrosis therapy. *Nat. Rev. Drug Discov.* **2020**, *19*, 57–75. [CrossRef]
55. Volkmann, E.R.; Hoffmann-Vold, A.-M.; Chang, Y.-L.; Jacobs, J.P.; Tillisch, K.; Mayer, E.A.; Clements, P.J.; Hov, J.R.; Kummen, M.; Midtvedt, Ø.; et al. Systemic sclerosis is associated with specific alterations in gastrointestinal microbiota in two independent cohorts. *BMJ Open Gastroenterol.* **2017**, *4*, 1–9. [CrossRef]
56. Visconti, A.; Le Roy, C.I.; Rosa, F.; Rossi, N.; Martin, T.C.; Mohney, R.P.; Li, W.; De Rinaldis, E.; Bell, J.T.; Venter, J.C.; et al. Interplay between the human gut microbiome and host metabolism. *Nat. Commun.* **2019**, *10*, 1–10. [CrossRef]

57. Bellocchi, C.; Fernández-Ochoa, Á.; Montanelli, G.; Vigone, B.; Santaniello, A.; Quirantes-Piné, R.; Borrás-Linares, I.; Gerosa, M.; Artusi, C.; Gualtierotti, R.; et al. Identification of a Shared Microbiomic and Metabolomic Profile in Systemic Autoimmune Diseases. *J. Clin. Med.* **2019**, *8*, 1291. [CrossRef] [PubMed]
58. Choi, S.C.; Brown, J.; Gong, M.; Ge, Y.; Zadeh, M.; Li, W.; Croker, B.P.; Michailidis, G.; Garrett, T.J.; Mohamadzadeh, M.; et al. Mansour Mohamadzadeh and Laurence Morel. Gut microbiota dysbiosis and altered tryptophan catabolism contribute to autoimmunity in lupus-susceptible mice. *Sci. Transl. Med.* **2020**, *12*, 1–15. [CrossRef] [PubMed]
59. Kumar, S.; Singh, J.; Rattan, S.; DiMarino, A.J.; Cohen, S.; Jimenez, S.A. Review article: Pathogenesis and clinical manifestations of gastrointestinal involvement in systemic sclerosis. *Aliment. Pharmacol. Ther.* **2017**, *45*, 883–898. [CrossRef]
60. He, Z.; Liu, Z.; Gong, L. Biomarker identification and pathway analysis of rheumatoid arthritis based on metabolomics in combination with ingenuity pathway analysis. *Proteomics* **2021**, *21*, 1–12. [CrossRef]
61. Sanchez-Lopez, E.; Cheng, A.; Guma, M. Can Metabolic Pathways Be Therapeutic Targets in Rheumatoid Arthritis? *J. Clin. Med.* **2019**, *8*, 753. [CrossRef] [PubMed]
62. Boros, F.; Vécsei, L. Progress in the development of kynurenine and quinoline-3-carboxamide-derived drugs. *Expert. Opin. Investig. Drugs.* **2020**, *29*, 1223–1247. [CrossRef]
63. Hesselstrand, R.; Distler, J.H.W.; Riemekasten, G.; Wuttge, D.M.; Törngren, M.; Nyhlén, H.C.; Andersson, F.; Eriksson, H.; Sparre, B.; Tuvesson, H.; et al. An open-label study to evaluate biomarkers and safety in systemic sclerosis patients treated with paquinimod. *Arthritis. Res. Ther.* **2021** *23*, 1–12. [CrossRef] [PubMed]
64. Mondanelli, G.; Iacono, A.; Carvalho, A.; Orabona, C.; Volpi, C.; Pallotta, M.T.; Matino, D.; Esposito, S.; Grohmann, U. Amino acid metabolism as drug target in autoimmune diseases. *Autoimmun. Rev.* **2019**, *18*, 334–348. [CrossRef] [PubMed]
65. Khanna, D.; Lin, C.J.F.; Furst, D.E.; Goldin, J.; Kim, G.; Kuwana, M.; Allanore, Y.; Matucci-Cerinic, M.; Distler, O.; Shima, Y.; et al. Tocilizumab in systemic sclerosis: A randomised, double-blind, placebo-controlled, phase 3 trial. *Lancet Respir. Med.* **2020**, *8*, 963–974. [CrossRef]
66. Azuma, A.; Nukiwa, T.; Tsuboi, E.; Suga, M.; Abe, S.; Nakata, K.; Taguchi, Y.; Nagai, S.; Itoh, H.; Ohi, M.; et al. Double-blind, placebo-controlled trial of pirfenidone in patients with idiopathic pulmonary fibrosis. *Am. J. Respir. Crit. Care. Med.* **2005**, *171*, 1040–1047. [CrossRef]
67. Hua-Huy, T.; Le-Dong, N.-N.; Duong-Quy, S.; Bei, Y.; Rivière, S.; Tiev, K.-P.; Nicco, C.; Chéreau, C.; Batteux, F.; Dinh-Xuan, A.T. Increased exhaled nitric oxide precedes lung fibrosis in two murine models of systemic sclerosis. *J. Breath. Res.* **2015**, *9*, 1–15. [CrossRef]

Review

Autoantibodies as Biomarker and Therapeutic Target in Systemic Sclerosis

Hanna Graßhoff *, Konstantinos Fourlakis, Sara Comdühr and Gabriela Riemekasten

Department of Rheumatology and Clinical Immunology, University of Lübeck, 23538 Lübeck, Germany
* Correspondence: hanna.grasshoff@uksh.de

Abstract: Systemic sclerosis (SSc) is a rare connective tissue disorder characterized by immune dysregulation evoking the pathophysiological triad of inflammation, fibrosis and vasculopathy. In SSc, several alterations in the B-cell compartment have been described, leading to polyclonal B-cell hyperreactivity, hypergammaglobulinemia and autoantibody production. Autoreactive B cells and autoantibodies promote and maintain pathologic mechanisms. In addition, autoantibodies in SSc are important biomarkers for predicting clinical phenotype and disease progression. Autoreactive B cells and autoantibodies represent potentially promising targets for therapeutic approaches including B-cell-targeting therapies, as well as strategies for unselective and selective removal of autoantibodies. In this review, we present mechanisms of the innate immune system leading to the generation of autoantibodies, alterations of the B-cell compartment in SSc, autoantibodies as biomarkers and autoantibody-mediated pathologies in SSc as well as potential therapeutic approaches to target these.

Keywords: systemic sclerosis; autoreactive B cells; autoantibodies; rituximab; aptamers; Seldeg

Citation: Graßhoff, H.; Fourlakis, K.; Comdühr, S.; Riemekasten, G. Autoantibodies as Biomarker and Therapeutic Target in Systemic Sclerosis. *Biomedicines* **2022**, *10*, 2150. https://doi.org/10.3390/biomedicines10092150

Academic Editor: Jun Lu

Received: 1 July 2022
Accepted: 23 August 2022
Published: 1 September 2022

Publisher's Note: MDPI stays neutral with regard to jurisdictional claims in published maps and institutional affiliations.

1. Introduction

Systemic sclerosis (SSc) is a rare connective tissue disorder characterized by immune dysregulation evoking the pathophysiological triad of inflammation, fibrosis and vasculopathy. This pathophysiological triad results in a heterogenous disease course involving the skin and internal organs such as the lung, the gastrointestinal tract, the heart, or the kidneys.

Several genes have been identified increasing disease susceptibility and several susceptibility haplotypes show an association to defined autoimmune profiles [1,2]. However, exposure to environmental agents and infectious pathogens is thought to play a major role in disease development and maintenance [3,4]. Furthermore, there are hints that the exposure to certain environmental factors leads to distinct clinical phenotypes and possibly also to a distinct autoantibody profile [5,6]. According to clinical and laboratory characteristics, SSc is divided into the limited cutaneous form (lcSSc) and the diffuse cutaneous form (dcSSc). In addition to this classification, further disease subsets have been distinguished, e.g., SSc-overlap syndromes, SSc sine scleroderma or paraneoplastic SSc. Beyond these classifications, however, the disease expression, the course of the disease, as well as the development of secondary disease complications and the mortality of the individual patient are heterogeneous. This heterogeneity has been a major obstacle in performing and analyzing clinical trials in SSc. For example, the risk of death varies greatly depending on the organ manifestations. Therefore, stratification of patients is essential in the therapy of SSc to predict disease progression and response to therapy.

In many other autoimmune diseases, but also in SSc, autoantibodies could be identified, which can be useful for stratification. These autoantibodies are not only valuable biomarkers, but together with autoreactive B cells, they are a crucial hallmark in the pathogenesis of SSc. Therefore, this review aims to address whether autoantibodies can be used as predictors of disease course and which therapies targeting autoantibody-mediated pathologies have been evaluated so far in SSc.

Thus, this review focuses on the following three questions:

1. Which mechanisms are involved in maturation of autoreactive B cells and secretion of autoantibodies in SSc?
2. Which autoantibodies might be useful as predictors of disease course and which role do autoantibodies play in the pathophysiology of SSc?
3. Which therapeutic approaches have been evaluated to target autoantibody-mediated pathologies in SSc?

By highlighting the topic of autoreactive B cells and autoantibodies in SSc, we present a disease-overarching pathomechanism mediating pathologies in several diseases and discuss potential therapeutic approaches [7].

2. Which Mechanisms Are Involved in Maturation of Autoreactive B Cells and Autoantibody Secretion in SSc?

B cells represent key cellular players in the pathophysiology of SSc. Accordingly, a gene expression study on SSc skin revealed an increase in gene clusters characteristic of B cells and plasma cells [8]. These data were confirmed by a recently published study also proving a B-cell characteristic signature by RNA sequencing in skin samples of SSc patients [9]. Moreover, CD20+ B-cell infiltrates were demonstrated in pulmonary tissue samples of SSc patients with interstitial lung disease [10].

In general, the B-cell compartment, in which short-lived plasmablasts and long-lived plasma cells secrete antibodies, is affected by a dichotomy between rapid and effective immune defense against pathogens and potentially harmful autoimmunity. To balance this dichotomy, maturation of the B-cell compartment and antibody secretion must be tightly regulated. Divergent regulation of these mechanisms is, therefore, a central feature of autoimmune diseases. In SSc, disease-specific autoantibodies are thought not only to play a role in disease maintenance but also to be involved in the development of the disease, thus representing an early pathomechanism. Evidence of this is that disease-specific autoantibodies in SSc are present before early clinical symptoms such as morning stiffness, Raynaud's phenomenon or swollen fingers [11]. Moreover, in patients with Raynaud's phenomenon, detection of autoantibodies predicts microvascular damage in nailfold capillary microscopy and subsequent diagnosis of SSc [12]. This aspect is taken into account in the criteria for "Very Early Diagnosis Of SSc" (VEDOSS), which include the presence of autoantibodies [13].

To understand the potential role of autoantibodies as biomarkers and in pathophysiology, as well as corresponding therapeutic approaches, we present the mechanisms involved in the generation of autoreactive B cells and the secretion of autoantibodies in SSc in the first section.

2.1. Tolerance Mechanisms in B-Cell Maturation

In humans, B cells are classified into three subclasses, namely B-1 cells, originating mainly from the fetal liver, B-2 cells, developing in the bone marrow, and regulatory B cells which prohibit the expansion of pro-inflammatory lymphocytes and, thus, contribute to immune homeostasis [14]. Maturation of the B-2 cells in the bone marrow is characterized by the formation of a functional B-cell receptor (BCR). In 2003, Wardemann et al. estimated that 55–75% of early immature B cells in the bone marrow display autoreactivity [15]. To ensure the balance between rapid and effective immune defense and autoimmunity, B cells undergo tolerance mechanisms during maturation (Figure 1).

In the bone marrow, high-affinity binding of endogenous antigens to the BCR of early immature B cells evokes receptor editing, induction of anergy or clonal deletion by apoptosis [16]. These mechanisms are comprised under the term "central tolerance mechanisms" and reduce frequency of autoreactive B cells from approximately 75% to 43% [15].

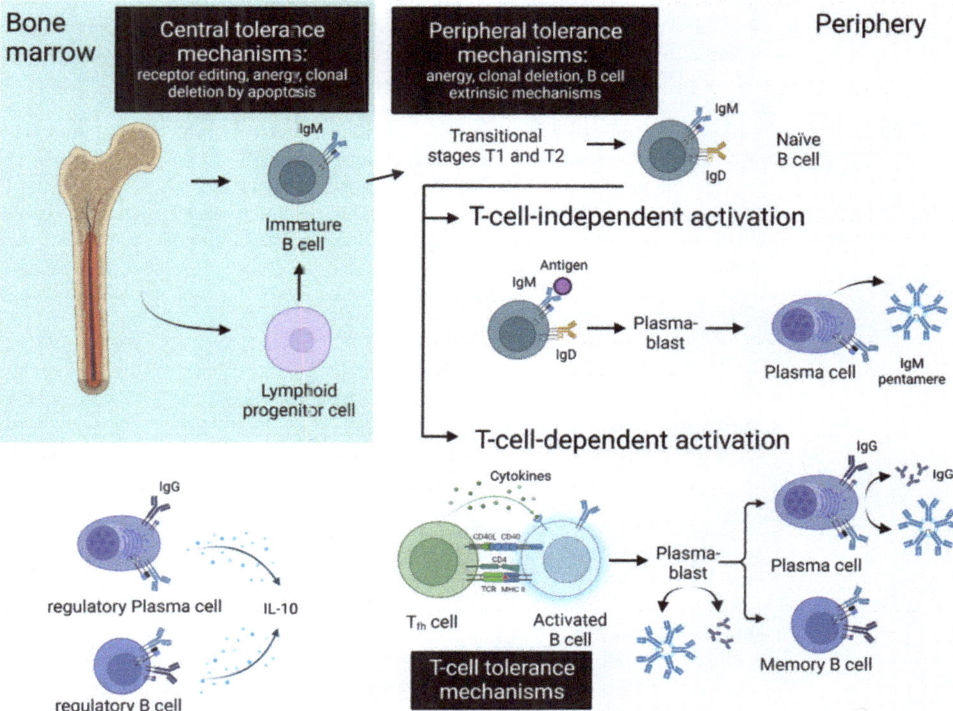

Figure 1. B-2 cell maturation in the bone marrow is characterized by the development of the BCR. Subsequently immature B cells leave the bone marrow. Naïve B cells can undergo T-cell-independent and T-cell-dependent activation. Tolerance mechanisms involving central tolerance mechanisms, peripheral tolerance mechanisms and T-cell tolerance mechanisms are marked in grey. The figure was created with 'Biorender.com'.

Subsequently, late immature B cells are released from the bone marrow to the blood stream. Peripheral tolerance mechanisms take effect between the transitional stages 1 and 2 in the spleen and secondary lymphoid organs. Peripheral tolerance mechanisms comprise B-cell-intrinsic mechanisms such as anergy or clonal deletion and B-cell-extrinsic mechanisms including ignorance and limited secretion of survival factors. These peripheral tolerance mechanisms reduce frequency of autoreactive B cells from approximately 42% to 20% [15].

Subsequently, mature B cells without antigen contact—also called naïve B cells—act as antigen-presenting cells and circulate in the blood and lymphatic organs. Here, the binding of an antigen to the BCR evokes a T-cell-independent or T-cell-dependent activation of the B cell. T-cell-independent activated B cells secrete antibodies of the IgM isotype and do not undergo class switch or formation of memory cells. T-cell-dependent activation of B cells promotes differentiation to plasma cells or to memory B cells. Therefore, a third and fourth important tolerance mechanism controlling activation of autoreactive B cells are the central and peripheral tolerance mechanisms of T cells, as these are more stringent than for B cells and most antigens induce a T-cell-dependent B-cell activation. Moreover, B cells require interaction with T cells for the germinal center reaction with class switch and somatic hypermutation [17].

Further mechanisms that can activate cells of the B-cell compartment are via Toll-like receptors (TLRs). TLRs belong to the innate immune system. With regard to their ligands, a distinction is made between pathogen-associated molecular patterns (PAMPs) and damage-associated molecular patterns (DAMPs). Whereas PAMPs are molecular motifs conserved within a class of microbes, DAMPs comprise molecules released from damaged cells due to trauma or infection. The expression of TLRs varies in the B-cell compartment. Naïve human B cells express only low levels of TLRs, whereas activated and memory B cells express increased levels of TLRs [18]. Moreover, increased gene expression of TLRs occurs physiologically in B cells after stimulation of BCR or CD40 [19]. Stimulation of TLRs on plasma cells has been shown to increase antibody secretion [20].

In addition, dendritic cells express TLRs and, subsequently, these cells could activate the B-cell compartment [21]. Regarding the role of TLRs in SSc, we refer to reviews by O'Reilly [22,23] as well as Frasca and Lande [24]. To date, the role of B-cell activation and the respective effects on antibody secretion via TLRs in the development and maintenance of SSc is poorly understood.

2.2. Natural and Pathogenic Autoantibodies

Despite the described tolerance mechanisms, autoreactive B cells and autoantibodies can be detected in the peripheral blood of humans and mice. Though there is a potentially increased risk for autoimmune diseases, in most cases, these diseases do not occur [15,25,26].

In fact, non-harmful autoantibodies represent a substantial proportion of antibodies [27]. This group of antibodies is termed natural autoantibodies. Natural autoantibodies are present from birth and mainly of the IgM isotype, less frequently of the IgA or IgG isotype [11,28]. They are encoded by unmutated V(D)J genes and have a moderate affinity to self-molecules. Natural autoantibodies are thought to play a key role in immune homeostasis. Polyreactive IgM autoantibodies are involved in early immune responses. Moreover, natural autoantibodies enhance phagocytic clearance of apoptotic cells and cell debris and, thus, prevent uncontrolled inflammation [29]. Furthermore, natural autoantibodies mask antigens and, as a result, prevent binding of pathologic autoantibodies which could promote autoimmunity. In addition, natural autoantibodies can suppress inflammatory responses to Toll-like receptor agonists [30]. In the peripheral immune system, binding of autoreactive T and B cells to endogenous antigens promotes both T- and B-cell survival. For further information on the function of natural autoantibodies, we refer to reviews published by Siloşi et al. [27], Silverman et al. [30] and Elkon and Casali [31]. Summarizing the described mechanisms, evidence suggests that natural autoantibodies might ameliorate risk and severity of autoimmune diseases [32–34] and may therefore have a therapeutic potential in autoimmune diseases.

Besides these natural autoantibodies, there are pathogenic somatically mutated autoantibodies—class-switched to IgG isotype—which are secreted by autoreactive B cells. Rarely, these pathogenic autoantibodies can also display the IgM or IgA isotype. These autoantibodies show a high affinity to their respective antigen and might be involved in autoimmune disease. In general, B cells secreting these autoantibodies can contribute to pathomechanisms in autoimmune diseases via autoantibody-dependent and autoantibody-independent pathways.

Autoantibody-independent mechanisms of B cells involve the secretion of proinflammatory cytokines [35], the formation of ectopic germinal centers in inflamed tissues [36,37] and the role of B cells as antigen-presenting cells. Especially via the latter mechanism, autoreactive B cells are involved in the pathophysiology of various autoimmune diseases. Pathologies in autoimmune diseases are often characterized as T-cell-mediated. However, this interpretation might underestimate the complex interactions of B and T cells and the role of B cells as antigen-presenting cells in CD4+ T-cell activation [38].

Autoantibody-dependent mechanisms involve complement activation and activation of neutrophils and NK cells by immune complexes composed of autoantibody and autoantigen. Activation of the classical complement pathway results in the release of C3a and C5a which promote the release of proinflammatory cytokines, migration of immune cells and upregulation of FcR on effector cells. The upregulation of FcR on effector cells augments the antibody-dependent cell-mediated cytotoxicity (ADCC). Autoantibodies facilitate antigen uptake by antigen-presenting cells such as monocytes and dendritic cells. By this mechanism, autoantibodies enhance T-cell responses to the respective antigens, which is of great importance in T-cell-mediated autoimmune diseases. Moreover, autoantibodies can stimulate and inhibit receptor function. Characteristics of natural and pathogenic autoantibodies are summarized in Table 1.

Table 1. Characteristics of natural and pathogenic autoantibodies.

	Natural Autoantibodies	**Pathogenic Autoantibodies**
Isotype	IgM, less frequently IgA or IgG	IgG, less frequently IgA or IgM
Generation of antibody diversity	Unmutated V(D)J recombination	V(D)J recombination, somatic hypermutation
Affinity	low	high
Mechanism of action	maintenance of immune homeostasis, amelioration of risk and severity of autoimmune diseases	contribution to autoimmune diseases via autoantibody-dependent and autoantibody-independent pathways

In the development of diagnostic methods for the detection of autoantibodies, it is crucial to differentiate natural autoantibodies and autoantibodies that mediate pathologies. Criteria for this distinction were first proposed in 1993 and further elaborated in subsequent years [39,40]. This is particularly important, as B-cell activation and hypergammaglobulinemia, which occur in the context of many chronic inflammatory diseases, could also increase the level of natural autoantibodies without inducing or maintaining pathologies. Recently, studies suggested the use of serum IgG levels to differentiate whether increased autoantibody levels are due to a non-specific B-cell activation or an antigen-specific autoimmune reaction [40,41].

2.3. *Autoreactive B Cells and Autoantibodies*

Several mechanisms have been described that may contribute to the maturation of autoreactive B cells and subsequent secretion of autoantibodies in SSc.

A physiological mechanism by which autoantibody formation occurs is the presence of polyreactive immunoglobulins that can bind diverse antigens. This reduced specificity can be advantageous in the immune defense, as a B cell can thus ward off different pathogens with similar antigens. However, a reduced specificity exhibits substantial cross-reactivity with endogenous antigens. Thus, polyreactive immunoglobulins are often autoreactive [42–44]. Along with this, the molecular mimicry hypothesis describes that T and B cells with specificity for an antigen of a pathogen also cross-react with self-antigens. This hypothesis was investigated in greater depth in SSc, as higher levels of antibodies to human cytomegalovirus proteins were detected in SSc than in healthy controls [45]. In addition, defined antigen-specific antibodies for human cytomegalovirus were found to be associated with autoantibody specificities of SSc [46,47]. Another interesting study investigating the molecular mimicry hypothesis was published in 2020 by Gourh et al. [48]. This study suggested a link between HLA alleles, peptides of viruses that infect amoebas or algae and anti-fibrillarin, anti-topoisomerase I and anti-centromere autoantibodies in African American and European American patients with SSc based on molecular mimicry.

A further mechanism that can predispose the formation of autoantibodies describes a loss of self-tolerance in the B-cell compartment, e.g., due to deficient negative selection or excessive stimulation, together with increased antigen expression or excessive antigen release due to cell damage. Exemplarily, the relationship between increased antigen expression and increased autoantibody formation could be shown for the AT1R. An increased expression of AT1R in peripheral blood mononuclear cells, skin and lungs corresponds to increased AT1R autoantibody levels in SSc [28].

Moreover, several aberrations promoting autoreactivity and autoantibody secretion were identified in SSc. Accordingly, Glauzy et al. demonstrated a deficiency in central and peripheral B-cell tolerance checkpoints in patients with SSc, promoting the development of autoreactive naïve B cells [49].

Moreover, SSc patients display higher levels of the B-cell survival factor BAFF and B cells of SSc patients exhibit increased levels of the BAFF receptor [50]. Complementarily, a genome-wide association study revealed an association between a variant in the BAFF gene (TNFS13B) and multiple sclerosis, as well as systemic lupus erythematosus. This variant is associated with increased levels of soluble BAFF, lymphocytes and immunoglobulins promoting humoral immunity [51].

Among others, these aberrations might contribute to the increased relative count of B cells and a disturbed composition of the B-cell compartment in SSc [52]. Several studies revealed distinct alterations of B-cell subsets that might promote autoreactivity. Firstly, SSc patients exhibit decreased levels of regulatory B cells and regulatory memory B cells [50,53]. This decrease is especially prominent in patients with pulmonary arterial hypertension [54]. Regulatory B cells secrete the anti-inflammatory cytokine IL-10 and they inhibit the induction of antigen-specific inflammatory reactions [52]. Accordingly, in SSc, levels of circulating regulatory B cells negatively correlated with anti-centromere and anti-topoisomerase I autoantibody levels as well as disease activity [50]. Moreover, regulatory B cells inhibit CD4+ Th1 and Th17 cell differentiation and cytokine secretion and induce regulatory T cells [55]. In addition, regulatory B cells are thought to participate in regulation of T_{fh} cells in SSc. Accordingly, levels of regulatory B cells and T_{fh} cells are negatively correlated [54].

Moreover, evidence indicates a decrease in plasmablasts and memory B cells due to an increased sensitivity to Fas-mediated apoptosis. Although the number of memory B cells (CD19+CD27+) decreases, these cells show an activated phenotype with increased expression of CD80 and CD86, which are co-stimulatory molecules of B cells [56]. Among memory B cells, distinct subsets are aberrated. Exemplarily, CD21low B cells were found to be increased in SSc patients, especially with visceral vascular manifestations, compared to healthy controls [57,58]. However, data on an increased prevalence of pulmonary arterial hypertension in patients with more than 10% of CD21low B cells are contradictory [54,58]. CD21low B cells are thought to have a high autoreactive potential as these cells express high levels of activation markers and act as antigen-presenting cells. An increase in the number of these cells has also already been described in other autoimmune diseases [59–62].

Probably, compensatory to the decrease in memory B cells, the number of naïve B cells (CD19+CD27-) is increased [56,63].

Furthermore, a change in the gene expression of B cells towards increased activity was detected [8]. Correspondingly, in SSc patients, expression of regulator molecules controlling B-cell responses are altered [64]. Increased CD19 expression was shown on naïve and memory B cells compared to healthy controls [56,65]. Experiments with CD19 transgenic mice showed that these mice produced elevated levels of autoantibodies, including SSc-specific autoantibodies, but without inducing a pathological phenotype [66–69].

To summarize, these aberrations in the B-cell compartment result in polyclonal B-cell hyperreactivity, hypergammaglobulinemia [70] and autoantibody production in SSc. However, in SSc, autoantibodies are detectable years before clinical disease manifestations [11]. Therefore, it is challenging to identify the mechanisms that lead to the loss of self-tolerance.

3. Which Autoantibodies Might Be Useful as Predictors of Disease Course and Which Role Do Autoantibodies Play in the Pathophysiology of SSc?

Because of the heterogeneity of SSc, biomarkers are essential for stratifying patients and predicting an individual disease course. Therefore, the field of biomarker research is an emerging field of investigation in SSc. Though the potential role of autoantibodies as biomarkers is frequently investigated in SSc, functional data on autoantibody-mediated pathomechanisms are rare and poorly understood. An improved understanding of autoantibody-mediated pathomechanisms is essential to identify appropriate targets for therapeutics that remove specific autoantibodies.

In the following section, we present clinical associations, roles as biomarkers and influences on pathomechanisms in SSc for autoantibodies against nuclear antigens (ANAs), anti-neutrophil cytoplasmic antibodies (ANCA), anti-phospholipid antibodies (aPL) and autoantibodies recognizing GPCRs.

3.1. Autoantibodies against Nuclear Antigens (ANAs)

A diagnostically important feature of SSc is the presence of circulating ANAs directed against nuclear or nucleolar proteins involved in transcription, splicing or cell proliferation. ANAs can be detected by indirect immunofluorescence. However, this technique is insufficient to identify specific ANAs except for anti-centromere antibodies. Therefore, additional techniques such as ELISA, immunodiffusion or Western immunoblotting can be used to determine the patients' individual specific antigenic targets [71].

The determination of ANAs as well as specific ANAs is well-established in diagnosing SSc and is part of the ACR/EULAR 2013 classification criteria for SSc [72]. ANAs can be detected in >90% of SSc patients [73,74]. The group of ANA-negative patients constitutes a distinct clinical subtype characterized by predominantly male patients with a severe disease course, involvement of the lower gastrointestinal tract with corresponding symptoms and less vasculopathy [75]. Another study revealed an association between ANA negativity and gastric antral vascular ectasia (GAVE) [76].

Until now, a variety of SSc-specific ANAs (anti-topoisomerase I, anti-centromere, anti-RNA polymerase III, anti-Th/To, anti-eukaryotic initiation factor 2B (anti-eIF2B), anti-U11/U12 RNP) and ANAs not exclusively specific for SSc (anti-Pm/Scl-100, anti-Pm/Scl-75, anti-Ro52, anti-Ku, anti-fibrillarin (U3-RNP), anti-U1 RNP, anti-NOR90/hUBF, anti-RuvBL1/2) have been identified. Each of these autoantibodies has been associated with a unique set of disease manifestations, enabling the prediction of disease course, development of organ manifestations and an individual prognosis [77]. To further explain clinical differences of ANA subspecificities, Clark et al. performed a transcriptional and proteomic analysis of blood and skin of SSc patients with anti-topoisomerase I and anti-RNA polymerase III antibody specificities, revealing pathogenetic differences between ANA-based subgroups [78]. For the association of SSc-specific ANAs with corresponding clinical phenotypes, we refer to recently published reviews, e.g., by Cavazzana et al. [71] or Stochmal et al. [79]. In addition, recently, further ANAs were identified in subsets of SSc patients. These autoantibodies target telomerase and shelterin proteins and show an association with severe interstitial lung disease [80]. Moreover, detection of ANA subspecificities in patients suffering from SSc-associated diseases such as primary biliary cholangitis can be used to identify patients at increased risk for developing SSc [81,82].

A remarkable observation is that the ANA titers and specific ANAs remain relatively stable over the disease course, which makes them a valuable diagnostic tool [83].

Moreover, in most patients, only few ANA subspecificities are detected in parallel, so mutual exclusivity is assumed. Against this background, however, the simultaneous occurrence of various SSc-specific autoantibodies has been detected in several studies in small subgroups of patients [84,85]. In particular, the joint occurrence of anti-centromere and anti-topoisomerase I antibodies was investigated. Depending on the ethnicity of the patients examined and the techniques used to determine the ANA subspecificities, divergent results were found. So far, no clear clinical cluster has been identified that is associated with the simultaneous presence of anti-centromere and anti-topoisomerase I autoantibodies. Currently, there is no proven pathophysiological concept that explains the relative mutual exclusivity of the autoantibodies. Hypotheses suggest that the different ANA subspecificities might be epiphenomena based on different environmental conditions or due to differences in antigen processing by B cells showing associations with particular HLA alleles [86–88].

Currently, due to increasing possibilities for quantitative measurements of ANA subspecificities, it has been investigated whether these measurements could provide additional information to predict disease progression. These analyses revealed evidence that levels of anti-topoisomerase I antibodies are associated with disease severity [89,90]. However, the results of large multicenter studies need to be awaited.

In addition to the well-established role of ANAs in diagnostics and predicting disease progression, data on a pathogenic role in disease onset or maintenance are rare in SSc. In 1996, Rudnicka et al. demonstrated an altered activity of the topoisomerase I enzyme in SSc and suggested to evaluate topoisomerase I inhibitors as a potential therapeutic approach [91]. Moreover, studies suggest a pathogenic role of anti-topoisomerase I autoantibodies. This assumption is based on the observation that anti-topoisomerase I autoantibodies bind to the cellular surface of fibroblasts [92]. Moreover, binding of anti-topoisomerase I autoantibodies to fibroblasts stimulates adhesion and activation of monocytes in vitro [93]. Another study brought further evidence for a pathogenicity of anti-topoisomerase I and anti-centromere autoantibodies by stimulating human dermal fibroblasts with these autoantibodies and, as a result, inducing an increase in pro-fibrotic markers and apoptosis resistance [94].

Although these findings do not necessarily prove a substantial role of ANA in disease pathogenesis, these data provide first evidence for the use of immunosuppressive treatment in early SSc.

3.2. Anti-Neutrophil Cytoplasmic Antibodies (ANCAs)

The term "anti-neutrophil cytoplasmic antibodies" (ANCAs) refers to autoantibodies against enzymes in primary granules of neutrophils or lysosomes of monocytes. ANCA subspecificities differ by indirect immunofluorescence, namely a cytoplasmic pattern (c-ANCA, e.g., autoantibodies against proteinase 3 (PR3)), a perinuclear pattern (p-ANCA, e.g., autoantibodies against myeloperoxidase (MPO)) and an atypical pattern showing aberrant patterns or a combination of c- and p-ANCA patterns (a-ANCA/x-ANCA). In addition to PR3 and MPO, ANCAs can be directed against further proteins such as elastase, cathepsin G, lactoferrin, α-enolase, catalase, azurocidin or actin [95]. The detection of ANCAs against PR3 and MPO is the hallmark of ANCA-associated vasculitis (AAV). In AAV, ANCAs lead to an activation of neutrophils primed by complement fragment C5a or cytokines (TNF-α, IL-1β), resulting in a translocation of PR3 or MPO to the cellular surface. After interaction between ANCAs with the target antigens PR3 or MPO and Fcγ receptors IIa or IIIb, intravascular near-wall degranulation of neutrophils and subsequent endothelial cell damage occurs [96–98]. ANCA-induced monocyte activation is mediated by a similar mechanism [99,100]. Subsequently, leukocyte migration and organ infiltration lead to secondary organ damage [101]. However, ANCAs can be detected in various other diseases (e.g., inflammatory bowel diseases, autoimmune hepatitis, primary sclerosing cholangitis), too [102].

Detection of ANCAs is a common phenomenon in SSc: the prevalence of ANCAs in SSc ranges up to 35% [103]. However, most studies report ANCA prevalence in 5–12% of SSc patients. The main antigenic targets are PR3 and MPO [104–106]. Although the presence of ANCAs is a common phenomenon in SSc, the development of systemic vasculitis in patients is rare [106,107]. The occurrence of systemic small-vessel vasculitis in SSc is grouped under the term SSc-AAV and is associated with a severe clinical phenotype involving the kidney, lung, peripheral and, rarely, the central nervous system, typically with a microscopic polyangiitis-like disease pattern [106]. In addition, patients may develop necrotizing vasculitis with critically reduced acral perfusion. Patients with renal manifestations typically present with pauci-immune glomerulonephritis or rapidly progressive glomerulonephritis. Patients with pulmonary involvement typically develop alveolar hemorrhage [106]. To date, there is limited evidence regarding promising treatment options for SSc-AAV. Randomized controlled trials are lacking. Possible treatment options that have been used so far include high-dose steroids, cyclophosphamide, rituximab, and plasma exchange. Before using high-dose steroids, the potential risk of developing a renal crisis should be considered in SSc [108]. In addition to autoantibodies against PR3 and MPO, autoantibodies with BPI and cathepsin G as antigenic targets have also been described in SSc [103]. Patients with ANCAs against BPI display lower skin scores.

3.3. *Anti-Phospholipid Antibodies (aPL)*

aPL comprise a group of antibodies directed against phospholipids and their cofactors, such as β2-glycoprotein I, prothrombin, annexin V and protein C or S. Phospholipids are components of the cell membrane and, together with the cofactors mentioned, play a role in hemostasis. In recent years, progress has been made in deciphering the pathomechanisms by which the binding of these autoantibodies to the protein–phospholipid complexes leads to increased blood coagulation, especially for the anti-β2-glycoprotein I antibodies.

aPL binding to β2GPI evoke a prothrombotic situation in endothelial cells through increased expression of adhesion molecules and tissue factor as well as reduced expression of the tissue factor pathway inhibitor [109,110]. In addition, complement activation is described [111]. Furthermore, in platelets, incubation with antibodies results in an activation of the glycoprotein IIa/IIIb (GPIIb/IIIa) receptor [112]. Moreover, activation of neutrophil granulocytes and monocytes by aPL has also been demonstrated [113–115].

aPL can be detected primarily without association to an underlying disease. In addition, aPL may be secondarily associated with autoimmune diseases such as systemic lupus erythematosus, rheumatoid arthritis, infections and cancers, but also with the use of medications such as oral contraceptives.

The detection of aPL also occurs frequently in SSc, with a prevalence of 9–20% [116]. In SSc, an association of anti-β2 glycoprotein 1 (β2GP1) positivity with active digital ulcerations has been demonstrated [117]. Moreover, a meta-analysis by Merashli et al. revealed an association between antibody positivity and pulmonary arterial hypertension, renal disease, thrombosis, miscarriage and digital ischemia compared to patients without aPL [118]. The association to venous thrombosis and miscarriage was confirmed in a meta-analysis by Sobanski et al. [116].

3.4. *Autoantibodies Recognizing G-Protein-Coupled Receptors, Growth Factors and Their Respective Receptors*

In addition to ANAs, ANCAs and aPL described in the previous sections, a new group of autoantibodies has recently been described which, in addition to their significance as biomarkers, are also becoming increasingly important in the pathophysiology of SSc. These comprise self-reactive autoantibodies recognizing GPCRs, growth factors and their respective receptors [26]. Anti-GPCR autoantibodies are present in healthy individuals and are thought to play a role in the regulation of immune cell homeostasis. These autoantibodies form a distinct, disease-specific network. It is hypothesized that this network reflects the patient's individual exposure to a specific environmental condition [28].

To investigate functional effects of the autoantibody network in a disease, several in vitro and in vivo technologies were established (Figure 2). In vitro stimulation of cell lines or isolated cells with purified IgG fractions of SSc patients and healthy donors is an established technology which led to the formation of autoantibody classes with the same cellular target, e.g., anti-endothelial cell autoantibodies (AECA) or anti-fibroblast autoantibodies (AFA). Since it has not yet been possible to purify target-specific autoantibodies, the effects of target-specific autoantibodies can only be studied by using receptor blockers. Applying this technology, Murthy et al. stimulated cells of the monocytic cell line THP-1 with IgG purified from patients with SSc. Stimulation with SSc–IgG induced a change to a pro-fibrotic and pro-inflammatory phenotype with IgG donor–individual alterations. Moreover, SSc–IgG induced pathways including AP-1, TAK/IKK-β/NFκB and ERK1/2, driving secretion of CCL18 and CXCL8 from stimulated cells [119].

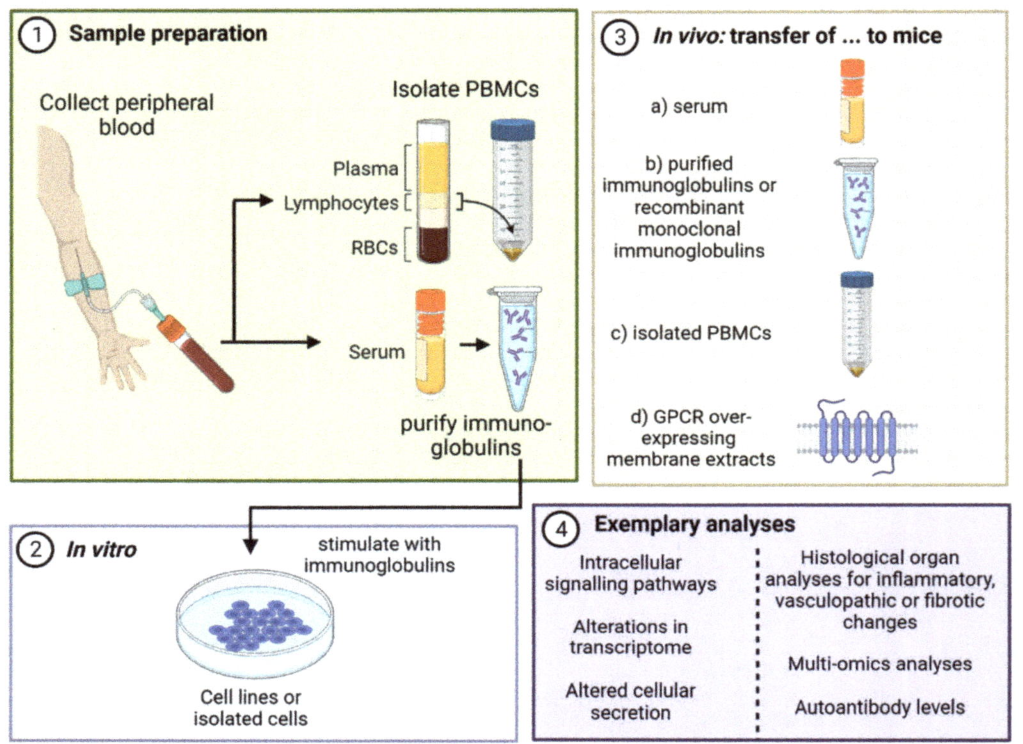

Figure 2. Established technologies to investigate functional effects of the autoantibodies in vitro and in vivo. The figure was created with 'Biorender.com'.

Moreover, several strategies have been established to transfer autoantibody-induced pathologies to animal models and build a platform to further investigate the respective pathomechanisms. These strategies involve the transfer of (a) serum [120], (b) IgG purified from serum [121] and (c) PBMC from diseased patients and healthy donors. A further strategy involves immunization with special agents, e.g., GPCR-overexpressing membrane extracts, (d), leading to the generation of autoantibodies and secondary to pathogenic effects [122,123]. These strategies were also applied to develop animal models of SSc mirroring autoantibody-mediated pathologies.

In 2014, Becker et al. transferred IgG purified from serum of SSc patients to C57BL/6J mice. These mice developed histological signs of an inflammatory pulmonary vasculopathy [124]. In addition, SSc–IgG positive for anti-AT1R and anti-ETAR autoantibodies induced increased neutrophil counts in bronchoalveolar fluid, pulmonary cellular infiltrations and cellular density after repeated passive transfer to C57BL/6J mice [125]. In a further study, a monoclonal AT1R autoantibody was injected into mice, which led to skin and lung inflammation. Interestingly, skin and lung inflammation did not occur in mice deficient in AT1Ra/b. This observation suggests a compelling involvement of autoantibody–receptor interaction in pathophysiological mechanisms [126].

In 2021, Yue et al. transferred PBMC of patients with SSc and granulomatosis with polyangiitis (GPA), as well as that of healthy donors, to Rag2$^{-/-}$/IL2rg$^{-/-}$ mice. Subsequently, mice engrafted with PBMC developed an ANA pattern similar to the respective donor [127]. In a subsequent study, performed by the same group, membrane-embedded human AT1R or empty membranes as controls were transferred to C57BL/6J mice [126]. Immunization with membrane-embedded human AT1R resulted in detectable levels of AT1R autoantibodies in mice and induced skin and lung inflammation as well as skin fibrosis.

These studies support the pathophysiological concepts of SSc described in the previous sections. Moreover, these studies provide animal models of SSc resembling human pathophysiology and, thus, these animal models enable evaluation of potential therapeutic approaches.

3.4.1. Functional Autoantibodies against GPCR

In the following, we summarize data on specific autoantibodies targeting GPCR. GPCRs form a large family of receptors in vertebrates and are characterized by seven transmembrane domains with intervening extracellular and intracellular amino acids. These receptors are named for their interaction with G-proteins, which mediate intracellular signal transduction. In addition to the G-proteins, G-protein-independent pathways for signal transduction have been described. In recent years, an increasing number of GPCRs that are recognized by autoantibodies have been identified. Studies have shown that the occurrence of these autoantibodies is not linked to diseases, but that physiological levels of autoantibodies that recognize GPCRs can also be detected in healthy controls [26]. Subsequently, altered levels of these autoantibodies were described for numerous diseases, e.g., solid organ or stem-cell transplantations, cardiovascular diseases, cancer, neurological, endocrine, pulmonary or rheumatic systemic diseases.

Anti-AT1R and Anti-ETAR Autoantibodies

Anti-AT1R Autoantibodies are well-investigated autoantibodies involved in the pathophysiology of several diseases, e.g., kidney transplant rejection, preeclampsia, diabetes mellitus [128], lupus nephritis [129] or COVID-19 disease [130]. Autoantibodies directed against AT1R were first described in 1999. Specifically, Wallukat et al. described the occurrence of autoantibodies that recognize AT1R in patients with preeclampsia [131]. AT1R is a GPCR whose binding of the endogenous ligand angiotensin results in Gq-mediated calcium release, β-arrestin-mediated cell signaling and the production of reactive oxygen species. Further research in 2005 demonstrated the role of these antibodies as biomarkers predicting rejection in kidney transplantation [132]. In the meantime, the measurement of AT1R autoantibodies in transplantation medicine is well-established in clinical routine. Furthermore, in addition to the role of AT1R autoantibodies as biomarkers, the pathophysiological mechanisms mediated by the antibodies in transplant rejection could be deciphered. Anti-AT1R autoantibodies bind to two different epitopes, namely AFHYESQ and ENTNIT, and act as allosteric agonists at the AT1R. Thus, they lead to sustained activation [132–134]. Interaction of the anti-AT1R autoantibody with the receptor results in vasoconstriction as well as the formation of proinflammatory, profibrotic and procoagulatory conditions in the microvascular circulation. The proinflammatory and procoagulatory processes are activated by the receptor activation-induced phosphorylation of extracellular signal-regulated kinase 1/2 (ERK1/2) and subsequent activation of the transcription factors activator protein 1 (AP-1) and nuclear factor kB (NFκB) in the endothelium and smooth muscle vascular cells. Expression of AT1R on the surface of immune cells, especially polymorphonuclear leukocytes, monocytes, T and B lymphocytes, results in pro-inflammatory gene expression (IL-1, IL-6, IL-8, IL-17, TNF-α, IFN-γ), which may even maintain AT1R expression on endothelial cells [135–137]. Moreover, microvascular inflammation and vasoconstriction induced by anti-AT1R autoantibodies promotes thrombosis, endarteritis and fibrinoid necrosis. Here, the interaction between receptor and autoantibody depends on the AT1R expression level. The AT1R expression level varies due to genetic and non-genetic factors. Genetic factors involve polymorphisms in the AT1R gene [138]. Non-genetic factors that increase AT1R expression are inflammation, endothelial damage or disturbances in microcirculation [136,139]. For further information on anti-AT1R autoantibodies in kidney transplant rejection, we refer to a review published by Sorohan et al. [140].

Interestingly, similar mechanisms are assumed to be involved in the pathogenesis of preeclampsia. Here, the epitope, targeted by anti-AT1R autoantibody, is AFHYESQ which has also been described in kidney transplant rejection [131]. A further mechanism proposed for mediating effects of anti-AT1R autoantibodies in preeclampsia is the long-term presence of anti-AT1R autoantibodies which reduces aldosterone production in vitro [141].

The similarity of the autoantibody binding sites and the secondary processes in kidney transplant rejection and preeclampsia suggest similar pathophysiologic processes in other diseases. However, the epitope to which anti-AT1R autoantibodies—which mediate pathologies—bind in SSc has not yet been identified.

In SSc, elevated levels of angiotensin II (AngII) and endothelin-1 (ET-1) were detected in blood and tissue samples [142,143]. Therefore, the corresponding receptors were discussed as potential therapeutic targets; endothelin-1 receptor blockers are recommended for treatment of pulmonary arterial hypertension [144]. Moreover, bosentan showed beneficial effects in the treatment of Raynaud's phenomenon and SSc-related digital ulcers [145–148].

In the majority of SSc patients, anti-AT1R and anti-ETAR autoantibodies are detectable. In addition, levels of anti-AT1R and anti-ETAR autoantibodies show a strong correlation with each other. Furthermore, both autoantibodies display the ability to exert functional activity at the respective receptor [28]. Regarding functional activity, in vitro experiments with IgG and receptor blockers for AT1R and ETAR proved that anti-AT1R and anti-ETAR autoantibodies bind to endothelial cells, exert phosphorylation of extracellular signal-regulated kinase 1/2 and increase TGFβ gene expression [149]. Moreover, Kill et al. demonstrated that anti-AT1R and anti-ETAR autoantibodies activate human microvascular endothelial cells in vitro, promoting secretion of proinflammatory chemokines and increased expression of adhesion molecules, enabling migration of neutrophils through an endothelial cell layer. Furthermore, using the same experimental setup, Kill et al. induced profibrotic processes in fibroblasts [125]. Günther et al. investigated the effects of PBMC stimulation with IgG from SSc patients and healthy controls, revealing an increased induction of IL-8 and CCL18 by SSc–IgG compared to healthy controls. These effects could be diminished by adding AT1R and ETAR blockers to the experimental setup [150].

In addition, anti-AT1R and anti-ETAR autoantibodies amplified vasoconstrictive effects of Ang II and ET-1 in small-vessel myography of intralobar pulmonary rat artery ring segments [124]. The amplification of Ang II- and ET-1-mediated effects by anti-AT1R and anti-ETAR autoantibodies was confirmed in in vitro analyses using the technology "dynamic mass redistribution" [126].

High levels of these autoantibodies showed an association with severe disease complications. Anti-ETAR autoantibodies—together with acute digital ulcers or ulcers in a patient's history—can be used to predict development of subsequent digital ulcers [151]. Moreover, anti-AT1R and anti-ETAR autoantibodies can predict the development of pulmonary arterial hypertension in SSc. Moreover, comparing levels of anti-AT1R and anti-ETAR autoantibodies in forms of pulmonary hypertension revealed highest levels for pulmonary arterial hypertension in SSc and connective tissue diseases. The same study revealed anti-AT1R and anti-ETAR autoantibodies as predictors of mortality in SSc [124]. Further information on the role of anti-AT1R and anti-ETAR autoantibodies are summarized in a review by Cabral-Marques and Riemekasten [152].

Anti-Muscarinic-3 Acetylcholine Receptor (M3R) Autoantibodies

Anti-M3R autoantibodies have been associated with intestinal dysmotility: a cardinal pathological condition and cause of the most gastrointestinal manifestations in patients with SSc [153]. Kumar et al. suggested a model of sequentially developing dysmotility: anti-M3R autoantibodies initially inhibit the release of acetylcholine (Ach) at the myenteric cholinergic neurons, inducing neuropathy through the blockage of cholinergic neurotransmission. Consequently, myopathy develops as a result of an anti-M3R autoantibody-related inhibition of Ach at the smooth muscle cells of the gastrointestinal tract. Smooth muscle fibrosis and atrophy ensue [154]. It can thus be hypothesized that an early and sustained elimination of anti-M3R autoantibodies could possibly lead to a reversal of SSc-associated dysmotility at the neuropathic and myopathic stages. Indeed, Kumar et al. presented evidence that application of IVIGs decreases binding intensity of anti-M3R autoantibodies and, thus, could probably decrease SSc-related gastrointestinal symptoms [154]. Accordingly, Raja et al. confirmed a significant improvement of gastrointestinal symptoms after repeated IVIG administration [155].

Anti-CXCR3 and Anti-CXCR4 Autoantibodies

Weigold et al. showed that autoantibody levels differ among subgroups of patients suffering from SSc, with dcSSc patients having the highest levels of autoantibodies directed to the N-terminal domain of CXCR3 and CXCR4. Comparable to anti-AT1R and anti-ETAR autoantibodies, anti-CXCR3 and anti-CXCR4 autoantibody levels also correlate with one another. Moreover, in SSc patients with interstitial lung disease, levels of autoantibodies directed to the N-terminus of CXCR3 and CXCR4 were lower in patients with progressive disease than in patients with stable disease [156]. Therefore, these autoantibodies might be a valuable tool to predict disease course of SSc patients with interstitial lung disease. A proposed hypothesis for an association between low autoantibody levels and disease progression is that the corresponding autoantibodies might be predominantly present in the tissues and, accordingly, detection of autoantibody levels in the blood may give lower levels [28]. So far, however, no studies have been conducted to bring evidence to this hypothesis. Currently, it is unclear whether autoantibodies directed to CXCR3 and CXCR4 exhibit functional activity. As shown for the AT1R autoantibody, functional activity of an autoantibody depends on the respective epitope. Therefore, Recke et al. applied a peptide-based epitope mapping for CXCR3. In this analysis, they could show differences in epitopes of anti-CXCR3 autoantibodies between SSc and healthy controls. Whereas autoantibodies from SSc patients preferentially bind to intracellular CXCR3 epitopes, autoantibodies of healthy controls bind to epitopes in the N-terminal domain [157].

Anti-PAR-1 Autoantibodies

Further potentially interesting autoantibodies in SSc are anti-PAR-1 autoantibodies. A first hint that anti-PAR-1 autoantibodies might exhibit functional activity in SSc was based on the observation that IL-6 release of HMECs stimulated with IgG from SSc patients could be reduced by a PAR-1 inhibitor. Moreover, stimulation of HMECs with IgG from SSc patients resulted in an increased expression of phosphorylated pAKT, p70S6K and pERK1/2, whereas this increase in expression was not observed after stimulation with IgG from healthy controls. Further experiments revealed an increased transcriptional activity of c-FOS and AP-1 after stimulation with SSc–IgG, which finally results in IL-6 mRNA expression and, subsequently, secretion of IL-6. To sum this up, anti-PAR-1 autoantibodies resemble the signaling pathway of thrombin, one of the natural ligands, which also leads to IL-6 secretion after binding to PAR-1 [158]. However, currently, the role of IL-6 in scleroderma renal crisis is obscure. Additional evidence for the pathomechanism described here provides a study showing an improvement in creatinine levels during therapy with tocilizumab in scleroderma renal crisis [159].

4. Which Therapeutic Approaches Have Been Evaluated to Target Autoantibody-Mediated Pathologies in SSc?

Various attempts have been conducted to prevent the pathogenic effects mediated by autoantibodies in diseases. As autoantibodies represent a cross-disease pathomechanism, therapeutic approaches developed in the field of other diseases might be transferrable to the treatment of SSc. In the following, we list possible therapeutic approaches that have already been investigated in SSc or could be applied to SSc in the future (Figure 3).

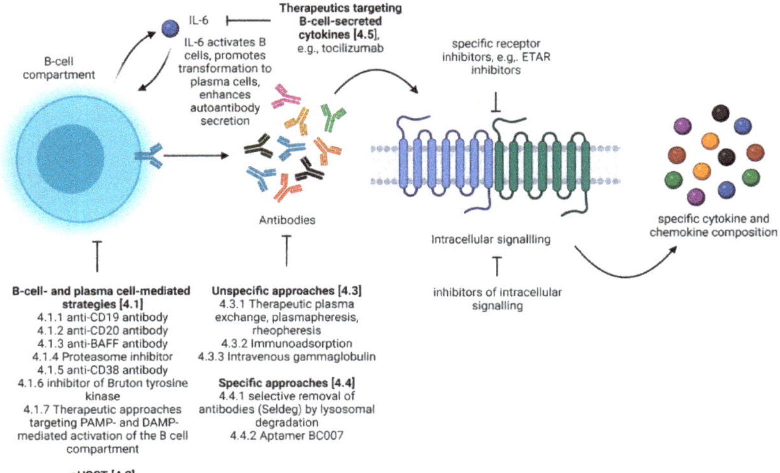

Figure 3. Potential therapeutic approaches for the treatment of autoantibody-induced pathologies in SSc. The figure was created with 'Biorender.com'.

4.1. B-Cell- and Plasma Cell-Mediated Strategies

As described, B cells play a critical role also in autoimmune diseases that are traditionally viewed as T-cell-mediated. Therefore, B-cell-targeting drugs are an emerging research area leading to the development of different therapeutic strategies. These include the elimination of defined cell subsets of the B-cell compartment, the neutralization of B-cell survival factors (e.g., BAFF and APRIL) and the prevention of the formation of ectopic germinal centers using antibodies against the lymphotoxin-β receptor.

4.1.1. Anti-CD19 Antibody

Cell subsets of the B-cell compartment express different surface proteins, so that a depletion of defined B-cell subsets can be achieved by targeting a specific protein. A potential target applying this mechanism of action is CD19, a B-cell surface antigen expressed from pre-B cells through plasmablasts and in some plasma cells. CD19 is targeted by inebilizumab/MEDI-551, a humanized monoclonal antibody that mediates ADCC. The application of MEDI-551 was investigated in a phase I multicenter, randomized, double-blind, placebo-controlled, single escalating dose study in SSc. One of the 24 study participants receiving treatment with MEDI-551 died during the course of the study from a cause which was not attributed to the drug. In general, MEDI-551 displayed a tolerable safety profile and achieved a dose-dependent depletion of circulating B cells and plasma cells. Assessments of the mRSS suggest a beneficial therapeutic effect of MEDI-551 on skin fibrosis [160].

4.1.2. Anti-CD20 Antibody

One of the best-studied B-cell-targeting therapeutics in SSc is rituximab. Rituximab is a chimeric anti-CD20 antibody. CD20 is expressed from cells of the pre-B cell stage to the pre-plasma cell stage. Thus, rituximab achieves a long-lasting, almost complete B-cell depletion in the blood and tissues. Six months after therapy initiation with rituximab, only a small number of CD19+ cells are detectable, which are predominantly IgA plasmablasts of mucosal origin [161]. Therapeutic efficacy of rituximab was assessed for several disease manifestations in SSc. Recently, Zamanian et al. investigated efficacy for the treatment of pulmonary arterial hypertension in SSc compared to placebo. Though the study failed to reach the primary endpoint, namely improvement in 6 min walk distance (6MWD) 24 weeks after treatment initiation, patients treated with rituximab had a significantly improved 6MWD after 48 weeks of treatment (n = 58, [162]). In addition, the therapeutic potential of rituximab on SSc-associated ILD was assessed in two small randomized controlled trials versus placebo (n = 16, [163]) and versus cyclophosphamide (n = 60, [164]) and several small non-controlled retrospective and prospective trials. The results of these two randomized controlled trials and 18 further studies or conference abstracts on rituximab in SSc-associated ILD were analyzed in a meta-analysis in 2021 by Goswami et al. (n = 575, [165]). This meta-analysis proves a significant improvement of FVC and DLCO under treatment with rituximab. Moreover, patients treated with rituximab exhibited less infectious complications than control patients. A key adverse event during treatment with rituximab remains a high potential for allergic reactions. Therefore, second-generation anti-CD20 antibodies (e.g., ocrelizumab, obinutuzumab, veltuzumab and ofatumumab) have been developed. These second-generation anti-CD20 antibodies are humanized or even fully human and have a higher therapeutic potential compared to rituximab in vitro [166]. Currently, none of these second-generation anti-CD20 antibodies have been evaluated in SSc.

A major obstacle in therapy with rituximab is the persistence of autoreactive long-lived plasma cells, which can produce autoantibodies and, thus, sustain the autoimmune disease. In addition, Mahévas et al. showed that B-cell depletion promotes differentiation from short-lived to long-lived autoimmune plasma cells by altering the splenic milieu [167].

4.1.3. Anti-BAFF Antibody

Rituximab treatment triggers the secretion of B-cell-activating factor (BAFF), which perpetuates autoreactive B cells in systemic lupus erythematosus [168]. Therefore, a possibility to prevent the persistence of long-lived plasma cells could be a combination therapy with a monoclonal anti-BAFF antibody (Belimumab). An interesting effect of therapy with BAFF inhibitors is that these depleted B effector cells secreting IL-6, but did not lead to a depletion of regulatory B cells [169]. Belimumab is already approved for the treatment of systemic lupus erythematosus. A phase II study investigating the effects of belimumab and mycophenolic acid versus placebo and mycophenolic acid was conducted in 20 patients with dcSSc. Patients who received belimumab did not develop significant improvement in skin thickness compared to the placebo group. Patients who responded to treatment with belimumab showed a decrease in the expression of profibrotic genes [170]. A phase II trial investigating the effects of a combination therapy of rituximab and belimumab is registered (NCT03844061).

4.1.4. Proteasome Inhibitor

An alternative therapeutic strategy that leads to a depletion of plasma cells is the use of proteasome inhibitors (e.g., bortezomib). Proteasome inhibitors are approved for the treatment of multiple myeloma. Early studies evaluating the use of proteasome inhibitors in mouse models of SSc were based on the premise of antifibrotic properties. This assumption was based on in vitro experiments on human fibroblasts, which showed a decrease in collagen production and an increase in collagen degradation when proteasome inhibitors were applied [171]. In a study conducted on the effects of proteasome inhibitors in mouse models of SSc, there was no effect on lung inflammation, lung fibrosis or skin fibrosis [172]. A phase II study was conducted comparing the effects on skin fibrosis and forced vital capacity in lung function as well as the tolerability of bortezomib and mycophenolate versus placebo with mycophenolate. The results were better for the combination therapy of bortezomib and mycophenolate (NCT02370693). In addition, a study is currently underway to investigate the safety and tolerability of oral ixazomib in scleroderma-related lung disease patients. This study is currently in recruitment status (NCT04837131). When applying proteasome inhibitors, however, it must be taken into account that depletion of the plasma cell compartment can lead to serious side effects.

4.1.5. Anti-CD38 Antibody

In 2019, Benfaremo et al. proposed an alternative treatment trial with an anti-CD38 antibody (e.g., daratumumab) for the treatment of SSc. CD38 is expressed on plasmablasts and plasma cells and might be useful in depletion of antibody-producing cells [173].

4.1.6. Inhibitor of Bruton Tyrosine Kinase

Ibrutinib is an irreversible inhibitor of Bruton tyrosine kinase, which is involved in intracellular signaling in B lymphocytes. Ibrutinib prevents signaling through the BCR, which promotes cell apoptosis and disrupts B-cell adhesion and B-cell migration. In addition, ibrutinib downregulates the expression of CD20 by targeting the CXCR4/SDF1 axis [174]. Ibrutinib is used to treat B-cell malignancies. In SSc, one study investigated possible effects of ibrutinib on B-cell pathologies in vitro. In this study, ibrutinib reduced the production of IL-6 and TNF-α by B effector cells. With the application of only small doses of ibrutinib, the function of regulatory B cells could be preserved [175].

4.1.7. Therapeutic Approaches Targeting PAMP- and DAMP-Mediated Activation of the B-Cell Compartment

Although the role of B-cell activation and the corresponding effects on antibody secretion via TLRs in the development and maintenance of SSc are poorly understood, this mechanism may also represent an interesting therapeutic strategy. In particular, this therapeutic strategy might be interesting because PAMP and DAMP contribute to the pathogenesis of SSc via further mechanisms. There are several therapeutic strategies that target PAMP- and DAMP-mediated activation of the B-cell compartment. These can either interfere with the interaction between ligand and TLR or with downstream signaling pathways. A summary of possible therapeutic strategies in SSc was provided by O'Reilly in 2018 [23].

4.2. Autologous Hematopoietic Stem-Cell Transplantation (aHSCT)

aHSCT is thought to eliminate autoreactive T and B cells by high-dose immunosuppression. Afterwards, reinfusion of autologous hematopoietic stem cells promotes reconstitution of a naïve, self-tolerant immune system. Thereby, aHSCT alters adaptive and innate immune systems [176]. Regarding the B-cell compartment, aHSCT induces a shift from memory B cells to naïve B cells and an increase in frequency of regulatory B cells [177,178]. Moreover, measurements of anti-topoisomerase I autoantibody levels after aHSCT reveal a decline [83].

aHSCT is recommended as a treatment option for patients with severe and rapidly progressive SSc refractory to immunosuppressive therapy. The first aHSCT was performed in 1997 in SSc [179]. Since then, around 500 aHSCTs have been reported [180] and more than 1000 SSc patients have been transplanted worldwide [181]. Furthermore, three randomized controlled trials have shown superiority of aHSCT versus intravenous cyclophosphamide therapy (ASSIST trial [182], ASTIS trial [183], SCOT trial [184]). In addition, a further non-interventional study (NISSC) confirmed the therapeutic efficacy of aHSCT [185]. SSc patients undergoing aHSCT yield significant improvements in survival, quality of life, skin fibrosis and lung function. However, patients undergoing aHSCT have an increased risk of infectious complications, especially for CMV reactivations and mycotic infections. A marker for the development of infectious complications is a lower number of B cells before aHSCT [186]. Further adverse events include the risk of engraftment syndrome and secondary autoimmune disorders [187]. Currently, different regimens have been compared to outweigh therapeutic efficacy and adverse events. So far, specific recommendations regarding transplant procedures are missing [181]. Recently, the German Society for Rheumatology (DGRh) suggested criteria for patient selection [188]. Moreover, optimal timing of aHSCT in SSc is still under discussion [180]. Therefore, timing of aHSCT is currently investigated in the UPSIDE study (NCT04464434).

4.3. Unspecific Approaches for the Removal of Antibodies

4.3.1. Therapeutic Plasma Exchange, Plasmapheresis and Rheopheresis

The term "therapeutic plasma exchange", also called "therapeutic apharesis", describes a procedure where the patient's blood is filtered and replaced, e.g., by albumin or fresh frozen plasma. Plasmapheresis is a related procedure removing less than 15% of blood volume and, thus, does not require fluid replacement. These technologies remove autoantibodies, immune complexes, cytokines or adhesion molecules from the blood [189]. Therapeutic plasma exchange removes approximately 65% of potential circulating pathogenic factors [190]. In SSc, therapeutic plasma exchange or plasmapheresis has been reported in more than 500 patients [189]. Evaluation of these technologies was conducted as case studies, small observational studies or in one of three prospective randomized clinical trials. Studies present mainly a favorable therapeutic effect on skin fibrosis, musculoskeletal symptoms, Raynaud's phenomenon, healing of digital ulcers and organ manifestation.

For the treatment of SSc, plasmapheresis is often combined with further therapeutics such as ACE inhibitors or immunosuppressive agents, e.g., with prednisolone alone or in a triple therapy with oral [191] or intravenous cyclophosphamide [192]. Further alternatives are combination therapies with IVIGs. This combination of therapies makes it difficult to compare study results and to estimate the therapeutic effects that can be achieved through plasma exchange or plasmapheresis.

Newer protocols suggest rheopheresis, a double-filtration plasmapheresis, which does not require replacement of fluid and, thus, reduces the risk of anaphylaxis. Rheopheresis is usually applied in conditions with microcirculatory alterations because of its beneficial effect on blood and plasma viscosity, erythrocyte deformability and aggregation. Therefore, rheopheresis was thought to have a beneficial effect on Raynaud's phenomenon and digital ulcers. However, data regarding effects of rheopheresis on microcirculation in SSc are contradictory [193,194].

Adverse events as a result of applying therapeutic plasma exchange, plasmapheresis or rheopheresis are rare [195,196]. Main adverse events to consider are complications due to venous access, hypocalcemia and hypovolemia. Excluding SSc, plasmapheresis reduced autoantibodies in serum and also in cerebrospinal fluid in several diseases [197].

4.3.2. Immunoadsorption

An alternative therapeutic procedure for the removal of autoantibodies from a patient's blood is a technology called immunoadsorption. In immunoadsorption, plasma and blood cells are separated in an extracorporeal circuit. The plasma is then passed through a high-affinity column. This procedure enables almost complete removal of human immunoglobulins and immune complexes from the patient's blood. Immunoadsorption is thus more effective than plasma exchange or plasma pheresis. However, this can also lead to a greater decrease in non-pathogenic immunoglobulins. So far, there are no studies that have investigated immunoadsorption in SSc. Evidence for the application of immunoadsorption in connective tissue diseases was reviewed by Hohenstein et al. [198].

4.3.3. Intravenous Gammaglobulin (IVIg)

IVIg is a blood product prepared from the serum of a great number of donors. Although the mechanism of action is not fully understood, a "high-dose" administration of 2 g/kg/month appears to have immunomodulatory effects, which is why the use of IVIGs in a number of autoimmune-mediated diseases has become widespread in recent decades [199,200]. The immunomodulatory mechanisms of action include inhibition of T-cell proliferation, modulation of apoptosis, inhibition of superantigen-mediated activation of T cells, inactivation of inflammatory factors such as TNF-a and IL-1α, effect on cytokine levels, inhibition of phagocytosis, enhancement of the catabolism of autoantibodies, effect on glucocorticoid receptor-binding affinity and inhibition of deposition of classical pathway components [119,201]. In SSc, in addition to the reduction of gastrointestinal symptoms, immunoglobulins can also help in decreasing skin thickness and reduce arthromyalgia and muscle weakness [202,203].

4.4. Specific Approaches for the Removal of Antibodies

4.4.1. Selective Removal of Autoantibodies by Lysosomal Degradation

Several mechanisms are involved in the regulation of antibody concentrations in the human body. An interesting mechanism involves the neonatal Fc receptor (FcRn) which has a structure similar to the major histocompatibility complex (MHC) class I and also associates with β2-microglobulin. The FcRn is involved in the transport of maternal IgG to the fetus. This receptor regulates the transport of not only antibodies, but also other serum proteins such as albumin, within and across cells, by binding IgG at pH < 6.5 in early endosomes and releasing IgG at neutral pH, enabling IgG recycling. As a result, FcRn reduces lysosomal IgG degradation and, thus, modulates IgG concentrations in serum and throughout the body. Whereas engineering of the variable regions of IgG is a widely used approach to develop therapeutically effective antibodies, modulation of the Fc region of IgG represents an emerging research area. E.g., the Fc part of tixagevimab and cilgavimab (Evusheld, formerly known as AZD7442), which is used for prophylaxis of developing symptomatic COVID-19 disease in patients with increased risk of insufficient response to vaccination, was mutated to extend half-life [204]. Moreover, several drugs were fused to Fc portions to increase half-lives through FcRn-mediated recycling, e.g., etanercept (Enbrel) or abatacept (Orencia) [205]. Regarding autoantibodies, new therapeutic strategies focus on the blocking of FcRn–IgG interaction to reduce autoantibody concentrations. Summarized under the term "Abdeg" technology, Fc fragments and antibodies were engineered to inhibit FcRn–IgG interaction [206]. In 2021, the first Fc-based inhibitor, namely efgartigimod (Vyvgart), was approved for treatment of generalized myasthenia gravis by the USA Food and Drug Administration (FDA). Moreover, this technology was evaluated in further diseases, e.g., primary immune thrombocytopenia or pemphigus vulgaris and foliaceus [207–211].

Another emerging therapeutic technology is summarized by the term "Seldeg", an abbreviation for "selective degradation". Seldegs enable the selective degradation of antigen-specific antibodies by binding, on one hand, to cell surface molecules via the targeting component and, on the other hand, to antigen-specific antibodies by the antigen component. After binding both components, the complex of the antigen-specific antibody and Seldeg is internalized and degraded in lysosomes. In animal models of several diseases, Seldegs were investigated. In an animal model of antibody-mediated exacerbation of experimental autoimmune encephalomyelitis, Seldegs binding to exposed phosphatidylserine or to FcRn were compared. Both Seldegs led to an amelioration of disease severity [212]. A key advantage of Seldegs compared to FcRn inhibitors is that Seldegs do not reduce antibody levels that are not target specific. Therefore, immunosuppressive effects of current treatments were not observed.

4.4.2. Selective Removal of Autoantibodies Using Aptamer BC007

Aptamers are artificially created, short, single-stranded DNA or RNA oligonucleotides that are capable of binding to a specific molecule with a high affinity [213]. The ability to digitally share their sequence information and produce them using a template enables fast and economical manufacturing [214]. Their heat resistance provides an additional advantage in processing, transporting and sterilizing the aptamers [215]. Aptamers can not only be useful diagnostic tools for recognizing pathogens and cancer or for tracking environmental contamination, but they also exhibit therapeutic uses [216]. Pegaptanib sodium (Macugen; Eyetech Pharma/Pfizer), a drug against age-related macular degeneration, is an RNA aptamer against vascular endothelial growth factor 165 (VEGF-165) and the only so-far-approved aptamer drug [217]. However, many other aptamers are currently being studied and their application could revolutionize the treatment of many diseases. BC007, which was originally developed as a thrombin inhibitor [218], is the only known aptamer that is able to neutralize functional antibodies against GPCRs. Haberland et al managed to neutralize antibodies against GPCRs that had been linked to cardiovascular pathologies such as dilated cardiomyopathy (DCM) and Chagas' cardiomyopathy in vivo through the application of BC007 [219]. Wallukat et al. demonstrated the neutralizing efficiency of BC007 in serum of patients with DCM, where this serum was treated with the aptamer ex vivo. Moreover, antibody neutralization was achieved in vivo in spontaneously hypertensive rats [220]. Antibodies against GPCRs have also recently been associated with COVID-19 severity [221]. Hohberger et al. published a case report of a patient with long-COVID syndrome and positivity for antibodies against GPCRs to whom BC007 was applied intravenously. After a single application and during the subsequent four-week observation, an inactivation of GPCRs and a sustained improvement of fatigue, taste and retinal capillary microcirculation was detected [222]. Moreover, in fibrosis, a key feature of SSc, aptamers targeting downstream TGF-β signaling have been investigated [223]. These aptamers could also be a potential therapeutic approach for the treatment of SSc.

4.5. Therapeutics Targeting B-Cell-Secreted Cytokines

Tocilizumab is a humanized monoclonal antibody inhibiting the binding of IL-6 to the membrane and soluble IL-6 receptor. Therapeutic efficacy of tocilizumab has been investigated in a phase II (faSScinated study, $n = 87$ [224]) and a phase III study (focuSSced study, $n = 212$ [225]). Both studies did not reach their primary endpoint: mRSS improved under therapy with tocilizumab without reaching statistical significance compared to placebo. However, in both studies, tocilizumab preserved lung functionality with a significantly smaller decline in FVC than with placebo. Patients with short disease duration and elevated inflammatory markers responded to therapy with tocilizumab. The most common adverse events under therapy with tocilizumab in both studies were infections. Based on these results, the FDA approved tocilizumab for treatment of SSc-associated ILD in March 2021. Though both trials could not demonstrate significant improvement in mRSS compared to placebo, in vitro treatment of cultured skin fibroblasts from SSc patients with tocilizumab revealed a normalization of the genetic profile resulting in an inactive molecular and functional fibroblast phenotype [226]. For further information on the role of IL-6 in the pathophysiology of SSc and the potential therapeutic efficacy of the IL-6 inhibitor tocilizumab in further studies, we refer to a recently published review by Cardoneanu et al. [227].

5. Conclusions

So far, the mechanisms leading to the development of autoreactivity in the B-cell compartment and, secondarily, to the formation of autoantibodies are insufficiently understood. Similarly, investigations regarding the mechanisms by which autoantibodies functionally intervene in the pathology of SSc are still in their infancy. We expect that research on autoreactive B cells and autoantibodies will identify further autoantibodies that might serve as biomarkers. In addition, the understanding of autoantibody-mediated pathologies will grow. As presented, autoreactive B cells and autoantibodies represent ubiquitous pathomechanisms; findings obtained in the context of a specific disease will be transferable to others. Moreover, an increased understanding of pathomechanisms will enable the identification of promising new therapeutic approaches, preventing harm through pathogenic autoantibodies, but preserving beneficial effects of natural autoantibodies. As autoantibody-mediated processes are among the mechanisms that occur early in SSc, it is crucial to identify patients in the early stages of the disease, treat them with the listed therapeutics, and evaluate the potential to halt the development of full-blown disease. Ultimately, future technologies could enable the application of precision medicine in the treatment of B-cell-mediated pathologies: patient-specific autoantibodies mediating pathogenic effects could be identified enabling a patient-specific removal of autoantibodies. However, autoreactive B cells and autoantibodies represent only one field of the pathogenesis of SSc. An investigation of the intertwining of autoantibody-mediated pathomechanisms with other pathogenetic factors in SSc will be a major field of research in the future.

Author Contributions: Conceptualization, H.G.; writing—original draft preparation, H.G.; writing—review and editing, K.F., S.C. and G.R.; visualization, H.G., K.F. and S.C.; supervision, G.R. All authors have read and agreed to the published version of the manuscript.

Funding: This research received funding by the excellence cluster 'Precision Medicine in Chronic Inflammation' and by the 'University of Lübeck'.

Institutional Review Board Statement: Not applicable.

Informed Consent Statement: Not applicable.

Conflicts of Interest: The authors declare no conflict of interest.

References

1. Zmorzyński, S.; Wojcierowska-Litwin, M.; Kowal, M.; Michalska-Jakubus, M.; Styk, W.; Filip, A.A.; Walecka, I.; Krasowska, D. NOTCH3 T6746C and TP53 P72R Polymorphisms Are Associated with the Susceptibility to Diffuse Cutaneous Systemic Sclerosis. *BioMed Res. Int.* **2020**, *2020*, 1–9. [CrossRef]
2. Hanson, A.L.; Sahhar, J.; Ngian, G.-S.; Roddy, J.; Walker, J.; Stevens, W.; Nikpour, M.; Assassi, S.; Proudman, S.; Mayes, M.D.; et al. Contribution of HLA and KIR Alleles to Systemic Sclerosis Susceptibility and Immunological and Clinical Disease Subtypes. *Front. Genet.* **2022**, *13*. [CrossRef]
3. Watad, A.; Rosenberg, V.; Tiosano, S.; Tervaert, J.W.C.; Yavne, Y.; Shoenfeld, Y.; Shalev, V.; Chodick, G.; Amital, H. Silicone breast implants and the risk of autoimmune/rheumatic disorders: A real-world analysis. *Int. J. Epidemiol.* **2018**, *47*, 1846–1854. [CrossRef]
4. Kassamali, B.; Kassamali, A.A.; Muntyanu, A.; Netchiporouk, E.; Vleugels, R.A.; LaChance, A. Geographic distribution and environmental triggers of systemic sclerosis cases from 2 large academic tertiary centers in Massachusetts. *J. Am. Acad. Dermatol.* **2021**, *86*, 925–927. [CrossRef]
5. Aguila, L.A.; da Silva, H.C.; Medeiros-Ribeiro, A.C.; Bunjes, B.G.; Luppino-Assad, A.P.; Sampaio-Barros, P.D. Is exposure to environmental factors associated with a characteristic clinical and laboratory profile in systemic sclerosis? A retrospective analysis. *Rheumatol. Int.* **2020**, *41*, 1143–1150. [CrossRef]
6. Lazzaroni, M.G.; Campochiaro, C.; Bertoldo, E.; De Luca, G.; Caimmi, C.; Tincani, A.; Franceschini, F.; Airò, P. Association of anti-RNA polymerase III antibody with silicone breast implants rupture in a multicentre series of Italian patients with systemic sclerosis. *Clin. Exp. Rheumatol.* **2021**, *39*, 25–28. [CrossRef]
7. Skiba, M.A.; Kruse, A.C. Autoantibodies as Endogenous Modulators of GPCR Signaling. *Trends Pharmacol. Sci.* **2020**, *42*, 135–150. [CrossRef]
8. Whitfield, M.L.; Finlay, D.R.; Murray, J.I.; Troyanskaya, O.G.; Chi, J.-T.; Pergamenschikov, A.; McCalmont, T.H.; Brown, P.O.; Botstein, D.; Connolly, M.K. Systemic and cell type-specific gene expression patterns in scleroderma skin. *Proc. Natl. Acad. Sci. USA* **2003**, *100*, 12319–12324. [CrossRef]
9. Roberson, E.D.; Carns, M.; Cao, L.; Aren, K.; Goldberg, I.A.; Morales-Heil, D.J.; Korman, B.D.; Atkinson, J.P.; Varga, J. RNA-Seq analysis identifies alterations of the primary cilia gene *SPAG17* and *SOX9* locus non-coding RNAs in systemic sclerosis. *Arthritis Rheumatol.* **2022**. [CrossRef]
10. Lafyatis, R.; O'Hara, C.; Feghali-Bostwick, C.A.; Matteson, E. B cell infiltration in systemic sclerosis–associated interstitial lung disease. *Arthritis Care Res.* **2007**, *56*, 3167–3168. [CrossRef]
11. Burbelo, P.D.; Gordon, S.M.; Waldman, M.; Edison, J.D.; Little, D.J.; Stitt, R.S.; Bailey, W.T.; Hughes, J.B.; Olson, S.W. Autoantibodies are present before the clinical diagnosis of systemic sclerosis. *PLoS ONE* **2019**, *14*, e0214202. [CrossRef]
12. Koenig, M.; Joyal, F.; Fritzler, M.J.; Roussin, A.; Abrahamowicz, M.; Boire, G.; Goulet, J.-R.; Rich, É.; Grodzicky, T.; Raymond, Y.; et al. Autoantibodies and microvascular damage are independent predictive factors for the progression of Raynaud's phenomenon to systemic sclerosis: A twenty-year prospective study of 586 patients, with validation of proposed criteria for early systemic sclerosis. *Arthritis Care Res.* **2008**, *58*, 3902–3912. [CrossRef]
13. Bellando-Randone, S.; Del Galdo, F.; Lepri, G.; Minier, T.; Huscher, D.; Furst, D.E.; Allanore, Y.; Distler, O.; Czirják, L.; Bruni, C.; et al. Progression of patients with Raynaud's phenomenon to systemic sclerosis: A five-year analysis of the European Scleroderma Trial and Research group multicentre, longitudinal registry study for Very Early Diagnosis of Systemic Sclerosis (VEDOSS). *Lancet Rheumatol.* **2021**, *3*, e834–e843. [CrossRef]
14. Rosser, E.C.; Mauri, C. Regulatory B Cells: Origin, Phenotype, and Function. *Immunity* **2015**, *42*, 607–612. [CrossRef]
15. Wardemann, H.; Yurasov, S.; Schaefer, A.; Young, J.W.; Meffre, E.; Nussenzweig, M.C. Predominant Autoantibody Production by Early Human B Cell Precursors. *Science* **2003**, *301*, 1374–1377. [CrossRef]
16. Nemazee, D. Mechanisms of central tolerance for B cells. *Nat. Rev. Immunol.* **2017**, *17*, 281–294. [CrossRef]
17. Getahun, A.; Smith, M.J.; Cambier, J.C. Mechanisms of Peripheral B Cell Tolerance. In *Encyclopedia of Immunobiology*; Academic Press: Oxford, UK, 2016; pp. 83–91. [CrossRef]
18. Hornung, V.; Rothenfusser, S.; Britsch, S.; Krug, A.; Jahrsdörfer, B.; Giese, T.; Endres, S.; Hartmann, G. Quantitative Expression of Toll-Like Receptor 1–10 mRNA in Cellular Subsets of Human Peripheral Blood Mononuclear Cells and Sensitivity to CpG Oligodeoxynucleotides. *J. Immunol.* **2002**, *168*, 4531–4537. [CrossRef]

19. Bourke, E.; Bosisio, D.; Golay, J.; Polentarutti, N.; Mantovani, A. The toll-like receptor repertoire of human B lymphocytes: Inducible and selective expression of TLR9 and TLR10 in normal and transformed cells. *Blood* **2003**, *102*, 956–963. [CrossRef]
20. Dorner, M.; Brandt, S.; Tinguely, M.; Zucol, F.; Bourquin, J.-P.; Zauner, L.; Berger, C.; Bernasconi, M.; Speck, R.; Nadal, D. Plasma cell toll-like receptor (TLR) expression differs from that of B cells, and plasma cell TLR triggering enhances immunoglobulin production. *Immunology* **2009**, *128*, 573–579. [CrossRef]
21. Browne, E.P. Regulation of B-cell responses by Toll-like receptors. *Immunology* **2012**, *136*, 370–379. [CrossRef]
22. O'Reilly, S. Toll-like receptor triggering in systemic sclerosis: Time to target. *Rheumatology* **2022**. [CrossRef] [PubMed]
23. O'Reilly, S. Toll Like Receptors in systemic sclerosis: An emerging target. *Immunol. Lett.* **2018**, *195*, 2–8. [CrossRef] [PubMed]
24. Frasca, L.; Lande, R. Toll-like receptors in mediating pathogenesis in systemic sclerosis. *Clin. Exp. Immunol.* **2020**, *201*, 14–24. [CrossRef] [PubMed]
25. Koelsch, K.; Zheng, N.-Y.; Zhang, Q.; Duty, A.; Helms, C.; Mathias, M.D.; Jared, M.; Smith, K.; Capra, J.D.; Wilson, P.C. Mature B cells class switched to IgD are autoreactive in healthy individuals. *J. Clin. Investig.* **2007**, *117*, 1558–1565. [CrossRef] [PubMed]
26. Marques, O.C.; Marques, A.; Giil, L.M.; De Vito, R.; Rademacher, J.; Günther, J.; Lange, T.; Humrich, J.Y.; Klapa, S.; Schinke, S.; et al. GPCR-specific autoantibody signatures are associated with physiological and pathological immune homeostasis. *Nat. Commun.* **2018**, *9*, 1–14. [CrossRef]
27. Siloşi, I.; Siloşi, C.A.; Boldeanu, M.V.; Cojocaru, M.; Biciuşcă, V.; Avrămescu, C.S.; Cojocaru, I.M.; Bogdan, M.; Folcuţi, R.M. The role of autoantibodies in health and disease. *Romanian J. Morphol. Embryol.* **2016**, *57*, 633–638.
28. Riemekasten, G.; Petersen, F.; Heidecke, H. What Makes Antibodies Against G Protein-Coupled Receptors so Special? A Novel Concept to Understand Chronic Diseases. *Front. Immunol.* **2020**, *11*, 564526. [CrossRef]
29. Devitt, A.; Marshall, L.J. The innate immune system and the clearance of apoptotic cells. *J. Leukoc. Biol.* **2011**, *90*, 447–457. [CrossRef]
30. Silverman, G.J.; Vas, J.; Grönwall, C. Protective autoantibodies in the rheumatic diseases: Lessons for therapy. *Nat. Rev. Rheumatol.* **2013**, *9*, 291–300. [CrossRef]
31. Elkon, K.; Casali, P. Nature and functions of autoantibodies. *Nat. Clin. Pract. Rheumatol.* **2008**, *4*, 491–498. [CrossRef]
32. Mehrani, T.; Petri, M. IgM Anti-ß$_2$ Glycoprotein I Is Protective Against Lupus Nephritis and Renal Damage in Systemic Lupus Erythematosus. *J. Rheumatol.* **2010**, *38*, 450–453. [CrossRef] [PubMed]
33. Mannoor, K.; Xu, Y.; Chen, C. Natural autoantibodies and associated B cells in immunity and autoimmunity. *Autoimmunity* **2013**, *46*, 138–147. [CrossRef]
34. Casali, P.; Schettino, E.W. Structure and Function of Natural Antibodies. *Immunol. Silicones* **1996**, *210*, 167–179. [CrossRef]
35. Duddy, M.E.; Alter, A.; Bar-Or, A. Distinct Profiles of Human B Cell Effector Cytokines: A Role in Immune Regulation? *J. Immunol.* **2004**, *172*, 3422–3427. [CrossRef] [PubMed]
36. Astorri, E.; Bombardieri, M.; Gabba, S.; Peakman, M.; Pozzilli, P.; Pitzalis, C. Evolution of Ectopic Lymphoid Neogenesis and In Situ Autoantibody Production in Autoimmune Nonobese Diabetic Mice: Cellular and Molecular Characterization of Tertiary Lymphoid Structures in Pancreatic Islets. *J. Immunol.* **2010**, *185*, 3359–3368. [CrossRef] [PubMed]
37. Nacionales, D.C.; Weinstein, J.S.; Yan, X.-J.; Albesiano, E.; Lee, P.Y.; Kelly-Scumpia, K.M.; Lyons, R.; Satoh, M.; Chiorazzi, N.; Reeves, W.H. B Cell Proliferation, Somatic Hypermutation, Class Switch Recombination, and Autoantibody Production in Ectopic Lymphoid Tissue in Murine Lupus. *J. Immunol.* **2009**, *182*, 4226–4236. [CrossRef]
38. Hampe, C.S. B Cells in Autoimmune Diseases. *Scientifica* **2012**, *2012*, 1–18. [CrossRef]
39. Naparstek, Y.; Plotz, P.H. The Role of Autoantibodies in Autoimmune Disease. *Annu. Rev. Immunol.* **1993**, *11*, 79–104. [CrossRef]
40. Senécal, J.-L.; Hoa, S.; Yang, R.; Koenig, M. Pathogenic roles of autoantibodies in systemic sclerosis: Current understandings in pathogenesis. *J. Scleroderma Relat. Disord.* **2019**, *5*, 103–129. [CrossRef]
41. Freitag, H.; Szklarski, M.; Lorenz, S.; Sotzny, F.; Bauer, S.; Philippe, A.; Kedor, C.; Grabowski, P.; Lange, T.; Riemekasten, G.; et al. Autoantibodies to Vasoregulative G-Protein-Coupled Receptors Correlate with Symptom Severity, Autonomic Dysfunction and Disability in Myalgic Encephalomyelitis/Chronic Fatigue Syndrome. *J. Clin. Med.* **2021**, *10*, 3675. [CrossRef]
42. Notkins, A.L. Polyreactivity of antibody molecules. *Trends Immunol.* **2004**, *25*, 174–179. [CrossRef] [PubMed]
43. Zhou, Z.-H.; Zhang, Y.; Hu, Y.-F.; Wahl, L.M.; Cisar, J.O.; Notkins, A.L. The Broad Antibacterial Activity of the Natural Antibody Repertoire Is Due to Polyreactive Antibodies. *Cell Host Microbe* **2007**, *1*, 51–61. [CrossRef]
44. Köhler, F.; Hug, E.; Eschbach, C.; Meixlsperger, S.; Hobeika, E.; Kofer, J.; Wardemann, H.; Jumaa, H. Autoreactive B Cell Receptors Mimic Autonomous Pre-B Cell Receptor Signaling and Induce Proliferation of Early B Cells. *Immunity* **2008**, *29*, 912–921. [CrossRef] [PubMed]
45. Neidhart, M.; Kuchen, S.; Distler, O.; Michel, B.A.; Gay, R.E.; Gay, S. Increased serum levels of antibodies against human cytomegalovirus and prevalence of autoantibodies in systemic sclerosis. *Arthritis Care Res.* **1999**, *42*, 389–392. [CrossRef]
46. Gkoutzourelas, A.; Liaskos, C.; Simopoulou, T.; Katsiari, C.; Efthymiou, G.; Scheper, T.; Meyer, W.; Tsirogianni, A.; Tsigalou, C.; Dardiotis, E.; et al. A study of antigen-specific anti-cytomegalovirus antibody reactivity in patients with systemic sclerosis and concomitant anti-Ro52 antibodies. *Rheumatol. Int.* **2020**, *40*, 1–11. [CrossRef]
47. Efthymiou, G.; Dardiotis, E.; Liaskos, C.; Marou, E.; Scheper, T.; Meyer, W.; Daponte, A.; Daoussis, D.; Hadjigeorgiou, G.; Bogdanos, D.P.; et al. A comprehensive analysis of antigen-specific antibody responses against human cytomegalovirus in patients with systemic sclerosis. *Clin. Immunol.* **2019**, *207*, 87–96. [CrossRef] [PubMed]

48. Gourh, P.; Safran, S.A.; Alexander, T.; Boyden, S.E.; Morgan, N.D.; Shah, A.A.; Mayes, M.D.; Doumatey, A.; Bentley, A.R.; Shriner, D.; et al. *HLA* and autoantibodies define scleroderma subtypes and risk in African and European Americans and suggest a role for molecular mimicry. *Proc. Natl. Acad. Sci. USA* **2019**, *117*, 552–562. [CrossRef] [PubMed]
49. Glauzy, S.; Olson, B.; May, C.K.; Parisi, D.; Massad, C.; Hansen, J.E.; Ryu, C.; Herzog, E.L.; Meffre, E. Defective Early B Cell Tolerance Checkpoints in Patients With Systemic Sclerosis Allow the Production of Self Antigen –Specific Clones. *Arthritis Rheumatol.* **2021**, *74*, 307–317. [CrossRef]
50. Matsushita, T.; Hamaguchi, Y.; Hasegawa, M.; Takehara, K.; Fujimoto, M. Decreased levels of regulatory B cells in patients with systemic sclerosis: Association with autoantibody production and disease activity. *Rheumatology* **2015**, *55*, 263–267. [CrossRef]
51. Steri, M.; Orrù, V.; Idda, M.L.; Pitzalis, M.; Pala, M.; Zara, I.; Sidore, C.; Faà, V.; Floris, M.; Deiana, M.; et al. Overexpression of the Cytokine BAFF and Autoimmunity Risk. *N. Engl. J. Med.* **2017**, *376*, 1615–1626. [CrossRef]
52. Yoshizaki, A. Pathogenic roles of B lymphocytes in systemic sclerosis. *Immunol. Lett.* **2018**, *195*, 76–82. [CrossRef] [PubMed]
53. Mavropoulos, A.; Simopoulou, T.; Varna, A.; Sakkas, L.I.; Liaskos, C.; Katsiari, C.G.; Bogdanos, D.P. Breg Cells Are Numerically Decreased and Functionally Impaired in Patients With Systemic Sclerosis. *Arthritis Rheumatol.* **2016**, *68*, 494–504. [CrossRef]
54. Ricard, L.; Malard, F.; Riviere, S.; Laurent, C.; Fain, O.; Mohty, M.; Gaugler, B.; Mekinian, A. Regulatory B cell imbalance correlates with Tfh expansion in systemic sclerosis. *Clin. Exp. Rheumatol.* **2021**, *39*, 20–24. [CrossRef]
55. Sakkas, L.I.; Daoussis, D.; Mavropoulos, A.; Liossis, S.-N.; Bogdanos, D.P. Regulatory B cells: New players in inflammatory and autoimmune rheumatic diseases. *Semin. Arthritis Rheum.* **2018**, *48*, 1133–1141. [CrossRef] [PubMed]
56. Sato, S.; Fujimoto, M.; Hasegawa, M.; Takehara, K. Altered blood B lymphocyte homeostasis in systemic sclerosis: Expanded naive B cells and diminished but activated memory B cells. *Arthritis Care Res.* **2004**, *50*, 1918–1927. [CrossRef]
57. Forestier, A.; Guerrier, T.; Jouvray, M.; Giovannelli, J.; Lefèvre, G.; Sobanski, V.; Hauspie, C.; Hachulla, E.; Hatron, P.-Y.; Zéphir, H.; et al. Altered B lymphocyte homeostasis and functions in systemic sclerosis. *Autoimmun. Rev.* **2018**, *17*, 244–255. [CrossRef]
58. Marrapodi, R.; Pellicano, C.; Radicchio, G.; Leodori, G.; Colantuono, S.; Iacolare, A.; Gigante, A.; Visentini, M.; Rosato, E. CD21low B cells in systemic sclerosis: A possible marker of vascular complications. *Clin. Immunol.* **2020**, *213*, 108364. [CrossRef]
59. Thorarinsdottir, K.; Camponeschi, A.; Gjertsson, I.; Mårtensson, I.-L. CD21$^{-/\text{low}}$B cells: A Snapshot of a Unique B Cell Subset in Health and Disease. *Scand. J. Immunol.* **2015**, *82*, 254–261. [CrossRef]
60. Thorarinsdottir, K.; Camponeschi, A.; Jonsson, C.; Granhagen Önnheim, K.; Nilsson, J.; Forslind, K.; Visentini, M.; Jacobsson, L.; Mårtensson, I.L.; Gjertsson, I. CD21$^{-/\text{low}}$ B cells associate with joint damage in rheumatoid arthritis patients. *Scand. J. Immunol.* **2019**, *90*, e12792. [CrossRef] [PubMed]
61. Saadoun, D.; Terrier, B.; Bannock, J.; Vazquez, T.; Massad, C.; Kang, I.; Joly, F.; Rosenzwajg, M.; Sene, D.; Benech, P.; et al. Expansion of Autoreactive Unresponsive CD21$^{-/\text{low}}$B Cells in Sjögren's Syndrome-Associated Lymphoproliferation. *Arthritis Care Res.* **2012**, *65*, 1085–1096. [CrossRef] [PubMed]
62. Wehr, C.; Eibel, H.; Masilamani, M.; Illges, H.; Schlesier, M.; Peter, H.-H.; Warnatz, K. A new CD21low B cell population in the peripheral blood of patients with SLE. *Clin. Immunol.* **2004**, *113*, 161–171. [CrossRef]
63. Wang, J.; Watanabe, T. Expression and function of Fas during differentiation and activation of B cells. *Int. Rev. Immunol.* **1999**, *18*, 367–379. [CrossRef]
64. Soto, L.; Ferrier, A.; Aravena, O.A.; Fonseca, E.; Berendsen, J.; Biere, A.; Bueno, D.; Ramos, V.; Aguillón, J.C.; Catalán, D.F. Systemic Sclerosis Patients Present Alterations in the Expression of Molecules Involved in B-Cell Regulation. *Front. Immunol.* **2015**, *6*, 496. [CrossRef] [PubMed]
65. Sato, S.; Hasegawa, M.; Fujimoto, M.; Tedder, T.F.; Takehara, K. Quantitative Genetic Variation in CD19 Expression Correlates with Autoimmunity. *J. Immunol.* **2000**, *165*, 6635–6643. [CrossRef]
66. Zhou, L.J.; Smith, H.M.; Waldschmidt, T.J.; Schwarting, R.; Daley, J.; Tedder, T.F. Tissue-specific expression of the human CD19 gene in transgenic mice inhibits antigen-independent B-lymphocyte development. *Mol. Cell. Biol.* **1994**, *14*. [CrossRef]
67. Asano, N.; Fujimoto, M.; Yazawa, N.; Shirasawa, S.; Hasegawa, M.; Okochi, H.; Tamaki, K.; Tedder, T.F.; Sato, S. B Lymphocyte Signaling Established by the CD19/CD22 Loop Regulates Autoimmunity in the Tight-Skin Mouse. *Am. J. Pathol.* **2004**, *165*, 641–650. [CrossRef]
68. Inaoki, M.; Sato, S.; Weintraub, B.C.; Goodnow, C.; Tedder, T.F. CD19-Regulated Signaling Thresholds Control Peripheral Tolerance and Autoantibody Production in B Lymphocytes. *J. Exp. Med.* **1997**, *186*, 1923–1931. [CrossRef]
69. Sato, S.; Ono, N.; Steeber, D.A.; Pisetsky, D.S.; Tedder, T.F. CD19 regulates B lymphocyte signaling thresholds critical for the development of B-1 lineage cells and autoimmunity. *J. Immunol.* **1996**, *157*, 4371–4378.
70. Komura, K.; Yanaba, K.; Ogawa, F.; Shimizu, K.; Takehara, K.; Sato, S. Elevation of IgG levels is a serological indicator for pulmonary fibrosis in systemic sclerosis with anti-topoisomerase I antibodies and those with anticentromere antibody. *Clin. Exp. Dermatol.* **2008**, *33*, 329–332. [CrossRef]
71. Cavazzana, I.; Vojinovic, T.; Airo', P.; Fredi, M.; Ceribelli, A.; Pedretti, E.; Lazzaroni, M.G.; Garrafa, E.; Franceschini, F. Systemic Sclerosis-Specific Antibodies: Novel and Classical Biomarkers. *Clin. Rev. Allergy Immunol.* **2022**, 1–19. [CrossRef]
72. Van den Hoogen, F.; Khanna, D.; Fransen, J.; Johnson, S.R.; Baron, M.; Tyndall, A.; Matucci-Cerinic, M.; Naden, R.P.; Medsger, T.A., Jr.; Carreira, P.E.; et al. 2013 classification criteria for systemic sclerosis: An American college of rheumatology/European league against rheumatism collaborative initiative. *Ann. Rheum. Dis.* **2013**, *72*, 1747–1755. [CrossRef]

73. Mierau, R.; Moinzadeh, P.; Riemekasten, G.; Melchers, I.; Meurer, M.; Reichenberger, F.; Buslau, M.; Worm, M.; Blank, N.; Hein, R.; et al. Frequency of disease-associated and other nuclear autoantibodies in patients of the German network for systemic scleroderma: Correlation with characteristic clinical features. *Arthritis Res. Ther.* **2011**, *13*, R172. [CrossRef]
74. Tan, E.M. Antinuclear Antibodies: Diagnostic Markers for Autoimmune Diseases and Probes for Cell Biology. *Adv. Immunol.* **1989**, *44*, 93–151. [CrossRef]
75. Salazar, G.A.; Assassi, S.; Wigley, F.; Hummers, L.; Varga, J.; Hinchcliff, M.; Khanna, D.; Schiopu, E.; Phillips, K.; Furst, D.E.; et al. Antinuclear antibody-negative systemic sclerosis. *Semin. Arthritis Rheum.* **2014**, *44*, 680–686. [CrossRef]
76. Serling-Boyd, N.; Chung, M.P.-S.; Li, S.; Becker, L.; Fernandez-Becker, N.; Clarke, J.; Fiorentino, D.; Chung, L. Gastric antral vascular ectasia in systemic sclerosis: Association with anti-RNA polymerase III and negative anti-nuclear antibodies. *Semin. Arthritis Rheum.* **2020**, *50*, 938–942. [CrossRef]
77. Patterson, K.A.; Roberts-Thomson, P.J.; Lester, S.; Tan, J.A.; Hakendorf, P.; Rischmueller, M.; Zochling, J.; Sahhar, J.; Nash, P.; Roddy, J.; et al. Interpretation of an Extended Autoantibody Profile in a Well-Characterized Australian Systemic Sclerosis (Scleroderma) Cohort Using Principal Components Analysis. *Arthritis Rheumatol.* **2015**, *67*, 3234–3244. [CrossRef]
78. Clark, K.E.N.; Campochiaro, C.; Csomor, E.; Taylor, A.; Nevin, K.; Galwey, N.; Morse, M.A.; Singh, J.; Teo, Y.V.; Ong, V.H.; et al. Molecular basis for clinical diversity between autoantibody subsets in diffuse cutaneous systemic sclerosis. *Ann. Rheum. Dis.* **2021**, *80*, 1584–1593. [CrossRef]
79. Stochmal, A.; Czuwara, J.; Trojanowska, M.; Rudnicka, L. Antinuclear Antibodies in Systemic Sclerosis: An Update. *Clin. Rev. Allergy Immunol.* **2019**, *58*, 40–51. [CrossRef]
80. Adler, B.L.; Boin, F.; Wolters, P.J.; Bingham, C.O.; Shah, A.A.; Greider, C.; Casciola-Rosen, L.; Rosen, A. Autoantibodies targeting telomere-associated proteins in systemic sclerosis. *Ann. Rheum. Dis.* **2021**, *80*, 912–919. [CrossRef]
81. Granito, A.; Muratori, P.; Pappas, G.; Cassani, F.; Worthington, J.; Ferri, S.; Quarneti, C.; Cipriano, V.; De Molo, C.; Lenzi, M.; et al. Antibodies to SS-A/Ro-52kD and centromere in autoimmune liver disease: A clue to diagnosis and prognosis of primary biliary cirrhosis. *Aliment. Pharmacol. Ther.* **2007**, *26*, 831–838. [CrossRef]
82. Zheng, B.; Vincent, C.; Fritzler, M.J.; Senécal, J.-L.; Koenig, M.; Joyal, F. Prevalence of Systemic Sclerosis in Primary Biliary Cholangitis Using the New ACR/EULAR Classification Criteria. *J. Rheumatol.* **2016**, *44*, 33–39. [CrossRef] [PubMed]
83. Henes, J.; Glaeser, L.; Kötter, I.; Vogel, W.; Kanz, L.; Klein, R. Analysis of anti–topoisomerase I antibodies in patients with systemic sclerosis before and after autologous stem cell transplantation. *Rheumatology* **2016**, *56*, 451–456. [CrossRef] [PubMed]
84. Jarzabek-Chorzelska, M.; Błaszczyk, M.; Kołacińska-Strasz, Z.; Jabłońska, S.; Chorzelski, T.; Maul, G. Are ACA and Scl 70 antibodies mutually exclusive? *Br. J. Dermatol.* **1990**, *122*, 201–208. [CrossRef] [PubMed]
85. Heijnen, I.A.F.M.; Foocharoen, C.; Bannert, B.; Carreira, P.E.; Caporali, R.F.; Smith, V.; Kumánovics, G.; Becker, M.O.; Vanthuyne, M.; Simsek, I.; et al. Clinical significance of coexisting antitopoisomerase I and anticentromere antibodies in patients with systemic sclerosis: A EUSTAR group-based study. *Clin. Exp. Rheumatol.* **2012**, *31*, S96–S102.
86. Harvey, G.R.; Butts, S.; Rands, A.L.; Patel, Y.; McHugh, N.J. Clinical and serological associations with anti-RNA polymerase antibodies in systemic sclerosis. *Clin. Exp. Immunol.* **1999**, *117*, 395–402. [CrossRef]
87. Harris, M.L.; Rosen, A. Autoimmunity in scleroderma: The origin, pathogenetic role, and clinical significance of autoantibodies. *Curr. Opin. Rheumatol.* **2003**, *15*, 778–784. [CrossRef]
88. Chen, M.; Dittmann, A.; Kuhn, A.; Ruzicka, T.; von Mikecz, A. Recruitment of topoisomerase I (Scl-70) to nucleoplasmic proteasomes in response to xenobiotics suggests a role for altered antigen processing in scleroderma. *Arthritis Care Res.* **2005**, *52*, 877–884. [CrossRef]
89. Perera, A.; Fertig, N.; Lucas, M.; Rodriguez-Reyna, T.S.; Hu, P.; Steen, V.D.; Medsger, T.A., Jr. Clinical subsets, skin thickness progression rate, and serum antibody levels in systemic sclerosis patients with anti–topoisomerase I antibody. *Arthritis Care Res.* **2007**, *56*, 2740–2746. [CrossRef]
90. Kuwana, M.; Kaburaki, J.; Mimori, T.; Kawakami, Y.; Tojo, T. Longitudinal analysis of autoantibody response to topoisomerase I in systemic sclerosis. *Arthritis Care Res.* **2000**, *43*, 1074–1084. [CrossRef]
91. Rudnicka, L.; Czuwara, J.; Barusińska, A.; Nowicka, U.; Makiela, B.; Jablonska, S. Implications for the Use of Topoisomerase I Inhibitors in Treatment of Patients with Systemic Sclerosis. *Ann. N. Y. Acad. Sci.* **1996**, *803*, 318–320. [CrossRef]
92. Hénault, J.; Tremblay, M.; Clément, I.; Raymond, Y.; Senécal, J.-L. Direct binding of anti-DNA topoisomerase I autoantibodies to the cell surface of fibroblasts in patients with systemic sclerosis. *Arthritis Care Res.* **2004**, *50*, 3265–3274. [CrossRef] [PubMed]
93. Hénault, J.; Robitaille, G.; Senécal, J.-L.; Raymond, Y. DNA topoisomerase I binding to fibroblasts induces monocyte adhesion and activation in the presence of anti–topoisomerase I autoantibodies from systemic sclerosis patients. *Arthritis Care Res.* **2006**, *54*, 963–973. [CrossRef] [PubMed]
94. Corallo, C.; Cheleschi, S.; Cutolo, M.; Soldano, S.; Fioravanti, A.; Volpi, N.; Franci, D.; Nuti, R.; Giordano, N. Antibodies against specific extractable nuclear antigens (ENAs) as diagnostic and prognostic tools and inducers of a profibrotic phenotype in cultured human skin fibroblasts: Are they functional? *Arthritis Res. Ther.* **2019**, *21*, 1–11. [CrossRef]
95. Suwanchote, S.; Rachayon, M.; Rodsaward, P.; Wongpiyabovorn, J.; Deekajorndech, T.; Wright, H.L.; Edwards, S.W.; Beresford, M.W.; Rerknimitr, P.; Chiewchengchol, D. Anti-neutrophil cytoplasmic antibodies and their clinical significance. *Clin. Rheumatol.* **2018**, *37*, 875–884. [CrossRef] [PubMed]
96. Walls, C.A.; Basu, N.; Hutcheon, G.; Erwig, L.P.; Little, M.A.; Kidder, D. A novel 4-dimensional live-cell imaging system to study leukocyte-endothelial dynamics in ANCA-associated vasculitis. *Autoimmunity* **2019**, *53*, 148–155. [CrossRef]

97. Kettritz, R. How anti-neutrophil cytoplasmic autoantibodies activate neutrophils. *Clin. Exp. Immunol.* **2012**, *169*, 220–228. [CrossRef]
98. Jerke, U.; Hernandez, D.P.; Beaudette, P.; Korkmaz, B.; Dittmar, G.; Kettritz, R. Neutrophil serine proteases exert proteolytic activity on endothelial cells. *Kidney Int.* **2015**, *88*, 764–775. [CrossRef]
99. Ralston, D.R.; Marsh, C.B.; Lowe, M.P.; Wewers, M.D. Antineutrophil cytoplasmic antibodies induce monocyte IL-8 release. Role of surface proteinase-3, alpha1-antitrypsin, and Fcgamma receptors. *J. Clin. Investig.* **1997**, *100*, 1416–1424. [CrossRef]
100. Jennette, J.C.; Falk, R.J. ANCAs Are Also Antimonocyte Cytoplasmic Autoantibodies. *Clin. J. Am. Soc. Nephrol.* **2014**, *10*, 4–6. [CrossRef]
101. Sibelius, U.; Hattar, K.; Schenkel, A.; Noll, T.; Csernok, E.; Gross, W.L.; Mayet, W.-J.; Piper, H.-M.; Seeger, W.; Grimminger, F. Wegener's Granulomatosis: Anti–proteinase 3 Antibodies Are Potent Inductors of Human Endothelial Cell Signaling and Leakage Response. *J. Exp. Med.* **1998**, *187*, 497–503. [CrossRef]
102. Granito, A.; Muratori, P.; Tovoli, F.; Muratori, L. Anti-neutrophil cytoplasm antibodies (ANCA) in autoimmune diseases: A matter of laboratory technique and clinical setting. *Autoimmun. Rev.* **2021**, *20*, 102787. [CrossRef]
103. Khanna, D.; Aggarwal, A.; Bhakuni, D.S.; Dayal, R.; Misra, R. Bactericidal/permeability-increasing protein and cathepsin G are the major antigenic targets of antineutrophil cytoplasmic autoantibodies in systemic sclerosis. *J. Rheumatol.* **2003**, *30*, 1248–1252. [PubMed]
104. Ruffatti, A.; Sinico, R.A.; Radice, A.; Ossi, E.; Cozzi, F.; Tonello, M.; Grypiotis, P.; Todesco, S. Autoantibodies to proteinase 3 and myeloperoxidase in systemic sclerosis. *J. Rheumatol.* **2002**, *29*, 918–923. [PubMed]
105. Caramaschi, P.; Biasi, D.; Tonolli, E.; Carletto, A.; Bambara, L.M. Antineutrophil cytoplasmic antibodies in scleroderma patients: First report of a case with anti-proteinase 3 antibodies and review of the literature. *Jt. Bone Spine* **2002**, *69*, 177–180. [CrossRef]
106. Quéméneur, T.; Mouthon, L.; Cacoub, P.; Meyer, O.; Michon-Pasturel, U.; Vanhille, P.; Hatron, P.-Y.; Guillevin, L.; Hachulla, E. Systemic Vasculitis During the Course of Systemic Sclerosis. *Medicine* **2013**, *92*, 1–9. [CrossRef] [PubMed]
107. Derrett-Smith, E.C.; Nihtyanova, S.I.; Harvey, J.; Salama, A.D.; Denton, C.P. Revisiting ANCA-associated vasculitis in systemic sclerosis: Clinical, serological and immunogenetic factors. *Rheumatology* **2013**, *52*, 1824–1831. [CrossRef]
108. Trang, G.; Steele, R.; Baron, M.; Hudson, M. Corticosteroids and the risk of scleroderma renal crisis: A systematic review. *Rheumatol. Int.* **2010**, *32*, 645–653. [CrossRef]
109. Vega-Ostertag, M.; Casper, K.; Swerlick, R.; Ferrara, D.; Harris, E.N.; Pierangeli, S.S. Involvement of p38 MAPK in the up-regulation of tissue factor on endothelial cells by antiphospholipid antibodies. *Arthritis Care Res.* **2005**, *52*, 1545–1554. [CrossRef]
110. Liestøl, S.; Sandset, P.M.; Jacobsen, E.M.; Mowinckel, M.-C.; Wisløff, F. Decreased anticoagulant response to tissue factor pathway inhibitor type 1 in plasmas from patients with lupus anticoagulants. *Br. J. Haematol.* **2007**, *136*, 131–137. [CrossRef]
111. Seed, P.; Parmar, K.; Moore, G.W.; Stuart-Smith, S.E.; Hunt, B.J.; Breen, K.A. Complement activation in patients with isolated antiphospholipid antibodies or primary antiphospholipid syndrome. *Thromb. Haemost.* **2012**, *107*, 423–429. [CrossRef]
112. Espinola, R.G.; Ghara, A.E.; Harris, E.N.; Pierangeli, S.S. Hydroxychloroquine Reverses Platelet Activation Induced by Human IgG Antiphospholipid Antibodies. *Thromb. Haemost.* **2002**, *87*, 518–522. [CrossRef] [PubMed]
113. Yalavarthi, S.; Gould, T.J.; Rao, A.N.; Mazza, L.F.; Morris, A.E.; Nunezalvarez, C.A.; Hernández-Ramírez, D.; Bockenstedt, P.L.; Liaw, P.C.; Cabral, A.R.; et al. Release of Neutrophil Extracellular Traps by Neutrophils Stimulated With Antiphospholipid Antibodies: A Newly Identified Mechanism of Thrombosis in the Antiphospholipid Syndrome. *Arthritis Rheumatol.* **2015**, *67*, 2990–3003. [CrossRef] [PubMed]
114. Meng, H.; Yalavarthi, S.; Kanthi, Y.; Mazza, L.F.; Elfline, M.A.; Luke, C.E.; Pinsky, D.J.; Henke, P.K.; Knight, J.S. In Vivo Role of Neutrophil Extracellular Traps in Antiphospholipid Antibody-Mediated Venous Thrombosis. *Arthritis Rheumatol.* **2016**, *69*, 655–667. [CrossRef] [PubMed]
115. López-Pedrera, C.; Buendía, P.; Cuadrado, M.J.; Siendones, E.; Aguirre, M.A.; Barbarroja, N.; Duarte, C.M.; Torres, A.; Khamashta, M.; Velasco, F. Antiphospholipid antibodies from patients with the antiphospholipid syndrome induce monocyte tissue factor expression through the simultaneous activation of NF-κB/Rel proteins via the p38 mitogen-activated protein kinase pathway, and of the MEK-1/ERK pathway. *Arthritis Care Res.* **2005**, *54*, 301–311. [CrossRef]
116. Sobanski, V.; Lemaire-Olivier, A.; Giovannelli, J.; Dauchet, L.; Simon, M.; Lopez, B.; Yelnik, C.; Lambert, M.; Hatron, P.-Y.; Hachulla, E.; et al. Prevalence and Clinical Associations of Antiphospholipid Antibodies in Systemic Sclerosis: New Data From a French Cross-Sectional Study, Systematic Review, and Meta-Analysis. *Front. Immunol.* **2018**, *9*, 2457. [CrossRef]
117. Martin, M.; Martinez, C.; Arnaud, L.; Weber, J.-C.; Poindron, V.; Blaison, G.; Kieffer, P.; Bonnotte, B.; Berthier, S.; Wahl, D.; et al. Association of antiphospholipid antibodies with active digital ulceration in systemic sclerosis. *RMD Open* **2019**, *5*, e001012. [CrossRef]
118. Merashli, M.; Alves, J.; Ames, P.R. Clinical relevance of antiphospholipid antibodies in systemic sclerosis: A systematic review and meta-analysis. *Semin. Arthritis Rheum.* **2017**, *46*, 615–624. [CrossRef]
119. Murthy, S.; Wannick, M.; Eleftheriadis, G.; Müller, A.; Luo, J.; Busch, H.; Dalmann, A.; Riemekasten, G.; Sadik, C.D. Immunoglobulin G of systemic sclerosis patients programs a pro-inflammatory and profibrotic phenotype in monocyte-like THP-1 cells. *Rheumatology* **2020**, *60*, 3012–3022. [CrossRef]
120. Rosetti, F.; Tsuboi, N.; Chen, K.; Nishi, H.; Ernandez, T.; Sethi, S.; Croce, K.; Stavrakis, G.; Alcocer-Varela, J.; Gómez-Martin, D.; et al. Human Lupus Serum Induces Neutrophil-Mediated Organ Damage in Mice That Is Enabled by Mac-1 Deficiency. *J. Immunol.* **2012**, *189*, 3714–3723. [CrossRef]

121. Stern, M.E.; Schaumburg, C.S.; Siemasko, K.F.; Gao, J.; Wheeler, L.A.; Grupe, D.A.; de Paiva, C.; Calder, V.L.; Calonge, M.; Niederkorn, J.Y.; et al. Autoantibodies Contribute to the Immunopathogenesis of Experimental Dry Eye Disease. *Investig. Opthalmol. Vis. Sci.* **2012**, *53*, 2062–2075. [CrossRef]
122. Scofield, R.H.; Asfa, S.; Obeso, D.; Jonsson, R.; Kurien, B.T. Immunization with Short Peptides from the 60-kDa Ro Antigen Recapitulates the Serological and Pathological Findings as well as the Salivary Gland Dysfunction of Sjögren's Syndrome. *J. Immunol.* **2005**, *175*, 8409–8414. [CrossRef]
123. Rosenberg, N.L.; Ringel, S.P.; Kotzin, B.L. Experimental autoimmune myositis in SJL/J mice. *Clin. Exp. Immunol.* **1987**, *68*, 117–129. [PubMed]
124. Becker, M.O.; Kill, A.; Kutsche, M.; Guenther, J.; Rose, A.; Tabeling, C.; Witzenrath, M.; Kühl, A.A.; Heidecke, H.; Ghofrani, H.A.; et al. Vascular Receptor Autoantibodies in Pulmonary Arterial Hypertension Associated with Systemic Sclerosis. *Am. J. Respir. Crit. Care Med.* **2014**, *190*, 808–817. [CrossRef] [PubMed]
125. Kill, A.; Tabeling, C.; Undeutsch, R.; Kühl, A.A.; Günther, J.; Radic, M.; Becker, M.O.; Heidecke, H.; Worm, M.; Witzenrath, M.; et al. Autoantibodies to angiotensin and endothelin receptors in systemic sclerosis induce cellular and systemic events associated with disease pathogenesis. *Arthritis Res. Ther.* **2014**, *16*, R29. [CrossRef] [PubMed]
126. Yue, X.; Yin, J.; Wang, X.; Heidecke, H.; Hackel, A.M.; Dong, X.; Kasper, B.; Wen, L.; Zhang, L.; Schulze-Forster, K.; et al. Induced antibodies directed to the angiotensin receptor type 1 provoke skin and lung inflammation, dermal fibrosis and act species overarching. *Ann. Rheum. Dis.* **2022**, *81*, 1281–1289. [CrossRef] [PubMed]
127. Yue, X.; Petersen, F.; Shu, Y.; Kasper, B.; Magatsin, J.D.T.; Ahmadi, M.; Yin, J.; Wax, J.; Wang, X.; Heidecke, H.; et al. Transfer of PBMC From SSc Patients Induces Autoantibodies and Systemic Inflammation in Rag2-/-/IL2rg-/- Mice. *Front. Immunol.* **2021**, *12*. [CrossRef]
128. Wang, J.; Li, D.; Zhang, Z.; Zhang, Y.; Lei, Z.; Jin, W.; Cao, J.; Jiao, X. Autoantibody against angiotensin II type I receptor induces pancreatic β-cell apoptosis via enhancing autophagy. *Acta Biochim. Biophys. Sin.* **2021**, *53*, 784–795. [CrossRef]
129. Mejia-Vilet, J.M.; López-Hernández, Y.J.; Santander-Vélez, J.I.; Trujeque-Matos, M.; Cruz, C.; Torre, C.A.C.D.L.; Espinosa-Cruz, V.; Espinosa-González, R.; Uribe-Uribe, N.O.; Morales-Buenrostro, L.E. Angiotensin II receptor agonist antibodies are associated with microvascular damage in lupus nephritis. *Lupus* **2020**, *29*, 371–378. [CrossRef]
130. Jiang, Y.; Duffy, F.; Hadlock, J.; Raappana, A.; Styrchak, S.; Beck, I.; Mast, F.D.; Miller, L.R.; Chour, W.; Houck, J.; et al. Angiotensin II receptor I auto-antibodies following SARS-CoV-2 infection. *PLoS ONE* **2021**, *16*, e0259902. [CrossRef]
131. Wallukat, G.; Homuth, V.; Fischer, T.; Lindschau, C.; Horstkamp, B.; Jüpner, A.; Baur, E.; Nissen, E.; Vetter, K.; Neichel, D.; et al. Patients with preeclampsia develop agonistic autoantibodies against the angiotensin AT1 receptor. *J. Clin. Investig.* **1999**, *103*, 945–952. [CrossRef]
132. Dragun, D.; Müller, D.N.; Bräsen, J.H.; Fritsche, L.; Nieminen-Kelhä, M.; Dechend, R.; Kintscher, U.; Rudolph, B.; Hoebeke, J.; Eckert, D.; et al. Angiotensin II Type 1–Receptor Activating Antibodies in Renal-Allograft Rejection. *N. Engl. J. Med.* **2005**, *352*, 558–569. [CrossRef] [PubMed]
133. Dragun, D.; Catar, R.; Philippe, A. Non-HLA antibodies against endothelial targets bridging allo- and autoimmunity. *Kidney Int.* **2016**, *90*, 280–288. [CrossRef] [PubMed]
134. Lukitsch, I.; Kehr, J.; Chaykovska, L.; Wallukat, G.; Nieminen-Kelhä, M.; Batuman, V.; Dragun, D.; Gollasch, M. Renal Ischemia and Transplantation Predispose to Vascular Constriction Mediated by Angiotensin II Type 1 Receptor-Activating Antibodies. *Transplantation* **2012**, *94*, 8–13. [CrossRef]
135. Rasini, E.; Cosentino, M.; Marino, F.; Legnaro, M.; Ferrari, M.; Guasti, L.; Venco, A.; Lecchini, S. Angiotensin II type 1 receptor expression on human leukocyte subsets: A flow cytometric and RT-PCR study. *Regul. Pept.* **2006**, *134*, 69–74. [CrossRef] [PubMed]
136. Wassmann, S.; Stumpf, M.; Strehlow, K.; Schmid, A.; Schieffer, B.; Böhm, M.; Nickenig, G. Interleukin-6 Induces Oxidative Stress and Endothelial Dysfunction by Overexpression of the Angiotensin II Type 1 Receptor. *Circ. Res.* **2004**, *94*, 534–541. [CrossRef] [PubMed]
137. Pearl, M.H.; Grotts, J.; Rossetti, M.; Zhang, Q.; Gjertson, D.W.; Weng, P.; Elashoff, D.; Reed, E.F.; Chambers, E.T. Cytokine Profiles Associated With Angiotensin II Type 1 Receptor Antibodies. *Kidney Int. Rep.* **2018**, *4*, 541–550. [CrossRef]
138. Ceolotto, G.; Papparella, I.; Bortoluzzi, A.; Strapazzon, G.; Ragazzo, F.; Bratti, P.; Fabricio, A.; Squarcina, E.; Gion, M.; Palatini, P.; et al. Interplay Between miR-155, AT1R A1166C Polymorphism, and AT1R Expression in Young Untreated Hypertensives. *Am. J. Hypertens.* **2011**, *24*, 241–246. [CrossRef]
139. Hrenak, J.; Simko, F. Renin–Angiotensin System: An Important Player in the Pathogenesis of Acute Respiratory Distress Syndrome. *Int. J. Mol. Sci.* **2020**, *21*, 8038. [CrossRef]
140. Sorohan, B.M.; Ismail, G.; Leca, N.; Tacu, D.; Obrișcă, B.; Constantinescu, I.; Baston, C.; Sinescu, I. Angiotensin II type 1 receptor antibodies in kidney transplantation: An evidence-based comprehensive review. *Transplant. Rev.* **2020**, *34*, 100573. [CrossRef]
141. Lei, J.; Zhang, S.; Wang, P.; Liao, Y.; Bian, J.; Yin, X.; Wu, Y.; Bai, L.; Wang, F.; Yang, X.; et al. Long-term presence of angiotensin II type 1 receptor autoantibody reduces aldosterone production by triggering Ca2+ overload in H295R cells. *Immunol. Res.* **2017**, *66*, 44–51. [CrossRef]
142. Kawaguchi, Y.; Takagi, K.; Hara, M.; Fukasawa, C.; Sugiura, T.; Nishimagi, E.; Harigai, M.; Kamatani, N. Angiotensin II in the lesional skin of systemic sclerosis patients contributes to tissue fibrosis via angiotensin II type 1 receptors. *Arthritis Care Res.* **2004**, *50*, 216–226. [CrossRef] [PubMed]

143. Yamane, K.; Miyauchi, T.; Suzuki, N.; Yuhara, T.; Akama, T.; Suzuki, H.; Kashiwagi, H. Significance of plasma endothelin-1 levels in patients with systemic sclerosis. *J. Rheumatol.* **1992**, *19*, 1566–1571. [PubMed]
144. Denton, C.P.; Pope, J.E.; Peter, H.-H.; Gabrielli, A.; Boonstra, A.; Hoogen, F.H.J.V.D.; Riemekasten, G.; De Vita, S.; Morganti, A.; Dolberg, M.; et al. Long-term effects of bosentan on quality of life, survival, safety and tolerability in pulmonary arterial hypertension related to connective tissue diseases. *Ann. Rheum. Dis.* **2007**, *67*, 1222–1228. [CrossRef] [PubMed]
145. Tillon, J.; Hervé, F.; Chevallier, D.; Muir, J.-F.; Levesque, H.; Marie, I. Successful treatment of systemic sclerosis-related digital ulcers and sarcoidosis with endothelin receptor antagonist (bosentan) therapy. *Br. J. Dermatol.* **2006**, *154*, 1000–1002. [CrossRef] [PubMed]
146. Arefiev, K.; Fiorentino, D.; Chung, L. Endothelin Receptor Antagonists for the Treatment of Raynaud's Phenomenon and Digital Ulcers in Systemic Sclerosis. *Int. J. Rheumatol.* **2011**, *2011*, 1–7. [CrossRef]
147. Dhillon, S. Bosentan. *Drugs* **2009**, *69*, 2005–2024. [CrossRef]
148. Korn, J.H.; Mayes, M.; Cerinic, M.M.; Rainisio, M.; Pope, J.; Hachulla, E.; Rich, E.; Carpentier, P.; Molitor, J.; Seibold, J.R.; et al. Digital ulcers in systemic sclerosis: Prevention by treatment with bosentan, an oral endothelin receptor antagonist. *Arthritis Care Res.* **2004**, *50*, 3985–3993. [CrossRef]
149. Riemekasten, G.; Philippe, A.; Näther, M.; Slowinski, T.; Muller, D.N.; Heidecke, H.; Matucci-Cerinic, M.; Czirják, L.; Lukitsch, I.; Becker, M.; et al. Involvement of functional autoantibodies against vascular receptors in systemic sclerosis. *Ann. Rheum. Dis.* **2010**, *70*, 530–536. [CrossRef]
150. Günther, J.; Kill, A.; Becker, M.O.; Heidecke, H.; Rademacher, J.; Siegert, E.; Radić, M.; Burmester, G.-R.; Dragun, D.; Riemekasten, G. Angiotensin receptor type 1 and endothelin receptor type A on immune cells mediate migration and the expression of IL-8 and CCL18 when stimulated by autoantibodies from systemic sclerosis patients. *Arthritis Res. Ther.* **2014**, *16*, R65. [CrossRef]
151. Avouac, J.; Riemekasten, G.; Meune, C.; Ruiz, B.; Kahan, A.; Allanore, Y. Autoantibodies against Endothelin 1 Type A Receptor Are Strong Predictors of Digital Ulcers in Systemic Sclerosis. *J. Rheumatol.* **2015**, *42*, 1801–1807. [CrossRef]
152. Cabral-Marques, O.; Riemekasten, G. Vascular hypothesis revisited: Role of stimulating antibodies against angiotensin and endothelin receptors in the pathogenesis of systemic sclerosis. *Autoimmun. Rev.* **2016**, *15*, 690–694. [CrossRef] [PubMed]
153. Kawaguchi, Y.; Nakamura, Y.; Matsumoto, I.; Nishimagi, E.; Satoh, T.; Kuwana, M.; Sumida, T.; Hara, M. Muscarinic-3 acetylcholine receptor autoantibody in patients with systemic sclerosis: Contribution to severe gastrointestinal tract dysmotility. *Ann. Rheum. Dis.* **2008**, *68*, 710–714. [CrossRef] [PubMed]
154. Kumar, S.; Singh, J.; Kedika, R.; Mendoza, F.; Jimenez, S.; Blomain, E.S.; Dimarino, A.J.; Cohen, S.; Rattan, S. Role of muscarinic-3 receptor antibody in systemic sclerosis: Correlation with disease duration and effects of IVIG. *Am. J. Physiol. Liver Physiol.* **2016**, *310*, G1052–G1060. [CrossRef] [PubMed]
155. Raja, J.; Nihtyanova, S.I.; Murray, C.; Denton, C.P.; Ong, V.H. Sustained benefit from intravenous immunoglobulin therapy for gastrointestinal involvement in systemic sclerosis. *Rheumatology* **2015**, *55*, 115–119. [CrossRef]
156. Weigold, F.; Günther, J.; Pfeiffenberger, M.; Marques, O.C.; Siegert, E.; Dragun, D.; Philippe, A.; Regensburger, A.-K.; Recke, A.; Yu, X.; et al. Antibodies against chemokine receptors CXCR3 and CXCR4 predict progressive deterioration of lung function in patients with systemic sclerosis. *Arthritis Res. Ther.* **2018**, *20*, 52. [CrossRef]
157. Recke, A.; Regensburger, A.-K.; Weigold, F.; Müller, A.; Heidecke, H.; Marschner, G.; Hammers, C.; Ludwig, R.; Riemekasten, G. Autoantibodies in Serum of Systemic Scleroderma Patients: Peptide-Based Epitope Mapping Indicates Increased Binding to Cytoplasmic Domains of CXCR3. *Front. Immunol.* **2018**, *9*, 428. [CrossRef]
158. Simon, M.; Lücht, C.; Hosp, I.; Zhao, H.; Wu, D.; Heidecke, H.; Witowski, J.; Budde, K.; Riemekasten, G.; Catar, R. Autoantibodies from Patients with Scleroderma Renal Crisis Promote PAR-1 Receptor Activation and IL-6 Production in Endothelial Cells. *Int. J. Mol. Sci.* **2021**, *22*, 11793. [CrossRef]
159. Shima, Y.; Kuwahara, Y.; Murota, H.; Kitaba, S.; Kawai, M.; Hirano, T.; Arimitsu, J.; Narazaki, M.; Hagihara, K.; Ogata, A.; et al. The skin of patients with systemic sclerosis softened during the treatment with anti-IL-6 receptor antibody tocilizumab. *Rheumatology* **2010**, *49*, 2408–2412. [CrossRef]
160. Schiopu, E.; Chatterjee, S.; Hsu, V.; Flor, A.; Cimbora, D.; Patra, K.; Yao, W.; Li, J.; Streicher, K.; McKeever, K.; et al. Safety and tolerability of an anti-CD19 monoclonal antibody, MEDI-551, in subjects with systemic sclerosis: A phase I, randomized, placebo-controlled, escalating single-dose study. *Arthritis Res. Ther.* **2016**, *18*, 131. [CrossRef]
161. Mei, H.E.; Frölich, D.; Giesecke, C.; Loddenkemper, C.; Reiter, K.; Schmidt, S.; Feist, E.; Daridon, C.; Tony, H.-P.; Radbruch, A.; et al. Steady-state generation of mucosal IgA+ plasmablasts is not abrogated by B-cell depletion therapy with rituximab. *Blood* **2010**, *116*, 5181–5190. [CrossRef]
162. Zamanian, R.T.; Badesch, D.; Chung, L.; Domsic, R.T.; Medsger, T.; Pinckney, A.; Keyes-Elstein, L.; D'Aveta, C.; Spychala, M.; White, R.J.; et al. Safety and Efficacy of B-Cell Depletion with Rituximab for the Treatment of Systemic Sclerosis–associated Pulmonary Arterial Hypertension: A Multicenter, Double-Blind, Randomized, Placebo-controlled Trial. *Am. J. Respir. Crit. Care Med.* **2021**, *204*, 209–221. [CrossRef] [PubMed]
163. Boonstra, M.; Meijs, J.; Dorjée, A.L.; Marsan, N.A.; Schouffoer, A.; Ninaber, M.K.; Quint, K.D.; Bonte-Mineur, F.; Huizinga, T.W.J.; Scherer, H.U.; et al. Rituximab in early systemic sclerosis. *RMD Open* **2017**, *3*, e000384. [CrossRef] [PubMed]
164. Sircar, G.; Goswami, R.P.; Sircar, D.; Ghosh, A.; Ghosh, P. Intravenous cyclophosphamide *vs* rituximab for the treatment of early diffuse scleroderma lung disease: Open label, randomized, controlled trial. *Rheumatology* **2018**, *57*, 2106–2113. [CrossRef]

165. Goswami, R.P.; Ray, A.; Chatterjee, M.; Mukherjee, A.; Sircar, G.; Ghosh, P. Rituximab in the treatment of systemic sclerosis–related interstitial lung disease: A systematic review and meta-analysis. *Rheumatology* **2020**, *60*, 557–567. [CrossRef]
166. Kaegi, C.; Wuest, B.; Crowley, C.; Boyman, O. Systematic Review of Safety and Efficacy of Second- and Third-Generation CD20-Targeting Biologics in Treating Immune-Mediated Disorders. *Front. Immunol.* **2022**, *12*. [CrossRef] [PubMed]
167. Mahévas, M.; Michel, M.; Weill, J.-C.; Reynaud, C.-A. Long-Lived Plasma Cells in Autoimmunity: Lessons from B-Cell Depleting Therapy. *Front. Immunol.* **2013**, *4*, 494. [CrossRef]
168. Ehrenstein, M.; Wing, M.R.E.C. The BAFFling effects of rituximab in lupus: Danger ahead? *Nat. Rev. Rheumatol.* **2016**, *12*, 367–372. [CrossRef]
169. Matsushita, T.; Kobayashi, T.; Mizumaki, K.; Kano, M.; Sawada, T.; Tennichi, M.; Okamura, A.; Hamaguchi, Y.; Iwakura, Y.; Hasegawa, M.; et al. BAFF inhibition attenuates fibrosis in scleroderma by modulating the regulatory and effector B cell balance. *Sci. Adv.* **2018**, *4*, eaas9944. [CrossRef]
170. Gordon, J.K.; Martyanov, V.; Franks, J.M.; Bernstein, E.J.; Szymonifka, J.; Magro, C.; Wildman, H.F.; Wood, T.A.; Whitfield, M.L.; Spiera, R.F. Belimumab for the Treatment of Early Diffuse Systemic Sclerosis. *Arthritis Rheumatol.* **2017**, *70*, 308–316. [CrossRef]
171. Fineschi, S.; Reith, W.; Guerne, P.A.; Dayer, J.; Chizzolini, C. Proteasome blockade exerts an antifibrotic activity by coordinately down-regulating type I collagen and tissue inhibitor of metalloproteinase-1 and up-regulating metalloproteinase-1 production in human dermal fibroblasts. *FASEB J.* **2006**, *20*, 562–564. [CrossRef]
172. Fineschi, S.; Bongiovanni, M.; Donati, Y.; Djaafar, S.; Naso, F.; Goffin, L.; Argiroffo, C.B.; Pache, J.-C.; Dayer, J.-M.; Ferrari-Lacraz, S.; et al. In Vivo Investigations on Anti-Fibrotic Potential of Proteasome Inhibition in Lung and Skin Fibrosis. *Am. J. Respir. Cell Mol. Biol.* **2008**, *39*, 458–465. [CrossRef] [PubMed]
173. Benfaremo, D.; Gabrielli, A. Is There a Future for Anti-CD38 Antibody Therapy in Systemic Autoimmune Diseases? *Cells* **2019**, *9*, 77. [CrossRef] [PubMed]
174. Pavlasova, G.; Borsky, M.; Seda, V.; Cerna, K.; Osickova, J.; Doubek, M.; Mayer, J.; Calogero, R.; Trbusek, M.; Pospisilova, S.; et al. Ibrutinib inhibits CD20 upregulation on CLL B cells mediated by the CXCR4/SDF-1 axis. *Blood* **2016**, *128*, 1609–1613. [CrossRef]
175. Einhaus, J.; Pecher, A.-C.; Asteriti, E.; Schmid, H.; Secker, K.-A.; Duerr-Stoerzer, S.; Keppeler, H.; Klein, R.; Schneidawind, C.; Henes, J.; et al. Inhibition of effector B cells by ibrutinib in systemic sclerosis. *Arthritis Res. Ther.* **2020**, *22*, 1–8. [CrossRef] [PubMed]
176. Servaas, N.H.; Spierings, J.; Pandit, A.; van Laar, J.M. The role of innate immune cells in systemic sclerosis in the context of autologous hematopoietic stem cell transplantation. *Clin. Exp. Immunol.* **2020**, *201*, 34–39. [CrossRef]
177. Gernert, M.; Tony, H.-P.; Schwaneck, E.C.; Gadeholt, O.; Schmalzing, M. Autologous hematopoietic stem cell transplantation in systemic sclerosis induces long-lasting changes in B cell homeostasis toward an anti-inflammatory B cell cytokine pattern. *Arthritis Res. Ther.* **2019**, *21*, 106. [CrossRef] [PubMed]
178. Arruda, L.C.M.; Lima-Júnior, J.R.; Clave, E.; Moraes, D.A.; Douay, C.; Fournier, I.; Moins-Teisserenc, H.; Covas, D.T.; Malmegrim, K.; Farge, D.; et al. Newly-Generated Regulatory B- and T-Cells Are Associated with Clinical Improvement and Reversal of Dermal Fibrosis in Systemic Sclerosis Patients after Autologous Hematopoietic Stem Cell Transplantation. *Blood* **2016**, *128*, 4625. [CrossRef]
179. Tyndall, A.; Black, C.; Finke, J.; Winkler, J.; Mertlesmann, R.; Peter, H.; Gratwohl, A. Treatment of systemic sclerosis with autologous haemopoietic stem cell transplantation. *Lancet* **1997**, *349*, 254. [CrossRef]
180. Shah, A.; Spierings, J.; van Laar, J.; Sullivan, K.M. Re-evaluating inclusion criteria for autologous hematopoietic stem cell transplantation in advanced systemic sclerosis: Three successful cases and review of the literature. *J. Scleroderma Relat. Disord.* **2021**, *6*, 199–205. [CrossRef]
181. Moraes, D.A.; Oliveira, M.C. Life after Autologous Hematopoietic Stem Cell Transplantation for Systemic Sclerosis. *J. Blood Med.* **2021**, *12*, 951–964. [CrossRef]
182. Burt, R.K.; Shah, S.J.; Dill, K.; Grant, T.; Gheorghiade, M.; Schroeder, J.; Craig, R.; Hirano, I.; Marshall, K.; Ruderman, E.; et al. Autologous non-myeloablative haemopoietic stem-cell transplantation compared with pulse cyclophosphamide once per month for systemic sclerosis (ASSIST): An open-label, randomised phase 2 trial. *Lancet* **2011**, *378*, 498–506. [CrossRef]
183. Van Laar, J.M.; Farge, D.; Sont, J.; Naraghi, K.; Marjanovic, Z.; Larghero, J.; Schuerwegh, A.J.; Marijt, E.W.A.; Vonk, M.C.; Schattenberg, A.V.; et al. Autologous Hematopoietic Stem Cell Transplantation vs Intravenous Pulse Cyclophosphamide in Diffuse Cutaneous Systemic Sclerosis. *JAMA* **2014**, *311*, 2490–2498. [CrossRef]
184. Sullivan, K.M.; Goldmuntz, E.A.; Keyes-Elstein, L.; McSweeney, P.A.; Pinckney, A.; Welch, B.; Mayes, M.D.; Nash, R.A.; Crofford, L.J.; Eggleston, B.; et al. Myeloablative Autologous Stem-Cell Transplantation for Severe Scleroderma. *N. Engl. J. Med.* **2018**, *378*, 35–47. [CrossRef] [PubMed]
185. Henes, J.; Oliveira, M.C.; Labopin, M.; Badoglio, M.; Scherer, H.U.; Del Papa, N.; Daikeler, T.; Schmalzing, M.; Schroers, R.; Martin, T.; et al. Autologous stem cell transplantation for progressive systemic sclerosis: A prospective non-interventional study from the European Society for Blood and Marrow Transplantation Autoimmune Disease Working Party. *Haematologica* **2020**, *106*, 375–383. [CrossRef] [PubMed]
186. Gernert, M.; Tony, H.-P.; Schwaneck, E.C.; Fröhlich, M.; Schmalzing, M. Low B cell counts as risk factor for infectious complications in systemic sclerosis after autologous hematopoietic stem cell transplantation. *Arthritis Res. Ther.* **2020**, *22*, 1–9. [CrossRef] [PubMed]

187. Strunz, P.-P.; Froehlich, M.; Gernert, M.; Schwaneck, E.C.; Fleischer, A.; Pecher, A.-C.; Tony, H.-P.; Henes, J.C.; Schmalzing, M. Immunological Adverse Events After Autologous Hematopoietic Stem Cell Transplantation in Systemic Sclerosis Patients. *Front. Immunol.* **2021**, *12*, 3526. [CrossRef]
188. Alexander, T.; Henes, J.; Distler, J.H.W.; Schmalzing, M.; Blank, N.; Kötter, I.; Hiepe, F. Autologe hämatopoetische Stammzelltransplantation bei systemischer Sklerose. *Z. Rheumatol.* **2020**, *79*, 429–436. [CrossRef]
189. Harris, E.S.; Meiselman, H.J.; Moriarty, P.M.; Metzger, A.; Malkovsky, M. Therapeutic plasma exchange for the treatment of systemic sclerosis: A comprehensive review and analysis. *J. Scleroderma Relat. Disord.* **2018**, *3*, 132–152. [CrossRef]
190. Patten, E.; Berkman, E.M. Therapeutic Plasmapheresis and Plasma Exchange. *CRC Crit. Rev. Clin. Lab. Sci.* **1986**, *23*, 147–175. [CrossRef]
191. Åkesson, A.; Wollheim, F.A.; Thysell, H.; Gustafson, T.; Forsberg, L.; Pahlm, O.; Wollmer, P.; Åkesson, B.; And, P.W. Visceral Improvement Following Combined Plasmapheresis and Immunosuppressive Drug Therapy in Progressive Systemic Sclerosis. *Scand. J. Rheumatol.* **1988**, *17*, 313–323. [CrossRef]
192. Suga, K.; Yamashita, H.; Takahashi, Y.; Katagiri, D.; Hinoshita, F.; Kaneko, H. Therapeutic efficacy of combined glucocorticoid, intravenous cyclophosphamide, and double-filtration plasmapheresis for skin sclerosis in diffuse systemic sclerosis. *Medicine* **2020**, *99*, e19301. [CrossRef] [PubMed]
193. Korsten, P.; Müller, G.A.; Rademacher, J.-G.; Zeisberg, M.; Tampe, B. Rheopheresis for Digital Ulcers and Raynaud's Phenomenon in Systemic Sclerosis Refractory to Conventional Treatments. *Front. Med.* **2019**, *6*, 208. [CrossRef] [PubMed]
194. Lutze, S.; Daeschlein, G.; Konschake, W.; Jünger, M. Rheopheresis as a causal therapy option for systemic scleroderma (SSc). *Clin. Hemorheol. Microcirc.* **2017**, *67*, 229–240. [CrossRef]
195. Cid, J.; Carbassé, G.; Andreu, B.; Baltanás, A.; Garcia-Carulla, A.; Lozano, M. Efficacy and safety of plasma exchange: An 11-year single-center experience of 2730 procedures in 317 patients. *Transfus. Apher. Sci.* **2014**, *51*, 209–214. [CrossRef] [PubMed]
196. Mokrzycki, M.H.; Balogun, R.A. Therapeutic apheresis: A review of complications and recommendations for Prevention and management. *J. Clin. Apher.* **2011**, *26*, 243–248. [CrossRef] [PubMed]
197. Graus, F.; Abos, J.; Roquer, J.; Mazzara, R.; Pereira, A. Effect of plasmapheresis on serum and CSF autoantibody levels in CNS paraneoplastic syndromes. *Neurology* **1990**, *40*, 1621. [CrossRef]
198. Hohenstein, B.; Bornstein, S.; Aringer, M. Immunoadsorption for connective tissue disease. *Atheroscler. Suppl.* **2013**, *14*, 185–189. [CrossRef]
199. Mulhearn, B.; Bruce, I.N. Indications for IVIG in rheumatic diseases. *Rheumatology* **2014**, *54*, 383–391. [CrossRef]
200. Jolles, S. A review of high-dose intravenous immunoglobulin (hdIVIg) in the treatment of the autoimmune blistering disorders. *Clin. Exp. Dermatol.* **2001**, *26*, 127–131. [CrossRef]
201. Sewell, W.A.C.; Jolles, S. Immunomodulatory action of intravenous immunoglobulin. *Immunology* **2002**, *107*, 387–393. [CrossRef]
202. Agostini, E.; De Luca, G.; Bruni, C.; Bartoli, F.; Tofani, L.; Campochiaro, C.; Pacini, G.; Moggi-Pignone, A.; Guiducci, S.; Bellando-Randone, S.; et al. Intravenous immunoglobulins reduce skin thickness in systemic sclerosis: Evidence from Systematic Literature Review and from real life experience. *Autoimmun. Rev.* **2021**, *20*, 102981. [CrossRef]
203. Sanges, S.; Rivière, S.; Mekinian, A.; Martin, T.; Le Quellec, A.; Chatelus, E.; Lescoat, A.; Jego, P.; Cazalets, C.; Quéméneur, T.; et al. Intravenous immunoglobulins in systemic sclerosis: Data from a French nationwide cohort of 46 patients and review of the literature. *Autoimmun. Rev.* **2017**, *16*, 377–384. [CrossRef]
204. Levin, M.J.; Ustianowski, A.; De Wit, S.; Launay, O.; Avila, M.; Templeton, A.; Yuan, Y.; Seegobin, S.; Ellery, A.; Levinson, D.J.; et al. Intramuscular AZD7442 (Tixagevimab–Cilgavimab) for Prevention of Covid-19. *N. Engl. J. Med.* **2022**, *386*, 2188–2200. [CrossRef] [PubMed]
205. Goldenberg, M.M. Etanercept, a novel drug for the treatment of patients with severe, active rheumatoid arthritis. *Clin. Ther.* **1999**, *21*, 75–87. [CrossRef]
206. Vaccaro, C.; Zhou, J.; Ober, R.J.; Ward, E.S. Engineering the Fc region of immunoglobulin G to modulate in vivo antibody levels. *Nat. Biotechnol.* **2005**, *23*, 1283–1288. [CrossRef] [PubMed]
207. Goebeler, M.; Bata-Csörgő, Z.; De Simone, C.; Didona, B.; Remenyik, E.; Reznichenko, N.; Stoevesandt, J.; Ward, E.; Parys, W.; de Haard, H.; et al. Treatment of pemphigus vulgaris and foliaceus with efgartigimod, a neonatal Fc receptor inhibitor: A phase II multicentre, open-label feasibility trial*. *Br. J. Dermatol.* **2021**, *186*, 429–439. [CrossRef]
208. Maho-Vaillant, M.; Sips, M.; Golinski, M.-L.; Vidarsson, G.; Goebeler, M.; Stoevesandt, J.; Bata-Csörgő, Z.; Balbino, B.; Verheesen, P.; Joly, P.; et al. FcRn Antagonism Leads to a Decrease of Desmoglein-Specific B Cells: Secondary Analysis of a Phase 2 Study of Efgartigimod in Pemphigus Vulgaris and Pemphigus Foliaceus. *Front. Immunol.* **2022**, *13*, 863095. [CrossRef]
209. Werth, V.P.; Culton, D.A.; Concha, J.S.; Graydon, J.S.; Blumberg, L.J.; Okawa, J.; Pyzik, M.; Blumberg, R.S.; Hall, R.P. Safety, Tolerability, and Activity of ALXN1830 Targeting the Neonatal Fc Receptor in Chronic Pemphigus. *J. Investig. Dermatol.* **2021**, *141*, 2858–2865.e4. [CrossRef]
210. Robak, T.; Kaźmierczak, M.; Jarque, I.; Musteata, V.; Treliński, J.; Cooper, N.; Kiessling, P.; Massow, U.; Woltering, F.; Snipes, R.; et al. Phase 2 multiple-dose study of an FcRn inhibitor, rozanolixizumab, in patients with primary immune thrombocytopenia. *Blood Adv.* **2020**, *4*, 4136–4146. [CrossRef]
211. Newland, A.C.; Sánchez-González, B.; Rejtő, L.; Egyed, M.; Romanyuk, N.; Godar, M.; Verschueren, K.; Gandini, D.; Ulrichts, P.; Beauchamp, J.; et al. Phase 2 study of efgartigimod, a novel FcRn antagonist, in adult patients with primary immune thrombocytopenia. *Am. J. Hematol.* **2019**, *95*, 178–187. [CrossRef]

212. Sun, W.; Khare, P.; Wang, X.; Challa, D.K.; Greenberg, B.M.; Ober, R.J.; Ward, E.S. Selective Depletion of Antigen-Specific Antibodies for the Treatment of Demyelinating Disease. *Mol. Ther.* **2021**, *29*, 1312–1323. [CrossRef] [PubMed]
213. Wang, P.; Yang, Y.; Hong, H.; Zhang, Y.; Cai, W.; Fang, D. Aptamers as therapeutics in cardiovascular diseases. *Curr. Med. Chem.* **2011**, *18*, 4169–4174. [CrossRef] [PubMed]
214. Famulok, M.; Mayer, G. Aptamers and SELEX in Chemistry & Biology. *Chem. Biol.* **2014**, *21*, 1055–1058. [CrossRef] [PubMed]
215. Haberland, A.; Wallukat, G.; Schimke, I. Aptamer Binding and Neutralization of β1-Adrenoceptor Autoantibodies: Basics and a Vision of Its Future in Cardiomyopathy Treatment☆ ☆This work was supported by the European Regional Development Fund (10141685; Berlin, Germany) and Stiftung Pathobiochemie, Deutsche Gesellschaft für Klinische Chemie und Laboratoriumsmedizin (66/2007) and 48/2011, Germany). *Trends Cardiovasc. Med.* **2011**, *21*, 177–182. [CrossRef]
216. Zhang, Y.; Lai, B.S.; Juhas, M. Recent Advances in Aptamer Discovery and Applications. *Molecules* **2019**, *24*, 941. [CrossRef]
217. Ng, E.W.M.; Shima, D.T.; Calias, P.; Cunningham, E.T.C., Jr.; Guyer, D.R.; Adamis, A.P. Pegaptanib, a targeted anti-VEGF aptamer for ocular vascular disease. *Nat. Rev. Drug Discov.* **2006**, *5*, 123–132. [CrossRef]
218. Paborsky, L.; McCurdy, S.; Griffin, L.; Toole, J.; Leung, L. The single-stranded DNA aptamer-binding site of human thrombin. *J. Biol. Chem.* **1993**, *268*, 20808–20811. [CrossRef]
219. Haberland, A.; Holtzhauer, M.; Schlichtiger, A.; Bartel, S.; Schimke, I.; Müller, J.; Dandel, M.; Luppa, P.B.; Wallukat, G. Aptamer BC 007—A broad spectrum neutralizer of pathogenic autoantibodies against G-protein-coupled receptors. *Eur. J. Pharmacol.* **2016**, *789*, 37–45. [CrossRef]
220. Wallukat, G.; Müller, J.; Haberland, A.; Berg, S.; Schulz, A.; Freyse, E.-J.; Vetter, R.; Salzsieder, E.; Kreutz, R.; Schimke, I. Aptamer BC007 for neutralization of pathogenic autoantibodies directed against G-protein coupled receptors: A vision of future treatment of patients with cardiomyopathies and positivity for those autoantibodies. *Atherosclerosis* **2015**, *244*, 44–47. [CrossRef]
221. Cabral-Marques, O.; Halpert, G.; Schimke, L.F.; Ostrinski, Y.; Vojdani, A.; Baiocchi, G.C.; Freire, P.P.; Filgueiras, I.S.; Zyskind, I.; Lattin, M.T.; et al. Autoantibodies targeting GPCRs and RAS-related molecules associate with COVID-19 severity. *Nat. Commun.* **2022**, *13*, 1–12. [CrossRef]
222. Hohberger, B.; Harrer, T.; Mardin, C.; Kruse, F.; Hoffmanns, J.; Rogge, L.; Heltmann, F.; Moritz, M.; Szewczykowski, C.; Schottenhamml, J.; et al. Case Report: Neutralization of Autoantibodies Targeting G-Protein-Coupled Receptors Improves Capillary Impairment and Fatigue Symptoms After COVID-19 Infection. *Front. Med.* **2021**, *8*, 2008. [CrossRef]
223. Zhu, X.; Li, L.; Zou, L.; Zhu, X.; Xian, G.; Li, H.; Tan, Y.; Xie, L. A Novel Aptamer Targeting TGF-β Receptor II Inhibits Transdifferentiation of Human Tenon's Fibroblasts into Myofibroblast. *Investig. Opthalmol. Vis. Sci.* **2012**, *53*, 6897–6903. [CrossRef]
224. Khanna, D.; Denton, C.P.; Jahreis, A.; van Laar, J.M.; Frech, T.M.; Anderson, M.E.; Baron, M.; Chung, L.; Fierlbeck, G.; Lakshminarayanan, S.; et al. Safety and efficacy of subcutaneous tocilizumab in adults with systemic sclerosis (faSScinate): A phase 2, randomised, controlled trial. *Lancet* **2016**, *387*, 2630–2640. [CrossRef] [PubMed]
225. Roofeh, D.; Lin, C.J.F.; Goldin, J.; Kim, G.H.; Furst, D.E.; Denton, C.P.; Huang, S.; Khanna, D.; the focuSSced Investigators. Tocilizumab Prevents Progression of Early Systemic Sclerosis–Associated Interstitial Lung Disease. *Arthritis Rheumatol.* **2021**, *73*, 1301–1310. [CrossRef]
226. Denton, C.P.; Ong, V.H.; Xu, S.; Chen-Harris, H.; Modrusan, Z.; Lafyatis, R.; Khanna, D.; Jahreis, A.; Siegel, J.; Sornasse, T. Therapeutic interleukin-6 blockade reverses transforming growth factor-beta pathway activation in dermal fibroblasts: Insights from the faSScinate clinical trial in systemic sclerosis. *Ann. Rheum. Dis.* **2018**, *77*, 1362–1371. [CrossRef]
227. Cardoneanu, A.; Burlui, A.M.; Macovei, L.A.; Bratoiu, I.; Richter, P.; Rezus, E. Targeting Systemic Sclerosis from Pathogenic Mechanisms to Clinical Manifestations: Why IL-6? *Biomedicines* **2022**, *10*, 318. [CrossRef]

Review

Emerging Evidence and Treatment Perspectives from Randomized Clinical Trials in Systemic Sclerosis: Focus on Interstitial Lung Disease

Caterina Oriana Aragona [1], Antonio Giovanni Versace [1], Carmelo Ioppolo [1], Daniela La Rosa [1], Rita Lauro [1], Maria Concetta Tringali [1], Simona Tomeo [1], Guido Ferlazzo [2], William Neal Roberts [3], Alessandra Bitto [1], Natasha Irrera [1,†] and Gianluca Bagnato [1,*,†]

1 Department of Clinical and Experimental Medicine, University of Messina, 98125 Messina, Italy; oriana.aragona@gmail.com (C.O.A.); antoniogiovanni.versace@polime.it (A.G.V.); carmelo.ioppolo@polime.it (C.I.); daniela.larosa@polime.it (D.L.R.); irritalauro@gmail.com (R.L.); mariaconcetta.tringali@polime.it (M.C.T.); simona.tomeo@polime.it (S.T.); abitto@unime.it (A.B.); nirrera@unime.it (N.I.)

2 Department of Human Pathology "G. Barresi", University of Messina, 98125 Messina, Italy; guido.ferlazzo@unime.it

3 Department of Medicine, University of Kentucky, Lexington, KY 40506, USA; neal.roberts@uky.edu

* Correspondence: gianbagnato@gmail.com

† These authors contributed equally to this work.

Citation: Aragona, C.O.; Versace, A.G.; Ioppolo, C.; La Rosa, D.; Lauro, R.; Tringali, M.C.; Tomeo, S.; Ferlazzo, G.; Roberts, W.N.; Bitto, A.; et al. Emerging Evidence and Treatment Perspectives from Randomized Clinical Trials in Systemic Sclerosis: Focus on Interstitial Lung Disease. *Biomedicines* **2022**, *10*, 504. https://doi.org/10.3390/biomedicines10020504

Academic Editor: Michal Tomcik

Received: 26 January 2022
Accepted: 17 February 2022
Published: 21 February 2022

Publisher's Note: MDPI stays neutral with regard to jurisdictional claims in published maps and institutional affiliations.

Abstract: Systemic sclerosis (SSc) is a complex rare autoimmune disease with heterogeneous clinical manifestations. Currently, interstitial lung disease (ILD) and cardiac involvement (including pulmonary arterial hypertension) are recognized as the leading causes of SSc-associated mortality. New molecular targets have been discovered and phase II and phase III clinical trials published in the last 5 years on SSc-ILD will be discussed in this review. Details on the study design; the drug tested and its dose; the inclusion and exclusion criteria of the study; the concomitant immunosuppression; the outcomes and the duration of the study were reviewed. The two most common drugs used for the treatment of SSc-ILD are cyclophosphamide and mycophenolate mofetil, both supported by randomized controlled trials. Additional drugs, such as nintedanib and tocilizumab, have been approved to slow pulmonary function decline in SSc-ILD. In this review, we discuss the therapeutic alternatives for SSc management, offering the option to customize the design of future studies to stratify SSc patients and provide a patient-specific treatment according to the new emerging pathogenic features of SSc-ILD.

Keywords: systemic sclerosis; interstitial lung disease; clinical trial

1. Introduction

Systemic sclerosis (SSc) is a complex rare autoimmune disease with heterogeneous clinical manifestations, including vasculopathy, immune dysfunction, musculoskeletal inflammation, and fibrosis of the skin and internal organs [1,2]. Currently, interstitial lung disease (ILD) and cardiac involvement (including pulmonary arterial hypertension) are recognized as the leading causes of SSc-associated mortality; in particular, ILD appears to be virtually always present in SSc patients according to autopsy studies, with up to 90% having evidence of pulmonary involvement on HRCT, with 40–75% showing reduced pulmonary function tests [3]. Indeed, interstitial lung disease accounts for up to a third of mortality causes [4,5], thus representing the leading cause of death in SSc. The discovery of targeted therapies is still an unmet clinical need, due to the complex multifactorial pathogenesis.

SSc-ILD progression is commonly characterized by a slow rate of progression, with about a quarter of SSc patients experiencing ILD progression in 1 year time and around a third in 5-year time [6]. A number of candidate therapies are under evaluation and, simultaneously, ongoing trials are also numerous, including phase I and II studies.

Cyclophosphamide (CYC) and mycophenolate mofetil (MMF) are the two most common drugs used for the treatment of SSc-ILD; their use is supported by randomized controlled trials (RCT) [7,8] that demonstrated similar effective results, although MMF has less risk to fertility, favorable ease of follow-up, with a reduced risk of secondary malignancies. Nevertheless, the latest SSc-ILD treatment guidelines recommend CYC and hematopoietic stem cell transplant considering these therapeutic approaches are supported by completed RCTs [9]. Additional drugs, such as nintedanib and tocilizumab, have been approved to slow down pulmonary function decline in SSc-ILD [10–12].

In this context, the aim of the present literature review is to analyze the phase II and phase III SSc-ILD randomized clinical trials published in the last 5 years.

Full peer-reviewed manuscripts reporting phase II, phase III, and head-to-head randomized clinical trials regarding SSc-ILD studies, published from 1 January 2016 to 31 December 2021 and describing outcomes following pharmaceutical-based interventions, were included. Studies with heterogeneous ILD populations (e.g., connective tissue disease-associated ILD (CTD-ILD)) as well as study designs, case reports, review articles, letters to the editor, editorials, preclinical studies, non-pharmacological interventions, and guidelines were excluded from the analysis.

Study design with sample size, treatment regimen details, participant baseline characteristics, study endpoints, assessment of pulmonary function, patient function or quality of life measures, survival and safety outcomes were reported in a descriptive table for phase II, phase III, and head-to-head randomized clinical trials.

2. Randomized Clinical Trials

2.1. Phase III

2.1.1. Nintedanib

Nintedanib, a multi-target tyrosine kinase inhibitor, is a small molecule designed as an ATP-competitive inhibitor of platelet-derived growth factor receptor (PDFGR), fibroblast growth factor receptor (FGFR), and vascular endothelial growth factor receptor (VEGFR), recently approved for the treatment of idiopathic pulmonary fibrosis (IPF) [13]. SSc-ILD and IPF share some similar fibrogenic mechanisms that include fibroblast activation, migration, proliferation, and differentiation into myofibroblasts, resulting in excessive collagen deposition [14]. The phase III, multi-center, randomized, double-blind, placebo-controlled SENSCIS (Safety and Efficacy of Nintedanib in Systemic SClerosIS) trial evaluated nintedanib safety and efficacy in patients affected by SSc-ILD (NCT02597933).

The primary endpoint was the annual rate of forced vital capacity (FVC) decline (ml/year) assessed over 52 weeks [15]. In the primary end-point analysis, the adjusted annual rate FVC change was −52.4 mL/year in the treated group and −93.3 mL/year in the placebo group (difference, 41.0 mL per year; 95% confidence interval (CI), 2.9 to 79.0; $p = 0.04$). The adjusted mean annual rate of FVC change as a percentage of the predicted value at week 52 was −1.4% in the nintedanib arm and −2.6% in the placebo arm (difference, 1.2 percentage points: 95% CI, 0.1 to 2.2). Patients in the arm treated with MMF at baseline had a better performance. Furthermore, the decrease in FVC differed in the placebo group and it is associated with MMF use. Although a randomization according to MMF administration was not used, these data suggest a potential additional benefit of MMF use with nintedanib on lung function [10]. However, additional studies are necessary to explore the combined use of MMF and nintedanib in SSc-ILD patients. In a post hoc analysis of the SENSCIS trial, the authors assessed the proportions of patients with categorical changes in % predicted FVC at week 52 and the time to absolute decline in FVC of ≥5% predicted or death and absolute decline in FVC of ≥10% predicted or death. Over 52 weeks, the hazard ratio (HR) for an absolute decline in FVC of ≥5% predicted or death with nintedanib versus placebo was 0.83 (95% confidence interval [95% CI] 0.66−1.06) ($p = 0.14$), and the HR for an absolute decline in FVC of ≥10% predicted was 0.64 (95% CI 0.43−0.95)

($p = 0.029$), confirming that nintedanib has a clinically significant advantage in reducing the progression of SSc-ILD [16].

2.1.2. Tocilizumab

Tocilizumab (TCZ) is a recombinant humanized monoclonal antibody of the immunoglobulin G1k subclass, that binds interleukin 6 (IL-6) receptor, thus blocking its signaling. The molecule is genetically engineered and is produced by grafting the complementarity determining region of mouse anti-human IL-6 R to human IgG1 [17]. TCZ is approved in Europe for the treatment of selected forms of rheumatoid arthritis and analogous diseases non-responsive to methotrexate or other disease-modifying anti-rheumatic drugs and it is under evaluation in other diseases sharing inflammatory pathogenesis [18]. The rationale for targeting IL-6 starts from the observation of high levels of IL-6 in skin and lung tissues of SSc-patients [19,20]; moreover, IL-6 levels tightly correlate with skin thickness scores supporting a causal relationship [21]. Starting from these data the faSScinate trial (NCT01532869) was designed. FaSScinate is a phase 2, randomized, double-blind, placebo-controlled trial enrolling 87 SSc patients [11]. Patients were assigned into different groups using randomization (1:1) to receive weekly subcutaneous treatment with TCZ 162 mg or placebo for 48 weeks followed by open-label weekly TCZ for additional 48 weeks. The primary efficacy endpoint was the difference in mean change from baseline in modified Rodnan skin score (mRSS) at week 24. Despite the primary endpoint was not achieved, with a difference of −2.70 mRSS units (95% CI: −5.85, 0.45) in favor of TCZ and in absence of a statistical significance ($p = 0.0915$), the study evidenced that fewer patients in the TCZ arm had a decline in % predicted FVC than in the placebo arm with respect to cumulative distribution (week 48, $p = 0.0373$). Overall, after 48 weeks of treatment, safety in faSScinate was coherent with the natural history of SSc and the known safety profile for TCZ [10]. Following these encouraging results, a phase 3 study, the focuSSced trial (NCT02453256), was designed [22]. The primary endpoint was the difference in change from baseline to week 48 in modified Rodnan skin score (mRSS), while % predicted FVC, time to treatment failure, and patient-reported and physician-reported outcomes were secondary endpoints. The change in mRSS was higher in the TCZ arm (−6.1 vs. −4.4) but it did not meet significance ($p = 0.1$). On the other hand, the TCZ group showed a significant stabilization of FVC compared to placebo, according to the subgroup analysis that considered only SSc participants with ILD (−14 mL vs. −255 mL at 48 weeks, $p < 0.001$). These findings were also associated with a reduced number of SSc-ILD patients having a ≥10% decline in % predicted FVC in the TCZ compared to placebo (5 vs. 14). Taken all together, these results support the use of TCZ in SSc patients with early ILD, as already approved by the Food and Drug Administration.

2.1.3. Lenabasum

Lenabasum is a synthetic cannabinoid receptor type-2 (CB2) agonist recognized as an inflammation-resolving drug candidate for the treatment of different diseases, thanks to its anti-inflammatory effects, and also for systemic sclerosis [23]. A randomized, double-blind, placebo-controlled, phase II study (NCT02465437) was designed to evaluate the safety and tolerability of lenabasum in patients with active SSc [24]. Lenabasum treatment was safe, well-tolerated, and improved multiple efficacy assessments of overall disease, such as skin involvement and patient-reported outcomes. FVC was used as a lung performance measure and was associated with numerical improvement from baseline in the lenabasum group compared to the placebo group starting at week 8, with a maximal but non-significant mean ± SEM treatment difference of 1.7 ± 1.6% observed at week 12. These data encouraged authors to provide other analyses and patients who had completed the 16-week phase 2 study were enrolled to continue lenabasum treatment at the dose of 20 mg twice a day. Thirty-six patients were enrolled and 26 patients were treated for >92 weeks. Predicted FVC values declined by 3.2% from the start of the study, but the trial design of this "open period" limits the conclusion. Despite the interesting results, the study shows some limitations, including the small number of patients, short-term observation period, and the subsequent open phase. RESOLVE-1 (NCT03398837) was the phase 3 study [25]. This study presents some novel aspects with respect to other studies: the American College of Rheumatology (ACR) Combined Response Index in diffuse cutaneous Systemic Sclerosis (CRISS) score was assigned as the primary outcome and interim analysis has been presented at the 2021 American College of Rheumatology meeting [26]. The CRISS score involves a two-step process: the first step is to identify relevant disease worsening or the occurrence of new-onset end-organ damage. The second step requires calculating the probability of patient improvement after 1 year of therapy based on a scale ranging from 0 to 1 point according to changes from baseline status in five domains: the mRSS, % predicted FVC, patient and physician global assessments, and the Health Assessment Questionnaire Disability Index (HAQ-DI). A responder is defined as a patient having a CRISS score of 0.6 or higher and no significant renal or cardiopulmonary worsening [27]. Despite in the RESOLVE-1 trial the primary outcome was not achieved, post hoc analysis suggests that patients taking lenabasum in association with background MMF for more than 2 years were more likely to have stable % predicted FVC over 1 year compared to MMF and placebo (64% vs. 35%). These findings, however, need to be confirmed in future studies.

2.2. *Phase II*

2.2.1. Pirfenidone

Pirfenidone is a small molecule comprising a modified phenyl pyridine with pleiotropic anti-fibrotic, antioxidant and anti-inflammatory properties, although its exact mechanism of action remains unclear. Pirfenidone was initially identified as an anti-inflammatory compound in animal models [28]. However, the unexpected identification of anti-fibrotic effects in animals treated with pirfenidone redefined the interest in this molecule [29]. Pirfenidone has been shown to attenuate fibrosis in different organs, including lung, liver, heart, and kidney [30]; it is actually approved for IPF treatment [31] and has also been tested in SSc-ILD patients in the LOTUSS study (an Open Label, RandOmized, Phase 2 STUdy of the Safety and Tolerability of Pirfenidone when Administered to Patients with Systemic Sclerosis-Related Interstitial Lung Disease) (NCT01933334) [32]. However, pirfenidone use is known to be associated with adverse events (AE) related to the liver, skin, and gastrointestinal (GI) system which are also frequently observed in SSc patients. Reasonably, the primary endpoint was the safety of this treatment in SSc-ILD patients, whereas the secondary endpoint was the effect of pirfenidone on predicted FVC and diffusion capacity of carbon monoxide (DLCO). Data showed an acceptable tolerability profile of pirfenidone in SSc-ILD and this tolerability was not affected by concomitant MMF use. However, FVC and DLCO values were basically unchanged throughout the observation period and no clinically relevant differences were observed in lung function variables between the groups or in any of the subgroup analyses. Lately, another double-blind, randomized, placebo-controlled, pilot study was conducted to evaluate the safety and efficacy of pirfenidone in SSc-ILD. Analogous to the LOTUSS trial, this study was unsuccessful in finding a relevant beneficial effect of pirfenidone over placebo in improving/stabilizing FVC, exercise capacity, symptoms, or skin disease. Nevertheless, the trial is underpowered to allow conclusive evidence [33].

2.2.2. Pomalidomide

Pomalidomide (POM) is an immunomodulatory agent structurally similar to thalidomide and lenalidomide. POM had shown immune-modulating activity on myeloid and lymphocyte cells and exhibited anti-fibrotic effects in pre-clinical models of dermal fibrosis [34]. Furthermore, it enhances T cell and natural killer (NK) cell-mediated immunity and inhibits pro-inflammatory cytokines release, such as tumor necrosis factor (TNF) or IL-6. In a study including 11 SSc patients, histologic comparison of skin biopsies showed changes in skin fibrosis and an increase in epidermal and dermal infiltrating CD8 (+) T cells following thalidomide treatment. Moreover, thalidomide reduced IL-12 and TNF plasma levels. These changes were associated with clinical effects, including dry skin, dermal edema, transient rashes, and healing of digital ulcers [35]. Starting from these data and from a similar structure, POM was tested in SSc patients. Due to difficulties in recruiting patients for the study owing to restrictive inclusion and exclusion criteria, the sponsor terminated enrollment. According to the interim analysis data, the study did not show a significant amelioration in any of the co-primary efficacy endpoints (changes from baseline in FVC and mRSS) for patients who concluded blinded treatment [36].

2.2.3. Romilkimab

Romilkimab is a bispecific monoclonal (immunoglobulin-G4) antibody that binds and neutralizes both IL-4 and IL-13 [37]. IL-4 and IL-13 are Th2-derived cytokines involved in the pro-fibrotic mechanisms of SSc; in fact, their increased levels have been detected both in serum and in skin biopsies of SSc patients [38]. A phase 2A, randomized, double-blind, placebo-controlled, 24-week trial was performed in SSc patients (NCT02921971) [39] that focused the primary endpoint on evaluating the change from baseline to week 24 in modified mRSS and the secondary endpoints on FVC and DLCO. Romilkimab resulted in a significant reduction in mRSS from baseline to week 24 versus placebo (−4.32 to −0.31; p = 0.0291. The least square mean (SE) change in FVC was −10 (40) mL for romilkimab versus −80 (40) mL for placebo at week 24 resulting in a non-significant mean (SE) difference (95% CI) of 70 (60) mL (−40 to 190; p = 0.10). In this study, romilkimab had a non-significant but favorable effect on lung outcomes, which might justify further assessment. Moreover, in patients treated with a placebo, the evidence of the loss of 80 mL for FVC between baseline and week 24 might supports the hypothesis that patients with early SSc may develop significant lung disease in a very short period of time.

2.2.4. Riociguat

The soluble guanylate cyclase (sGC) stimulator riociguat increases intracellular cyclic guanosine monophosphate (cGMP) and consequently activates protein kinases G, with an effect on the regulation of vascular tone and remodeling [40]. Riociguat was approved for the treatment of pulmonary arterial hypertension [41]. A 52 week, double-blind, placebo-controlled, multi-center, randomized phase 2 study (RISE-SSc, NCT0228376219) was performed in recently diagnosed SSc patients (disease duration $\leq$ 18 months) in order to investigate the potential effects of riociguat. The primary endpoint was the change in mRSS from baseline to week 52 whereas among secondary endpoints, change in FVC% was evaluated. Among secondary endpoints, DLCO% and FVC% were evaluated overall and (post hoc) in patients with ILD according to medical history and restrictive lung disease (FVC% 50–75% at baseline). The modification in FVC% between baseline and week 52 was −2.38% (SD 7.52) with riociguat and −2.95% (SD 9.73) with placebo (difference of LS means −0.20 (SE 1.61); 95% CI −3.40 to 3.00; nominal p = 0.901). Depending on the diagnosis (patients with ILD or patients with restrictive disease) the mean change in FVC% from baseline to week 52 was −7.6 to −8.7% with placebo and +0.7 to −2.7% with riociguat. Some measures of mRSS, lung function in patients with evidence for pre-existing ILD and the prevention of new Raynaud's phenomenon and digital ulcers symptoms suggest potential signals for efficacy. It is important to highlight that the results of descriptive analyses of secondary and exploratory endpoints should not be interpreted as the efficacy of riociguat, but as a potential indicator to be examined in further studies [42].

2.2.5. Rituximab

Rituximab (RTX) is a chimeric murine/human monoclonal antibody targeted against CD20, a specific pan-B-cell marker and the only binding site for RTX. RTX binding to the cell surface results in the destruction of lymphocytes through several mechanisms including apoptosis activation, complement-dependent cytotoxicity, or antibody-dependent cytotoxicity.

B cells have been found in the skin and lungs of patients with SSc-ILD, and an increased expression of B cell-related genes has also been demonstrated in the skin [11]. Moreover, in SSc patients, the B cell population is predominantly represented by naïve B cells compared to memory B cells, although the latter are highly active [43].

A multi-center, open-label, comparative study was performed on a total of 51 patients with SSc-ILD and treated with rituximab (RTX) or conventional treatment (azathioprine, methotrexate, and MMF) [44]. Patients treated with RTX showed an increase in FVC at 2 years (mean ± SD of FVC: 80.60 ± 21.21 vs. 86.90 ± 20.56 at baseline vs. 2 years, respectively, $p = 0.041$ compared to baseline). Patients in the control group had no significant change in FVC during the first 2 years of follow-up. At the 7th year, the remaining patients in the RTX group had higher FVC compared to baseline (mean ± SD of FVC: 91.60 ± 14.81, $p = 0.158$ compared to baseline) in contrast to patients in the control group that showed an FVC deterioration ($p < 0.01$, compared to baseline). Direct comparison between the 2 groups showed a significant benefit for the RTX group in FVC ($p = 0.013$).

Of note, a recent trial analyzed the efficacy and safety of RTX in SSc patients. This was a phase II, multi-center, double-blind, parallel-group, investigator-initiated, randomized, placebo-controlled trial (DESIRES) conducted in four Japanese hospitals. The results, despite the limited number of patients, are relevant demonstrating that RTX at the dose of 375 mg/m^2 every week for 4 weeks produces a significant improvement in mRSS compared to placebo. Additionally, among the secondary endpoints included, FVC% appears to be stable in RTX-treated overall compared to the reported reduction of the placebo, with promising results reported in the subgroup analysis regarding SSc-ILD patients [45].

2.2.6. Abatacept

Based on growing evidence suggesting a key role for T cells in the development of both skin and internal organ damage in systemic sclerosis [46,47], a phase II investigator-initiated, multi-center, double-blind randomized placebo-controlled trial was designed to assess the safety and efficacy of abatacept in 88 early diffuse SSc [48]. mRSS change at 12 months, which was the primary endpoint of the study, was not statistically significant; neither was % predicted FVC. Nonetheless, a significant difference was observed in HAQ-DI and ACR CRISS between abatacept and placebo. These results, together with the superior safety profile of abatacept compared to placebo and the higher number of SSc patients requiring rescue therapy in the placebo group, led to the extension of this trial with an additional 6-month open-label, double-blind, randomized trial [49]. Overall, the authors concluded that both mRSS and FVC% predicted showed numerical improvement in patients assigned abatacept compared with those assigned to placebo which was not statistically significant; moreover, substantial individual heterogeneity should be considered.

2.3. *Head-to-Head*

2.3.1. Rituximab vs. Cyclophosphamide

RTX has been used in patients with SSc with pulmonary and renal involvement and has shown efficacy in patients refractory to CYC [44,50,51]. A prospective, randomized, open-label, parallel-group trial was performed including SSc patients with skin and lung involvement [52]. This trial offers the unique feature to provide a head-to-head study, including CYC as a comparator for the assessment of endpoints. The primary outcome was the % predicted FVC value at 6 months. There was a significant improvement in the predicted FVC in the RTX group [from 61.30 (11.28) at baseline to 67.52 (13.59) at the end of the study; $p = 0.002$]. The mean difference in predicted FVC was 9.46 (95% CI: 3.01, 15.90; $p = 0.003$) in favor of the RTX group. This study met its primary endpoint, demonstrating that RTX improved the % predicted FVC, while CYC was not associated with a % precited FVC stability after 6 months. In fact, % predicted FVC in the RTX group significantly increased from 61.30 at baseline to 67.52 at the end of the study, while in the CYC group FVC showed a non-significant decrease from 59.25 at baseline to 58.06. In conclusion, the authors described a greater increase in % predicted FVC value as well as decreased mRSS in the RTX arm compared with the CYC arm. Since the mean difference in % predicted FVC was in favor of the RTX arm (9.46; 95% CI: 3.01, 15.90; $p = 0.003$) and the lower limit of 95% CI of the mean difference of % predicted FVC was 3.01, this study met the non-inferiority criterion vs. CYC (margin 2%). Furthermore, RTX therapy was associated with a good safety profile.

2.3.2. Mycophenolate Mofetil vs. Cyclophosphamide

Because of its immunosuppressive activity and an acceptable safety and tolerability profile, MMF has been widely employed in uncontrolled studies for the treatment of SSc-ILD in the last years. In fact, Tashkin et al. designed a multi-center randomized, double-blind, clinical trial in order to assess the efficacy and safety of MMF using CYC as a comparator drug in the Scleroderma Lung Study II (SLS II) [8], as the natural evolution of the SLS I [7]. In this trial, 69 SSc-ILD symptomatic patients were treated with MMF for 24 months and 73 with CYC for 12 months followed by placebo for additional 12 months. This preliminary hypothesis was based on the results of SLS I, thus assuming that MMF would be effective as CYC at 18 months, while the 12 months treatment with CYC would be associated with a fall in %-predicted FVC back to untreated values after additional 12 months of placebo treatment. While this trial failed to demonstrate MMF superiority, the results showed that MMF is also not inferior to CYC in producing an improvement in lung function, though modest and measured in an increase in % predicted FVC of $3.0 \pm 1.2\%$ and $3.3 \pm 1.1\%$ within the CYC and MMF treatment arms, respectively. Additionally, MMF induced significant between-treatment differences in DLCO %-predicted and DL/VA %-predicted supporting a slower decline during the MMF treatment than CYC.

Regarding safety, in the CYC arm, hematopoietic suppression occurred more frequently. On the other hand, the number of pneumonias, infections, systemic adverse events, or deaths, was not different between the treatment arms. Nevertheless, a greater percentage of subjects in the CYC arm prematurely discontinued the drug and the tolerated dose of CYC decreased over time to approximately 75% of the target dose.

3. Discussion

SSc exhibits a complex heterogeneity in multisystem organ involvement due to multiple cellular and molecular interactions. A growing body of data identified several potential targets, leading to a better definition of the pathogenesis of the disease. In the last 5 years, SSc-ILD, as the leading cause of mortality among SSc patients, attracted numerous interventional trials with a design improved for stratification, screening, timing, and evaluation [53,54].

In the present critical review of published phase II, phase III, and head-to-head clinical trials from the last 5 years, we found high-quality trials regarding mycophenolate mofetil, rituximab, riociguat, romilkimab, pomalidomide, pirfenidone, abatacept, lenabasum, nintedanib, and tocilizumab (Tables 1–3). These molecules act at different levels in the fibrotic and immunologic pathways of SSc-ILD (Figure 1).

From the comparison of these clinical trials, it appears evident that most of the results related to SSc-ILD and pulmonary function are extrapolated from clinical trials where the primary endpoint is the progression of skin disease, thus patients' selection criteria are not defined specifically for SSc-ILD [55]. A very elegant score, attempting to identify a core set of variables to include in each clinical trial in SSc, is the ACR CRISS and its revised version for response (rCRISS), which has been used recently and encompasses, apart from mRSS and FVC, also patient-and physician-reported outcomes in a single score [56].

As evidenced in Table 2, which shows the trials in the most advanced phase, the primary endpoint is always different and it does not allow to perform an ideal comparison. The nintedanib trial offers the ideal patients' selection, restricted to SSc-ILD, without needing to provide sub-group analysis, such as for tocilizumab and lenabasum.

Table 1. Phase II randomized double-blind clinical trials in SSc-ILD.

Molecule and Trial	Study Design	Endpoints	Efficacy	Safety
Tocilizumab (FaSScinate) (NCT01532869)	Randomization 1:1 TCZ (n = 43) 162 mg (s.c. weekly) Placebo (n = 44) Duration 24 weeks followed by open-label weekly TCZ for additional 24 weeks.	Primary: mRSS Secondary: patient-reported and physician-reported outcomes % predicted FVC, % predicted DLCO	The primary endpoint not reached (treatment difference of −2.70 mRSS units (95% CI: −5.85, 0.45) in favour of TCZ, with no statistical significance (p = 0.0915). Fewer patients in the TCZ arm had a decline in FVC% than in the placebo arm (week 48, p = 0.0373).	42/43 (98%) patients in the TCZ group vs. 40/44 (91%) in the placebo group had adverse events. 14 (33%) vs. 15 (34%) had serious adverse events. Serious infections were more common in the TCZ group (7/43 (16%)) than in the placebo group (2/44 (5%)). One patient died in the TCZ group
Lenabasum (NCT02465437)	Randomization: 2:1 Lenabasum (n = 27) (5 mg once daily, 20 mg once daily, or 20 mg twice daily for 4 weeks, followed by 20 mg twice daily for 8 weeks) Placebo (n = 15) Duration: 12 weeks	Primary: CRISS score	The median CRISS score increased in the lenabasum group during the study, reaching 0.33, versus 0.00 in the placebo group, at week 16 (p = 0.07 by 2-sided mixed-effects model repeated-measures analysis).	9/15 (60%) of the placebo-treated subjects and 17/27 (63%) of the lenabasum-treated subjects had adverse events (AEs). No serious or severe AEs related to lenabasum were observed. No deaths.
Pirfenidone (LOTUSS) (NCT01933334)	Randomization: 1:1 Pirfenidone starting dose: 801 mg/day-maintenance dose of 2403 mg/day 2 week titration (n = 32) 4 week titration (n = 31) Duration: 16 weeks	Safety	Data showed an acceptable tolerability profile of pirfenidone in SSc-ILD, and tolerability was not affected by concomitant MMF use. FVC% and DLCO values were unchanged throughout the observation period, and no clinically relevant differences were observed in lung function variables between the groups or in any of the subgroup analyses	96.8% experienced an adverse event 6 patients discontinued early for treatment-related adverse events in the pirfenidone group.
Pirfenidone (NCT03856853)	Randomization: 1:1 Pirfenidone (n = 17) 2400 mg/day Placebo (n = 17) Duration: 24 weeks	Primary: stabilisation or improvement in FVC Secondary: absolute change in the % predicted FVC, Mahler's dyspnoea index, 6MWD, MRSS and TNF and TGF-β serum levels	Stabilisation/improvement in FVC was seen in 94.1% and 76.5% subjects in the pirfenidone and placebo groups, respectively (p = 0.33). The median (range) absolute change in % predicted FVC was −0.55 (−9 to 7%) and 1.0 (−42 to 11.5%) in the treatment and control groups, respectively (p = 0.51). The changes in 6MWD, dyspnoea scores, MRSS, and levels of TNF and TGF-β were not significantly different between groups.	Adverse events were common among the groups (96.1% in the prifenidone and 100% in the placebo group) Gastrointestinal intolerance led to discontinuation of the drug in two patients in the pirfenidone group and one of these subjects required hospitalisation

Table 1. *Cont.*

Molecule and Trial	Study Design	Endpoints	Efficacy	Safety
Pomalidomide (NCT01559129)	Randmization 1:1 POM (n = 11) 1 mg QD Placebo (n = 12) Duration: 52 weeks (discontinued early)	Primary: mRSS, % predicted FVC and UCLA Scleroderma Clinical Trial Consortium Gastrointestinal Tract	Because of recruitment challenges, subject enrollment was discontinued early. Interim analysis showed that primary endpoints were not met	POM was generally well tolerated, with an adverse event profile consistent with previous studies regaridng POM
Romilkimab (NCT02921971)	Randmization 1:1 Romilkimab (n = 48) 200 mg s.c. weekly Placebo (n = 49) Duration: 24 weeks	Primary: mRSS. Secondary: HAQ-DI, observed FVC/observed DLCO.	Romilkimab resulted in a statistically significant decrease in mRSS versus placebo. FVC and DLCO show a positive trend for romilkimab without reaching statistical significance	Overall incidence of treatment-emergent AEs (TEAEs) was high (>80% in both groups. Mild or moderate in intensity (40%) and severe (2%) for romilkimab; and 76% mild or moderate and severe (8%) for placebo
Riociguat (RISE-SSc) (NCT0228376219)	Randomization 1:1 Riociguat (n = 60) individually adjusted every 2 weeks from 0.5 mg to 2.5 mg orally three times daily over 10 weeks Placebo (n = 61) Duration: 52 weeks	Primary: mRSS Secondary: ACR CRISS, HAQ-DI, patient's global assessment, physician's global assessment and change in FVC%.	Riociguat did not show significantly benefit in mRSS versus placebo at the predefined $p < 0.05$. Overall, the change in FVC% was not significant. Subgroup analysis suggests potential signals for efficacy	Overall, 96.7% patients in the riociguat group and 55 90.2% in the placebo group experienced an AE. Severe adverse events were less common with riociguat than with placebo
Abituzumab (STRATUS) NCT02745145)	Randomization: 2:2:1 24 SSc-ILD patients on stable mycophenolate Abituzumab 1500 mg, abituzumab 500 mg, or placebo every 4 weeks Duration: 104 weeks	Annual rate of change in absolute forced vital capacity	terminated prematurely	Well tolerated No new safety signals were detected.
Abatacept (ASSET) (NCT02161406)	Randomization: 1:1 Abatacept (n = 44) 125 mg s.c. weekly Placebo (n = 44) no background immunomodulatory therapies were allowed apart steroids	Primary: mRSS Secondary: 28-swollen and tender joint count, patient global assessment for overall disease, physician global assessment for overall disease, PROMIS-29 v2 Profile, HAQ-DI, Scleroderma-HAQ-DI VAS pain, burden of digital ulcers and Raynaud's, UCLA GIT 2.0; FVC% predicted, ACR-CRISS	The primary outcome measure was not statistically significant. According to intrinsic gene expression subset based on a machine learning classifier. The fibroproliferative subset showed a numerical increase in FVC% in the abatacept arm (p = 0.19) while all other groups showed decreases in FVC%.	Abatacept was found to be generally safe with no new safety signals, with lower numbers of participants experiencing AEs, infectious AEs, and SAEs compared to the placebo group

Table 1. *Cont.*

Molecule and Trial	Study Design	Endpoints	Efficacy	Safety
Rituximab (DESIRES) (NCT04274257)	Randomization 1:1 Rituximab (n = 28) 375 mg/m^2 once a week for 4 weeks Placebo (n = 28) Duration: 24 weeks	Primary: mRSS Secondary: % predicted FVC, predicted DLCO%, TLC, SF-36, HAQ-DI.	The absolute change in mRSS was significantly lower in the rituximab group than in the placebo group (−6.30 vs. 2.14; difference −8.44 (95% CI −11.00 to −5.88); $p < 0.0001$). Absolute change in % predicted FVC was statistical signficant (0.09% in the rituximab group vs. 2.87% in the placebo group)	Adverse events were almost similar in both groups (100% in the rituximab and 88% in the placebo group) One serious adverse event leading to treatment discontinuation occurred in one patient in each group. Upper respiratory infections occurred in 39% rituximab-treated patients and in 38% of the placebo-treated patients There were no deaths during follow-up.

CRISS: American College of Rheumatology Combined Response Index in diffuse cutaneous Systemic Sclerosis (CRISS) score: MRSS, Health Assessment Questionnaire (HAQ) disability index (DI), physician global assessment of overall patient health, patient global assessment of health, FVC%.

Table 2. Phase III randomized double-blind clinical trials in SSc-ILD.

Molecule and Trial	Study Design	Endpoints	Efficacy	Safety
Tocilizumab (FocuSSced) (NCT02453256)	Ranzomization 1:1 with IL-6 stratification TCZ (n = 104) 162 mg (s.c. weekly) Placebo (n = 106) Duration: 48 weeks	Primary: mean change difference in mRSS. Secondary: % predicted FVC, time to treatment failure, patient-reported and physician-reported outcomes	The change in mRSS was higher in TCZ arm (−6.1 vs. −4.4) but not significant (p = 0.1). TCZ group showed instead a significant slower reduction of FVC% compared to placebo, and the results appear more significant in the subset with SSc-ILD with a stabilization of FVC% contrasting to the decline in the placebo group (−20 mL vs. −257 mL, $p < 0.001$)	Infections were the most common adverse events (52% in the TCZ group vs. 50% in the placebo group). Serious adverse events were reported in 13 participants treated with TCZ and 18 with placebo, primarily infections and cardiac events.
Nintedanib (SENSCIS) (NCT02597933)	Randomization: 1:1 with anti-topoisomerase stratification Nintedanib (n = 288) 150 mg (orally twice daily) Placebo (n = 288), SSc-ILD patients receiving prednisone (≤10 mg/day) and/or a stable dose of MMF or methotrexate, were considered eligible. Duration: 52 weeks	Primary: annual rate of decline in FVC (ml/year) Secondary: absolute changes from baseline in the modified Rodnan skin score and the total score on the St. George's Respiratory Questionnaire (SGRQ)	The adjusted annual rate of change in FVC was −52.4 mL/year in the treated group and −93.3 mL/year in the placebo group (difference, 41.0 mL per year; 95% confidence interval (CI), 2.9 to 79.0; p = 0.04). The adjusted mean annual rate of change in % precited FVC at week 52 was −1.4% in the nintedanib arm and −2.6% in the placebo arm (difference, 1.2 percentage points; 95% CI, 0.1 to 2.2).	Discontinuation was higher in the nintedanib group than in the placebo group (16% vs. 8.7%) Diarrhea, the most common adverse event was reported in 75.7% of the patients in the nintedanib group and in 31.6% of those in the placebo group. 3.5% in the nintedanib group and 3.1% in the placebo group died

Table 2. *Cont.*

Molecule and Trial	Study Design	Endpoints	Efficacy	Safety
Lenabasum (RESOLVE-1) (NCT03398837) Data from interim analysis	Randomization 1:1:1 Lenabasum ($n = 100$) 20 mg orally twice daily Lenabasum ($n = 113$) 5 mg orally twice daily Placebo ($n = 115$) Background immunosuppressive therapies (bIST) were allowed if doses were stable for ≥8 weeks before screening Duration: 52 weeks	Primary: ACR CRISS score—rCRISS Secondary: mRSS, HAQ-DI, FVC	Primary endpoint was not achieved Stable FVC in subjects treated with lenabasum 20 mg and MMF for more than 2 years compared to placebo	Less severe side effects in patients treated with lenabasum 20 mg compared to placebo

CRISS: American College of Rheumatology Combined Response Index in diffuse cutaneous Systemic Sclerosis (CRISS) score: MRSS, Health Assessment Questionnaire (HAQ) disability index (DI), physician global assessment of overall patient health, patient global assessment of health, FVC%.

Table 3. Head-to-head randomized double-blind clinical trials in SSc-ILD.

Molecule and Trial	Study Design	Endpoints	Efficacy	Safety
Rituximab vs. Cyclophosphamide (CTRI/2017/07/009152)	Randomization 1:1 RTX ($n = 30$): 1000 mg × 2 doses at 0, 15 days CYC ($n = 30$): pulses of CYC 500 mg/m^2 Duration: 24 weeks	Primary: % predicted FVC Secondary: variables considered were: absolute change in liters (FVC-l); modified Rodnan skin scores, 6-min walk test, Medsgers score and new onset or worsening of existing pulmonary hypertension by echocardiographic criteria	A significantly higher percentage of patients experienced an improvement of FVC (%) in the RTX group vs. the CYC group (26.7% vs. 6.7%, respectively, $p = 0.038$). However, the rate of worsening of FVC (%) was similar in the RTX and CYC treated patients (3.3% vs. 3% $p = 0.612$)	The total number of patients having an adverse event was lower in the RTX group (30%) vs. the CYC group (70%) ($p = 0.02$) One patient developed severe pulmonary arterial hypertension 5 months after the completion of the trial (in the RTX group) and died.
Mycophenolate mofetil vs. cyclophophamide (NCT00883129)	Randomization 1 1 MMF ($n = 69$): 1500 mg orally twice daily for 104 weeks CYC ($n = 73$) 18 to 2.3 mg/kg orally once daily for 52 weeks + placebo for 52 weeks Duration: 104 weeks	Primary: % predicted FVC superior to CYC Secondary: the course from 3 to 24 months of the DLCO %-predicted, TDI and mRSS scores, and the change from baseline in quantitative HRCT scores for lung fibrosis and total ILD at 24 months	No significant difference between groups, with non-inferiority of the MMF arm vs. CYC arm.	Treatment related adverse events were more frequent in the CYC vs. MMF Time to withdrawal from the study medication or treatment failure was significantly shorter in the CYC arm. Sixteen deaths (11.3% of randomized patients) occurred during the 2-year course of the trial (11 CYC; 5 MMF)

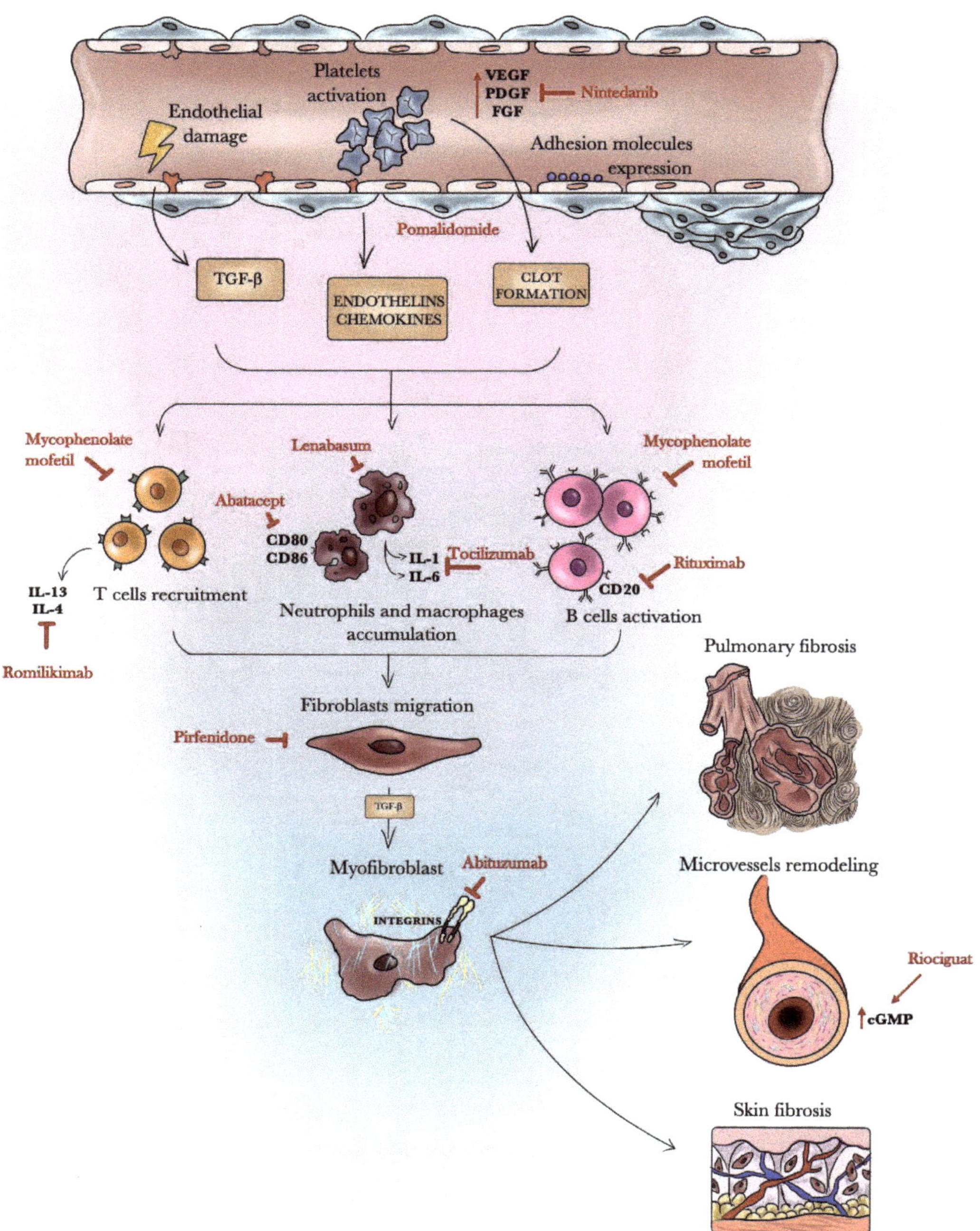

Figure 1. Schematic representation of the mechanism of action of nintedanib, pomalidomide, tocilizumab, rituximab, romilikimab, pirfenidone, lenabasum, and riociguat.

In addition, the enrollment of patients exposed to background immunosuppressive therapy, allowed in the nintedanib and lenabasum trials, and the addition of rescue therapy during the tocilizumab trials at week 16 if needed, further complicates the analysis of the results.

Ideally, the trial design of a study regarding SSc-ILD should provide the additional or cumulative effect of a drug over MMF, based on the results of previous trials confirming its non-inferior efficacy to CYC, already modest [7], and its superiority to CYC in terms of tolerability and treatment discontinuation [8]. Indeed, after the results obtained by MMF in the phase II trial, and the promising data from pirfenidone trial in SSc [33], a randomized clinical trial, namely the Scleroderma Lung Study III, has been designed (NCT03221257), aiming at assessing the efficacy and safety of the combination of the anti-fibrotic effects of pirfenidone with mycophenolate for the treatment of SSc-ILD.

However, it remains difficult to identify and stratify SSc-ILD patients according to the most appropriate target or combination therapy, as supporting evidence to employ anti-fibrotic vs. immunosuppressant medications is lacking. An elegant attempt to suggest a patient stratification strategy, based on autoantibodies profile and skin gene profiling could prove useful in driving the most appropriate therapeutic combination [57].

Rituximab could be another candidate for future trials in SSc-ILD employing a combination therapy design, as this drug showed a better safety profile in a head-to-head study vs. CYC and the stabilization of pulmonary function in a randomized clinical trial compared to placebo [58]. In fact, recent reports support RTX use in association with MMF [59,60]. Similarly, a subgroup analysis of the nintedanib trial has been published reporting the post hoc analysis performed to estimate the proportion of patients with an absolute decrease in FVC of at least 3.3% predicted at week 52, considered as the minimal clinically important difference estimate for worsening of FVC in patients with SSc-ILD, by concomitant MMF use at baseline [61]. This study suggests that combination therapy might have additional beneficial effects for the treatment of SSc-ILD.

This emerging evidence suggests that the default consensus reference drug for SSc-ILD is MMF [10,62,63], as supported by the recent post hoc analysis of the nintedanib trial [61].

Since the sequencing of immunosuppressants versus anti-fibrotic treatments remains to be tested, further combination studies are needed, as already proposed for idiopathic pulmonary fibrosis [31], to clarify if the combination is superior to either one alone.

Revised ACR CRISS (rCRISS) responses: the proportion of participants that improve in $\geq$3/5 core items by certain percentages (30%, except $\geq$5% in FVC%).

4. Conclusions

The primary outcome in ILD studies is to halt disease progression and avoid irreversible lung damage and pulmonary function deterioration. Among the most recent phase III trials, nintedanib and tocilizumab achieved this goal with different endpoints, while the results of the lenabasum study are not available [22,61]. Furthermore, the promising results reported for rituximab, according to the recent phase II trial and the results of the study in comparison with CYC, warrant a confirmatory long-term phase III trial for extensive use. While several published [10,16,62] and upcoming (NCT02370693-NCT03221257) trials confirm the use of MMF as the primary drug in SSc-ILD combination therapy, several additional cellular and molecular targets have been investigated in placebo-controlled trials.

On one hand, the trials regarding pomalidomide, an analog of thalidomide, and abituzumab, a pan-αv integrin inhibiting monoclonal antibody, were interrupted for recruiting issues. On the other hand, subgroup analyses show promising effects on pulmonary function for romilkimab, an anti-interleukin-4/interleukin-13 monoclonal antibody, abatacept, an antibody against costimulatory molecules CD80 and CD86, and riociguat, a cGMP inhibitor. Taken together, these phase II trials confirm that an essential unmet need in SSc-ILD studies is a better definition of the eligible population and a definitive identification of treatment arms for combination therapy and duration of treatment exposure.

5. Future Directions

While there is a general consensus based upon recommendations on the use of immunosuppressants in early SSc to treat cutaneous progression [9], the choice of the most appropriate drug for SSc-ILD between antifibrotic and immunosuppressants remains challenging.

Hopefully, ongoing or upcoming trials, such as the combination of bortezomib and MMF (NCT02370693) or the combination of pirfenidone and MMF (Scleroderma Lung Study III—NCT03221257), with a trial design dedicated to SSc-ILD, will provide new alternatives, increasing the quality of evidence already available, with a precise definition of patients' selection criteria, background immunosuppressive therapy, and treatment duration.

We now have growing therapeutic alternatives for SSc management, offering the option to customize the design of future studies to stratify SSc patients and provide a patient-specific treatment according to the new emerging pathogenic features of SSc-ILD.

Further research is required to identify the therapeutic algorithm to support combination therapy, improve criteria for patient enrollment in clinical trials, and provide the optimal timing for treatment initiation to achieve the ambitious endpoint of improving and stabilizing over time pulmonary function.

Funding: This research received no external funding.

Institutional Review Board Statement: Not applicable.

Informed Consent Statement: Not applicable.

Data Availability Statement: Not applicable.

Conflicts of Interest: The authors declare no conflict of interest.

Abbreviations

Adverse events, AEs; American college of rheumatology, ACR; cannabinoid receptor type-2, CB2; connective tissue disease-associated ILD, CTD-ILD; cluster of differentiation, CD; cyclic guanosine monophosphate, cGMP; combined response index for systemic sclerosis, CRISS; connective tissue growth factor, CTGF; cyclophosphamide, CYC; diffusing capacity of the lungs for carbon monoxide, DLCO; fibroblast growth factor receptor, FGFR; forced vital capacity, FVC; health assessment questionnaire disability index, HAQ-DI; high-resolution computed tomography, HRCT; idiopathic pulmonary fibrosis, IPF; interleukin, IL; interstitial lung disease, ILD; modified Rodnan skin score, mRSS; mycophenolate mofetil, MMF; natural killer, NK; platelet-derived growth factor receptor, PDFGR; pomalidomide, POM; randomized controlled trials, RCT; rituximab, RTX; soluble guanylate cyclase, sGC; systemic sclerosis, SSc; tocilizumab, TCZ; transforming growth factor, TGF; tumor necrosis factor, TNF; vascular endothelial growth factor receptor, VEGFR.

References

1. Denton, C.P.; Khanna, D. Systemic sclerosis. *Lancet* **2017**, *390*, 1685–1699. [CrossRef]
2. Khanna, D.; Tashkin, D.P.; Denton, C.P.; Lubell, M.W.; Vazquez-Mateo, C.; Wax, S. Ongoing clinical trials and treatment options for patients with systemic sclerosis–associated interstitial lung disease. *Rheumatology* **2019**, *58*, 567–579. [CrossRef] [PubMed]
3. Solomon, J.J.; Olson, A.L.; Fischer, A.; Bull, T.; Brown, K.K.; Raghu, G. Scleroderma lung disease. *Eur. Respir. Rev.* **2013**, *22*, 6–19. [CrossRef] [PubMed]
4. Elhai, M.; Meune, C.; Boubaya, M.; Avouac, J.; Hachulla, E.; Balbir-Gurman, A.; Riemekasten, G.; Airò, P.; Joven, B.; Vettori, S.; et al. Mapping and predicting mortality from systemic sclerosis. *Ann. Rheum. Dis.* **2017**, *76*, 1897–1905. [CrossRef]
5. Hoffmann-Vold, A.-M.; Fretheim, H.; Halse, A.-K.; Seip, M.; Bitter, H.; Wallenius, M.; Garen, T.; Salberg, A.; Brunborg, C.; Midtvedt, Ø.; et al. Tracking Impact of Interstitial Lung Disease in Systemic Sclerosis in a Complete Nationwide Cohort. *Am. J. Respir. Crit. Care Med.* **2019**, *200*, 1258–1266. [CrossRef] [PubMed]
6. Hoffmann-Vold, A.-M.; Allanore, Y.; Alves, M.; Brunborg, C.; Airó, P.; Ananieva, L.P.; Czirják, L.; Guiducci, S.; Hachulla, E.; Li, M.; et al. Progressive interstitial lung disease in patients with systemic sclerosis-associated interstitial lung disease in the EUSTAR database. *Ann. Rheum. Dis.* **2021**, *80*, 219–227. [CrossRef] [PubMed]

7. Tashkin, D.P.; Elashoff, R.; Clements, P.J.; Goldin, J.; Roth, M.D.; Furst, D.E.; Arriola, E.; Silver, R.; Strange, C.; Bolster, M.; et al. Cyclophosphamide versus Placebo in Scleroderma Lung Disease. *N. Engl. J. Med.* **2006**, *354*, 2655–2666. [CrossRef]
8. Tashkin, D.P.; Roth, M.D.; Clements, P.J.; Furst, D.E.; Khanna, D.; Kleerup, E.C.; Goldin, J.; Arriola, E.; Volkmann, E.R.; Kafaja, S.; et al. Mycophenolate mofetil versus oral cyclophosphamide in scleroderma-related interstitial lung disease (SLS II): A randomised controlled, double-blind, parallel group trial. *Lancet Respir. Med.* **2016**, *4*, 708–719. [CrossRef]
9. Kowal-Bielecka, O.; Fransen, J.; Avouac, J.; Becker, M.; Kulak, A.; Allanore, Y.; Distler, O.; Clements, P.; Cutolo, M.; Czirjak, L.; et al. Update of EULAR recommendations for the treatment of systemic sclerosis. *Ann. Rheum. Dis.* **2017**, *76*, 1327–1339. [CrossRef]
10. Distler, O.; Highland, K.B.; Gahlemann, M.; Azuma, A.; Fischer, A.; Mayes, M.D.; Raghu, G.; Sauter, W.; Girard, M.; Alves, M.; et al. Nintedanib for Systemic Sclerosis–Associated Interstitial Lung Disease. *N. Engl. J. Med.* **2019**, *380*, 2518–2528. [CrossRef]
11. Khanna, D.; Denton, C.P.; Jahreis, A.; van Laar, J.M.; Frech, T.M.; Anderson, M.E.; Baron, M.; Chung, L.; Fierlbeck, G.; Lakshminarayanan, S.; et al. Safety and efficacy of subcutaneous tocilizumab in adults with systemic sclerosis (faSScinate): A phase 2, randomised, controlled trial. *Lancet* **2016**, *387*, 2630–2640. [CrossRef]
12. Boleto, G.; Avouac, J.; Allanore, Y. An update on recent randomized clinical trials in systemic sclerosis. *Jt. Bone Spine* **2021**, *88*, 105184. [CrossRef] [PubMed]
13. Richeldi, L.; Du Bois, R.M.; Raghu, G.; Azuma, A.; Brown, K.K.; Costabel, U.; Cottin, V.; Flaherty, K.R.; Hansell, D.M.; Inoue, Y.; et al. Efficacy and Safety of Nintedanib in Idiopathic Pulmonary Fibrosis. *N. Engl. J. Med.* **2014**, *370*, 2071–2082. [CrossRef]
14. Bagnato, G.; Harari, S. Cellular interactions in the pathogenesis of interstitial lung diseases. *Eur. Respir. Rev.* **2015**, *24*, 102–114. [CrossRef] [PubMed]
15. Distler, O.; Brown, K.K.; Distler, J.H.W.; Assassi, S.; Maher, T.M.; Cottin, V.; Varga, J.; Coeck, C.; Gahlemann, M.; Sauter, W.; et al. Design of a randomised, placebo-controlled clinical trial of nintedanib in patients with systemic sclerosis-associated interstitial lung disease (SENSCIS™). *Clin. Exp. Rheumatol.* **2017**, *35* (Suppl. 106), 75–81. [PubMed]
16. Maher, T.M.; Mayes, M.D.; Kreuter, M.; Volkmann, E.R.; Aringer, M.; Castellvi, I.; Cutolo, M.; Stock, C.; Schoof, N.; Alves, M.; et al. Effect of Nintedanib on Lung Function in Patients with Systemic Sclerosis—Associated Interstitial Lung Disease: Further Analyses of a Randomized, Double-Blind, Placebo-Controlled Trial. *Arthritis Rheumatol.* **2021**, *73*, 671–676. [CrossRef] [PubMed]
17. Sheppard, M.; Laskou, F.; Stapleton, P.P.; Hadavi, S.; Dasgupta, B. Tocilizumab (Actemra). *Hum. Vaccines Immunother.* **2017**, *13*, 1972–1988. [CrossRef]
18. Shetty, A.; Hanson, R.; Korsten, P.; Shawagfeh, M.; Arami, S.; Volkov, S.; Vila, O.; Swedler, W.; Shunaigat, A.N.; Smadi, S.; et al. Tocilizumab in the treatment of rheumatoid arthritis and beyond. *Drug Des. Dev. Ther.* **2014**, *8*, 349–364. [CrossRef]
19. De Lauretis, A.; Sestini, P.; Pantelidis, P.; Hoyles, R.; Hansell, D.M.; Goh, N.S.; Zappala, C.J.; Visca, D.; Maher, T.M.; Denton, C.P.; et al. Serum Interleukin 6 Is Predictive of Early Functional Decline and Mortality in Interstitial Lung Disease Associated with Systemic Sclerosis. *J. Rheumatol.* **2013**, *40*, 435–446. [CrossRef]
20. Scala, E.; Pallotta, S.; Frezzolini, A.; Abeni, D.; Barbieri, C.; Sampogna, F.; De Pità, O.; Puddu, P.; Paganelli, R.; Russo, G. Cytokine and chemokine levels in systemic sclerosis: Relationship with cutaneous and internal organ involvement. *Clin. Exp. Immunol.* **2004**, *138*, 540–546. [CrossRef]
21. O'Reilly, S.; Cant, R.; Ciechomska, M.; Van Laar, J.M. Interleukin-6: A new therapeutic target in systemic sclerosis? *Clin. Transl. Immunol.* **2013**, *2*, e4. [CrossRef] [PubMed]
22. Khanna, D.; Lin, C.J.F.; Furst, D.E.; Goldin, J.; Kim, G.; Kuwana, M.; Allanore, Y.; Matucci-Cerinic, M.; Distler, O.; Shima, Y.; et al. Tocilizumab in systemic sclerosis: A randomised, double-blind, placebo-controlled, phase 3 trial. *Lancet Respir. Med.* **2020**, *8*, 963–974. [CrossRef]
23. Burstein, S. Molecular Mechanisms for the Inflammation-Resolving Actions of Lenabasum. *Mol. Pharmacol.* **2021**, *99*, 125–132. [CrossRef] [PubMed]
24. Spiera, R.; Hummers, L.; Chung, L.; Frech, T.M.; Domsic, R.; Hsu, V.; Furst, D.E.; Gordon, J.; Mayes, M.; Simms, R.; et al. Safety and Efficacy of Lenabasum in a Phase II, Randomized, Placebo-Controlled Trial in Adults with Systemic Sclerosis. *Arthritis Rheumatol.* **2020**, *72*, 1350–1360. [CrossRef] [PubMed]
25. Spiera, R.; Khanna, D.; Kuwana, M.; Furst, D.E.; Frech, T.M.; Hummers, L.; Stevens, W.; Matucci-Cerinic, M.; Baron, M.; Distler, O.; et al. A randomised, double-blind, placebo-controlled phase 3 study of lenabasum in diffuse cutaneous systemic sclerosis: RESOLVE-1 design and rationale. *Clin. Exp. Rheumatol.* **2021**, *39* (Suppl. 131), 124–133. [PubMed]
26. Abstract Supplement ACR Convergence 2021. *Arthritis Rheumatol.* **2021**, *73*, 1–4259. [CrossRef]
27. Khanna, D.; Berrocal, V.J.; Giannini, E.H.; Seibold, J.R.; Merkel, P.A.; Mayes, M.D.; Baron, M.; Clements, P.J.; Steen, V.; Assassi, S.; et al. The American College of Rheumatology Provisional Composite Response Index for Clinical Trials in Early Diffuse Cutaneous Systemic Sclerosis. *Arthritis Rheumatol.* **2016**, *68*, 299–311. [CrossRef]
28. Iyer, S.N.; Hyde, D.M.; Giri, S.N. Anti-Inflammatory Effect of Pirfenidone in the Bleomycin-Hamster Model of Lung Inflammation. *Inflammation* **2000**, *24*, 477–491. [CrossRef]
29. Azuma, A. Pirfenidone: Antifibrotic agent for idiopathic pulmonary fibrosis. *Expert Rev. Respir. Med.* **2010**, *4*, 301–310. [CrossRef]
30. Schaefer, C.J.; Ruhrmund, D.W.; Pan, L.; Seiwert, S.D.; Kossen, K. Antifibrotic activities of pirfenidone in animal models. *Eur. Respir. Rev.* **2011**, *20*, 85–97. [CrossRef]
31. Wuyts, W.A.; Antoniou, K.M.; Borensztajn, K.; Costabel, U.; Cottin, V.; Crestani, B.; Grutters, J.C.; Maher, T.M.; Poletti, V.; Richeldi, L.; et al. Combination therapy: The future of management for idiopathic pulmonary fibrosis? *Lancet Respir. Med.* **2014**, *2*, 933–942. [CrossRef]

32. Khanna, D.; Albera, C.; Fischer, A.; Khalidi, N.; Raghu, G.; Chung, L.; Chen, D.; Schiopu, E.; Tagliaferri, M.; Seibold, J.R.; et al. An Open-label, Phase II Study of the Safety and Tolerability of Pirfenidone in Patients with Scleroderma-associated Interstitial Lung Disease: The LOTUSS Trial. *J. Rheumatol.* **2016**, *43*, 1672–1679. [CrossRef] [PubMed]
33. Acharya, N.; Sharma, S.K.; Mishra, D.; Dhooria, S.; Dhir, V.; Jain, S. Efficacy and safety of pirfenidone in systemic sclerosis-related interstitial lung disease—A randomised controlled trial. *Rheumatol. Int.* **2020**, *40*, 703–710. [CrossRef] [PubMed]
34. Weingärtner, S.; Zerr, P.; Tomcik, M.; Palumbo-Zerr, K.; Distler, A.; Dees, C.; Beyer, C.; Shankar, S.L.; Cedzik, D.; Schafer, P.H.; et al. Pomalidomide is effective for prevention and treatment of experimental skin fibrosis. *Ann. Rheum. Dis.* **2012**, *71*, 1895–1899. [CrossRef] [PubMed]
35. Oliver, S.J.; Moreira, A.; Kaplan, G. Immune Stimulation in Scleroderma Patients Treated with Thalidomide. *Clin. Immunol.* **2000**, *97*, 109–120. [CrossRef] [PubMed]
36. Hsu, V.M.; Denton, C.P.; Domsic, R.T.; Furst, D.E.; Rischmueller, M.; Stanislav, M.; Steen, V.D.; Distler, J.H.W.; Korish, S.; Cooper, A.; et al. Pomalidomide in Patients with Interstitial Lung Disease due to Systemic Sclerosis: A Phase II, Multicenter, Randomized, Double-blind, Placebo-controlled, Parallel-group Study. *J. Rheumatol.* **2018**, *45*, 405–410. [CrossRef]
37. Zhao, Q. Bispecific Antibodies for Autoimmune and Inflammatory Diseases: Clinical Progress to Date. *BioDrugs* **2020**, *34*, 111–119. [CrossRef]
38. Huang, X.-L.; Wang, Y.-J.; Yan, J.-W.; Wan, Y.-N.; Chen, B.; Li, B.-Z.; Yang, G.-J.; Wang, J. Role of anti-inflammatory cytokines IL-4 and IL-13 in systemic sclerosis. *Inflamm. Res.* **2015**, *64*, 151–159. [CrossRef]
39. Allanore, Y.; Wung, P.; Soubrane, C.; Esperet, C.; Marrache, F.; Bejuit, R.; Lahmar, A.; Khanna, D.; Denton, C.P.; Investigators. A randomised, double-blind, placebo-controlled, 24-week, phase II, proof-of-concept study of romilkimab (SAR156597) in early diffuse cutaneous systemic sclerosis. *Ann. Rheum. Dis.* **2020**, *79*, 1600–1607. [CrossRef]
40. Stasch, J.-P.; Pacher, P.; Evgenov, O.V. Soluble Guanylate Cyclase as an Emerging Therapeutic Target in Cardiopulmonary Disease. *Circulation* **2011**, *123*, 2263–2273. [CrossRef]
41. Ghofrani, H.-A.; Galie, N.; Grimminger, F.; Grünig, E.; Humbert, M.; Jing, Z.-C.; Keogh, A.M.; Langleben, D.; Kilama, M.O.; Fritsch, A.; et al. Riociguat for the Treatment of Pulmonary Arterial Hypertension. *N. Engl. J. Med.* **2013**, *369*, 330–340. [CrossRef] [PubMed]
42. Khanna, D.; Allanore, Y.; Denton, C.P.; Kuwana, M.; Matucci-Cerinic, M.; Pope, J.E.; Atsumi, T.; Bečvář, R.; Czirják, L.; Hachulla, E.; et al. Riociguat in patients with early diffuse cutaneous systemic sclerosis (RISE-SSc): Randomised, double-blind, placebo-controlled multicentre trial. *Ann. Rheum. Dis.* **2020**, *79*, 618–625. [CrossRef] [PubMed]
43. Sato, S.; Fujimoto, M.; Hasegawa, M.; Takehara, K. Altered blood B lymphocyte homeostasis in systemic sclerosis: Expanded naive B cells and diminished but activated memory B cells. *Arthritis Rheum.* **2004**, *50*, 1918–1927. [CrossRef] [PubMed]
44. Daoussis, D.; Melissaropoulos, K.; Sakellaropoulos, G.; Antonopoulos, I.; Markatseli, T.E.; Simopoulou, T.; Georgiou, P.; Andonopoulos, A.P.; Drosos, A.A.; Sakkas, L.; et al. A multicenter, open-label, comparative study of B-cell depletion therapy with Rituximab for systemic sclerosis-associated interstitial lung disease. *Semin. Arthritis Rheum.* **2017**, *46*, 625–631. [CrossRef]
45. Ebata, S.; Yoshizaki, A.; Oba, K.; Kashiwabara, K.; Ueda, K.; Uemura, Y.; Watadani, T.; Fukasawa, T.; Miura, S.; Yoshizaki-Ogawa, A.; et al. Safety and efficacy of rituximab in systemic sclerosis (DESIRES): A double-blind, investigator-initiated, randomised, placebo-controlled trial. *Lancet Rheumatol.* **2021**, *3*, e489–e497. [CrossRef]
46. Ponsoye, M.; Frantz, C.; Ruzehaji, N.; Nicco, C.; Elhai, M.; Ruiz, B.; Cauvet, A.; Pezet, S.; Brandely, M.L.; Batteux, F.; et al. Treatment with abatacept prevents experimental dermal fibrosis and induces regression of established inflammation-driven fibrosis. *Ann. Rheum. Dis.* **2016**, *75*, 2142–2149. [CrossRef]
47. Kalogerou, A.; Gelou, E.; Mountantonakis, S.; Settas, L.; Zafiriou, E.; Sakkas, L. Early T cell activation in the skin from patients with systemic sclerosis. *Ann. Rheum. Dis.* **2005**, *64*, 1233–1235. [CrossRef]
48. Khanna, D.; Spino, C.; Johnson, S.; Chung, L.; Whitfield, M.L.; Denton, C.P.; Berrocal, V.; Franks, J.; Mehta, B.; Molitor, J.; et al. Abatacept in Early Diffuse Cutaneous Systemic Sclerosis: Results of a Phase II Investigator-Initiated, Multicenter, Double-Blind, Randomized, Placebo-Controlled Trial. *Arthritis Rheumatol.* **2020**, *72*, 125–136. [CrossRef]
49. Chung, L.; Spino, C.; McLain, R.; Johnson, S.R.; Denton, C.P.; Molitor, J.A.; Steen, V.D.; Lafyatis, R.; Simms, R.W.; Kafaja, S.; et al. Safety and efficacy of abatacept in early diffuse cutaneous systemic sclerosis (ASSET): Open-label extension of a phase 2, double-blind randomised trial. *Lancet Rheumatol.* **2020**, *2*, e743–e753. [CrossRef]
50. Haroon, M.; McLaughlin, P.; Henry, M.; Harney, S. Cyclophosphamide-refractory scleroderma-associated interstitial lung disease: Remarkable clinical and radiological response to a single course of rituximab combined with high-dose corticosteroids. *Ther. Adv. Respir. Dis.* **2011**, *5*, 299–304. [CrossRef]
51. Braun-Moscovici, Y.; Butbul-Aviel, Y.; Guralnik, L.; Toledano, K.; Markovits, D.; Rozin, A.; Nahir, M.A.; Balbir-Gurman, A. Rituximab: Rescue therapy in life-threatening complications or refractory autoimmune diseases: A single center experience. *Rheumatol. Int.* **2013**, *33*, 1495–1504. [CrossRef] [PubMed]
52. Sircar, G.; Goswami, R.P.; Sircar, D.; Ghosh, A.; Ghosh, P. Intravenous cyclophosphamide vs rituximab for the treatment of early diffuse scleroderma lung disease: Open label, randomized, controlled trial. *Rheumatology* **2018**, *57*, 2106–2113. [CrossRef] [PubMed]
53. Del Galdo, F.; Hartley, C.; Allanore, Y. Randomised controlled trials in systemic sclerosis: Patient selection and endpoints for next generation trials. *Lancet Rheumatol.* **2020**, *2*, e173–e184. [CrossRef]
54. Denton, C.P. Challenges in systemic sclerosis trial design. *Semin. Arthritis Rheum.* **2019**, *49*, S3–S7. [CrossRef] [PubMed]

55. Pope, J.E. The future of treatment in systemic sclerosis: Can we design better trials? *Lancet Rheumatol.* **2020**, *2*, e185–e194. [CrossRef]
56. Khanna, D.; Huang, S.; Lin, C.J.F.; Spino, C. New composite endpoint in early diffuse cutaneous systemic sclerosis: Revisiting the provisional American College of Rheumatology Composite Response Index in Systemic Sclerosis. *Ann. Rheum. Dis.* **2021**, *80*, 641–650. [CrossRef]
57. Quinlivan, A.; Ross, L.; Proudman, S. Systemic sclerosis: Advances towards stratified medicine. *Best Pract. Res. Clin. Rheumatol.* **2020**, *34*, 101469. [CrossRef]
58. Goswami, R.P.; Ray, A.; Chatterjee, M.; Mukherjee, A.; Sircar, G.; Ghosh, P. Rituximab in the treatment of systemic sclerosis–related interstitial lung disease: A systematic review and meta-analysis. *Rheumatology* **2021**, *60*, 557–567. [CrossRef]
59. Fraticelli, P.; Fischetti, C.; Salaffi, F.; Carotti, M.; Mattioli, M.; Pomponio, G.; Gabrielli, A. Combination therapy with rituximab and mycophenolate mofetil in systemic sclerosis. A single-centre case series study. *Clin. Exp. Rheumatol.* **2018**, *36* (Suppl. 113), 142–145.
60. Rimar, D.; Rosner, I.; Slobodin, G. Upfront Combination Therapy with Rituximab and Mycophenolate Mofetil for Progressive Systemic Sclerosis. *J. Rheumatol.* **2021**, *48*, 304–305. [CrossRef]
61. Highland, K.B.; Distler, O.; Kuwana, M.; Allanore, Y.; Assassi, S.; Azuma, A.; Bourdin, A.; Denton, C.P.; Distler, J.H.W.; Hoffmann-Vold, A.M.; et al. Efficacy and safety of nintedanib in patients with systemic sclerosis-associated interstitial lung disease treated with mycophenolate: A subgroup analysis of the SENSCIS trial. *Lancet Respir. Med.* **2021**, *9*, 96–106. [CrossRef]
62. Naidu, G.S.R.S.N.K.; Sharma, S.K.; Adarsh, M.B.; Dhir, V.; Sinha, A.; Dhooria, S.; Jain, S. Effect of mycophenolate mofetil (MMF) on systemic sclerosis-related interstitial lung disease with mildly impaired lung function: A double-blind, placebo-controlled, randomized trial. *Rheumatol. Int.* **2020**, *40*, 207–216. [CrossRef] [PubMed]
63. Volkmann, E.R.; Tashkin, D.P.; Li, N.; Roth, M.D.; Khanna, D.; Hoffmann-Vold, A.-M.; Kim, G.; Goldin, J.; Clements, P.J.; Furst, D.E.; et al. Mycophenolate Mofetil Versus Placebo for Systemic Sclerosis-Related Interstitial Lung Disease: An Analysis of Scleroderma Lung Studies I and II. *Arthritis Rheumatol.* **2017**, *69*, 1451–1460. [CrossRef] [PubMed]

 biomedicines

Article

Clinical Predictors of Lung-Function Decline in Systemic-Sclerosis-Associated Interstitial Lung Disease Patients with Normal Spirometry

Tamas Nagy [1], Nora Melinda Toth [1], Erik Palmer [1], Lorinc Polivka [1], Balazs Csoma [1], Alexandra Nagy [1], Noémi Eszes [1], Krisztina Vincze [1], Enikő Bárczi [1], Anikó Bohács [1], Ádám Domonkos Tárnoki [2], Dávid László Tárnoki [2], György Nagy [3,4], Emese Kiss [5,6], Pál Maurovich-Horvát [2] and Veronika Müller [1,*]

1 Department of Pulmonology, Semmelweis University, 1083 Budapest, Hungary
2 Medical Imaging Centre, Semmelweis University, 1082 Budapest, Hungary
3 Department of Genetics, Cell- and Immunobiology, Semmelweis University, 1082 Budapest, Hungary
4 Department of Rheumatology and Clinical Immunology, Department of Internal Medicine and Oncology, Semmelweis University, 1083 Budapest, Hungary
5 Department of Clinical Immunology, Adult and Pediatric Rheumatology, National Institute of Locomotor Diseases and Disabilities, 1023 Budapest, Hungary
6 Department of Haematology and Internal Medicine, Semmelweis University, 1088 Budapest, Hungary
* Correspondence: muller.veronika@med.semmelweis-univ.hu

Citation: Nagy, T.; Toth, N.M.; Palmer, E.; Polivka, L.; Csoma, B.; Nagy, A.; Eszes, N.; Vincze, K.; Bárczi, E.; Bohács, A.; et al. Clinical Predictors of Lung-Function Decline in Systemic-Sclerosis-Associated Interstitial Lung Disease Patients with Normal Spirometry. *Biomedicines* **2022**, *10*, 2129. https://doi.org/10.3390/biomedicines10092129

Academic Editors: Michal Tomcik and Alberto Ricci

Received: 30 June 2022
Accepted: 26 August 2022
Published: 31 August 2022

Publisher's Note: MDPI stays neutral with regard to jurisdictional claims in published maps and institutional affiliations.

Abstract: Interstitial lung disease (ILD) is the leading cause of mortality in systemic sclerosis (SSc). Progressive pulmonary fibrosis (PPF) is defined as progression in 2 domains including clinical, radiological or lung-function parameters. Our aim was to assess predictors of functional decline in SSc-ILD patients and compare disease behavior to that in idiopathic pulmonary fibrosis (IPF) patients. Patients with normal forced vital capacity (FVC > 80% predicted; SSc-ILD: n = 31; IPF: n = 53) were followed for at least 1 year. Predictors of functional decline including clinical symptoms, comorbidities, lung-function values, high-resolution CT pattern, and treatment data were analyzed. SSc-ILD patents were significantly younger (59.8 ± 13.1) and more often women (93 %) than IPF patients. The median yearly FVC decline was similar in both groups (SSc-ILD = −67.5 and IPF = −65.3 mL/year). A total of 11 SSc-ILD patients met the PPF criteria for functional deterioration, presenting an FVC decline of −153.9 mL/year. Cough and pulmonary hypertension were significant prognostic factors for SSc-ILD functional progression. SSc-ILD patients with normal initial spirometry presenting with cough and PH are at higher risk for showing progressive functional decline.

Keywords: systemic sclerosis; interstitial lung disease; cough; pulmonary hypertension; predictors of treatment response

1. Introduction

Systemic sclerosis (SSc) is a chronic connective tissue disease with multiple organ involvement [1]. Skin and internal organ manifestations are common; however, SSc-associated interstitial lung disease (SSc-ILD) is the leading cause of mortality in SSc [2]. Another significant cardiopulmonary complication accounting for additional high mortality is pulmonary hypertension (PH) [3,4]. The clinical course of this disease shows a high variability in progression [5].

Goh criteria are widely used to assess the severity of pulmonary involvement in SSc based on the extent of lung fibrosis on high-resolution computed tomography (HRCT) scan and forced vital capacity (FVC) % predicted value [6]. According to these criteria, SSc-ILD patients are classified into limited or extensive disease subgroups; however, functional decline in patients with normal spirometry, progression of symptoms, or HRCT is not included in the disease stratification [4]. The SSc-ILD Safety and Efficacy of Nintedanib in Systemic Sclerosis (SENSCIS) clinical trial confirmed that the annual rate of FVC decline in placebo-treated patients was on average −93.3 mL/year, and 46% of participants without specific treatment lost over 5% of FVC during a 1-year follow-up [7]. Furthermore, recent studies additionally suggested that SSc-ILD patients might lose lung function even in the physiological range and should be followed more closely to prevent functional deterioration [8].

Idiopathic pulmonary fibrosis (IPF) is a progressive fibrosing ILD of unknown origin and is associated with a poor outcome. IPF is characterized by a progressive functional decline that is mainly assessed using FVC and diffusing capacity of the lung for carbon monoxide (DL_{CO}) [9]. FVC is currently the most extensively studied marker of disease progression in IPF, as this functional value is the only parameter accepted by regulatory bodies in predicting mortality [10].

Progressive pulmonary fibrosis (PPF) is defined by two parameters of the following three criteria: (1) worsening of respiratory symptoms; (2) physiological criteria of PFT decline (FVC $\geq$ 5%/year and/or $DL_{CO} \geq$ 10%/year); and (3) radiological signs of disease progression within a 1-year follow-up [11]. PPF might show similar clinical, lung functional, and radiological features over long term with IPF, the prototype of rapidly progressive ILD.

Our goal was to assess the PPF criteria on lung functional decline and possible clinical predictors of functional progression in SSc-ILD patients with normal spirometry and compare data with that of IPF patients to identify possible common factors defining functional progression.

2. Materials and Methods

2.1. Study Design and Parameters

Our study was a retrospective longitudinal observational study. All patients referred between February 2015 and January 2021 to the ILD multidisciplinary team (MDT) of the Department of Pulmonology, Semmelweis University, Budapest, Hungary, were screened. We enrolled subjects in our study who had been diagnosed with either SSc-ILD or IPF, had physiologic spirometric lung-function parameters (forced vital capacity (FVC) > 80% of predicted value), and had at least 12 months of follow-up data. SSc-ILD patients (classified with both J84 and M34 codes according to the 10th edition of the International Classification of Diseases (ICD10)) were diagnosed between January 2017 and July 2019. The diagnosis of SSc was established previously based on the American College of Rheumatology/European League Against Rheumatism Collaborative Initiative (EULAR-ACR) criteria [12] by immunology and rheumatology specialists; the disease-specific therapy was coordinated at immunological–rheumatological centers in central Hungary. Patients diagnosed with IPF between February 2015 and January 2021 according to the 2011 American Thoracic Society/European Respiratory Society (ATS/ERS) IPF guideline included in the EMPIRE registry were reviewed and all entered into the analysis if they met the inclusion criteria of having lung-function follow-up of at least one year with a baseline FVC > 80% predicted [10,13]. The study's functional and follow-up criteria were met by 53 IPF and 31 SSc-ILD patients (Figure 1).

Baseline characteristics including smoking history, symptoms, detailed pulmonary function test (PFT) values (FVC, forced expiratory volume in 1st second (FEV_1), total lung capacity (TLC), DL_{CO}, carbon monoxide transfer coefficient (KL_{CO})). HRCT pattern and treatment data were analyzed and compared between the study populations. Arterial blood gas (ABG) analysis, body mass index (BMI), and 6-minute walk test (6MWT) results were examined. The gender, age, and physiology (GAP) index was calculated as a prognostic staging system [14]. At each follow-up visit, routine PFTs, ABGs, and the 6MWT were documented. Methods of the measurements (PFTs, ABGs, 6MWT, and HRCT) were described in detail in our previous study [8]. The median follow-up was 34 months and the functional decline of PPF was established throughout the follow-up according to the criteria of $\geq$5% FVC decline and/or $\geq$10% DL_{CO} decline within 1 year. A PPF diagnosis was not made because no additional second criterion of worsening symptoms and/or progression on HRCT was assessed [11].

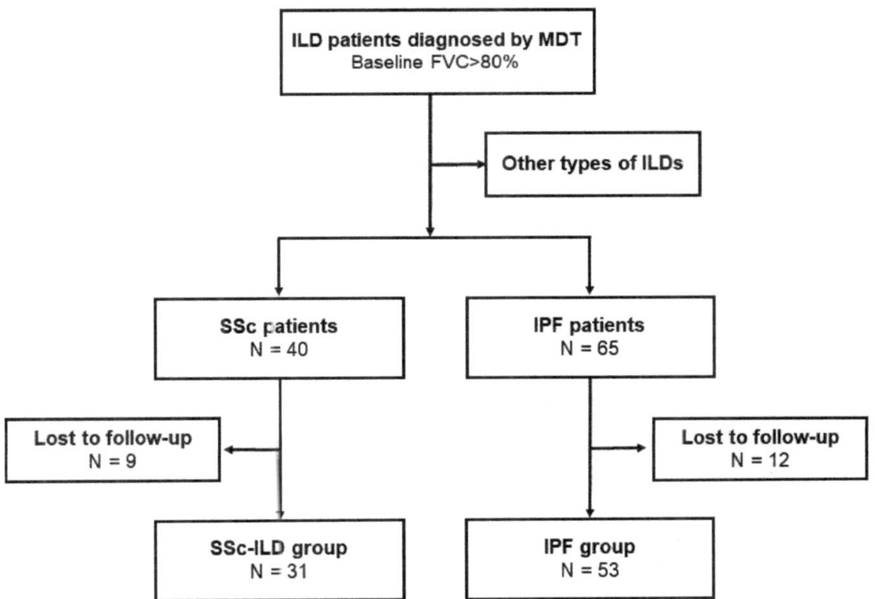

Figure 1. Patient selection for analysis. FVC, forced vital capacity; ILD, interstitial lung disease; IPF, idiopathic pulmonary fibrosis; MDT, multidisciplinary team; SSc-ILD, systemic-sclerosis-associated interstitial lung disease.

2.2. Statistical Analysis

Data were analyzed using Graph Pad software (GraphPad Prism 5.0 Software, Inc., La Jolla, CA, USA) and SPSS v25 (IBM Corporation, Armonk, NY, USA). Continuous variables were expressed as the mean ± standard deviation (SD) or median (interquartile range (IQR)) and were compared using a *t*-test or a Mann–Whitney U-test according to the distribution of the variable. The test for normality was performed using the Kolmogorov–Smirnov test. Categorical variables are presented as percentages (%) expressed for the entire study population (all patients) or respective subgroups as indicated and were compared using a chi-squared test or two-tailed Fisher's exact test. Multiple logistic regression analysis was used to assess predictors of possible functional progression including age (as a continuous variable), sex (male/female), smoking history (present/absent), cough (present/absent), PH (present/absent), baseline FVC, TLC, DL_{CO}, KL_{CO} (all functional parameters in % predicted as continuous variables), and therapy (applied/none). A p-value < 0.05 was defined as statistically significant.

3. Results

The characteristics of the patients are summarized in Table 1. The mean age of the SSc-ILD patients was significantly lower than in IPF patients, with female predominance. IPF patients had been more frequently smokers and a higher proportion of patients qualified as overweight. A majority of the patients in both groups were in GAP stage I at baseline.

Table 1. Patient characteristics.

Characteristics	IPF (n = 53)	SSc-ILD (n = 31)	p-Value
Age (years)	68.9 ± 8.5	59.8 ± 13.1	**0.001**
Sex (male:female)	28:25	2:29	**<0.001**
Smoking history			
Ever smoker	33 (62.3)	7 (22.6)	**0.001**
Non-smoker	20 (37.7)	24 (77.4)	
BMI (kg/m^2)	27.7 ± 4.4	25.2 ± 4.4	**0.006**
Overweight (25.0–29.9 kg/m^2)	21 (39.6%)	5 (16.2%)	**0.025**
Signs and symptoms			
Dyspnea	52 (98.1)	11 (35.5)	**<0.001**
Cough	26 (49.1)	9 (29.0)	0.108
Finger clubbing	12 (22.6)	0	NA
Crackles	47 (88.7)	12 (38.7)	**<0.001**
Raynaud phenomenon	0	23 (74.2)	NA
GAP score			
Stage I	50 (94.3)	31 (100.0)	NA
Stage II	3 (5.7)	0	NA
Stage III	0	0	NA
Specific comorbidities			
PH	7 (13.2)	5 (16.1)	0.712
GERD	6 (11.3)	6 (19.4)	0.345
HRCT pattern			
UIP/pUIP	26/25 (96.2)	0/3 (9.7)	**<0.001**
NSIP	0	26 (83.9)	NA
Other	2 (3.8)	2 (6.5)	0.578
Therapy			
Nintedanib	39 (73.6)	0	NA
Pirfenidone	8 (15.0)	0	NA
ISU	0	26 (83.9)	NA
Biological treatment	0	9 (29.0)	NA
None	14 (26.4)	3 (9.7)	NA

Data are presented as n (%) or mean ± SD. BMI, body mass index; GAP, Gender–Age–Physiology index; GERD, gastro-esophageal reflux disease; HRCT, high-resolution computed tomography; IPF, idiopathic pulmonary fibrosis; NA, not assessed; NSIP, non-specific interstitial pneumonia; PH, pulmonary hypertension; pUIP, probable usual interstitial pneumonia; SSc-ILD, systemic-sclerosis-associated interstitial lung disease; UIP, usual interstitial pneumonia; ISU, immunosuppressive therapy. GAP index: stage I = 0–3 points; stage II = 4–5 points; stage III = 6–8 points. Statistically significant values are highlighted in bold.

ILD-associated symptoms showed a significant difference between the two groups. More patients presented with respiratory symptoms such as dyspnea, cough, and crackles at the time of the diagnosis in the IPF group. On the other hand, the Raynaud phenomenon was only observed in the SSc-ILD group. No difference between groups was observed regarding the presence of PH and gastroesophageal reflux disease The Scl-70 antibody was present in 48% of SSc-ILD patients. The HRCT pattern was in line with actual guidelines and the literature, as usual interstitial pneumonia (UIP) or probable (p)UIP was the predominant pattern in IPF, while non-specific interstitial pneumonia (NSIP) predominated in SSc-ILD.

SSc therapy in 26 patients' was conventional immunosuppressive therapy (ISU), while biological treatment was given in 9 cases, including 7 subjects with combined ISU and biological therapy. Antifibrotic agents (nintedanib or pirfenidone) were applied in 39 cases in the IPF group, while 14 patients did not receive IPF-specific treatment.

The functional parameters for the patients are summarized in Table 2. The values for FVC and FEV_1 were similar in both groups, while TLC was significantly impaired and below the normal range in IPF patients. Baseline CO diffusion parameters showed a significant difference; DL_{CO} was decreased in the IPF group and SSc-ILD patients had lower KL_{CO} values. In the ABG, IPF patients had significantly lower pO_2 values compared to SSc-ILD patients. The 6MWT distance was similar for both groups, representing a diminished functional capacity in SSc-ILD patients since normal percentile values are age-dependent [15,16]. Post-exercise desaturation in the 6MWT was present in both groups to a similar degree.

Table 2. Baseline lung function, ABG, and 6MWT functional parameters.

Values	IPF (n = 53)	SSc-ILD (n = 31)	p-Value
Lung-function parameters			
FVC (mL)	3035.7 ± 836.2	2725.5 ± 655.6	0.080
FVC (% pred)	96.4 ± 13.9	98.7 ± 12.2	0.263
FEV_1 (mL)	2488.5 ± 696.4	2301.6 ± 569.2	0.209
FEV_1 (% pred)	98.6 ± 16.2	99.7 ± 13.3	0.748
FEV_1/FVC (%)	82.6 ± 8.2	84.5 ± 5.2	0.329
TLC (mL)	4648.9 ± 1358.4	4263.9 ± 823.3	0.157
TLC (% pred)	79.5 ± 14.4	88.4 ± 15.4	**0.022**
Diffusion parameters			
DL_{CO} (mmol/min/kPa)	5.9 ± 1.8	6.4 ± 1.6	0.201
DL_{CO} (% pred)	74.1 ± 17.6	83.7 ± 18.3	**0.020**
KL_{CO} (mmol/min/kPa/L)	1.3 ± 0.3	1.4 ± 0.3	**0.042**
KL_{CO} (% pred)	88.8 ± 24.2	71.2 ± 16.4	**<0.001**
ABGs			
pH	7.4 ±0.0	7.4 ±0.0	0.655
pCO_2 (mmHg)	37.7 ± 5.5	37.1 ± 2.3	0.119
pO_2 (mmHg)	67.8 ± 11.0	78.6 ± 8.6	**<0.001**
6MWT			
Distance (m)	454.4 ± 103.1	449.3 ± 70.8	0.502
Initial SpO_2 (%)	95.3 ± 2.9	94.9 ± 3.0	0.463
Final SpO_2 (%)	88.8 ± 9.0	89.2 ± 10.8	0.407
Initial HR (1/min)	81.0 ± 13.9	84.4 ± 14.3	0.425
Final HR (1/min)	111.2 ± 19.7	106.5 ± 20.2	0.443
Initial Borg score (0–10)	0 (0–0)	0 (0–0)	0.885
Final Borg score (0–10)	2 (0–4)	1.5 (1–3)	0.924

Data are presented as mean ± SD, median (IQR). 6MWT, 6-minute walk test; ABGs, arterialized capillary blood gases; DL_{CO}, diffusing capacity of the lungs for carbon monoxide; FEV_1, forced expiratory volume in 1 s; FVC, forced vital capacity; HR, heart rate; IPF, idiopathic pulmonary fibrosis; KL_{CO}, transfer coefficient of the lung for carbon monoxide; pCO_2; partial pressure of carbon dioxide; pO_2, partial pressure of oxygen; SpO_2, oxygen saturation; SSc-ILD, systemic-sclerosis-associated interstitial lung disease; TLC, total lung capacity. Statistically significant values are highlighted in bold.

The annual median decline in FVC for SSc-ILD was −67.5 (−146.0 to −4.0) mL/year and −65.3 (−173.8 to −65.3) mL/year in the IPF group. During the follow-up period, 11 (35%) SSc-ILD patients met the criteria of PPF for functional decline: 7 patients presented with ≥5% FVC and 6 patients presented with ≥10% DL_{CO} predicted value decline, including 3 patients that met both criteria. In IPF patients, FVC and/or DL_{CO} decline, as defined for PFF, was present in 16 (30.2%) cases: 14 patients presented with ≥5% FVC and 7 patients presented with ≥10% DL_{CO} predicted value decline, including 5 patients meeting both criteria.

The annual median decline in FVC for the functionally progressive SSc-ILD subgroup was −153.9 (−278.3 to −121.4) mL/year, which was significantly higher compared to −26.2 (−75.4 to −1.6) mL/year in the stable/improved SSc-ILD subgroup (p = 0.017). In the functionally progressive IPF and stable/improved IPF subgroups, the respective data were −264.7 (−404.9 to −204.6) vs. −39.2 (−85.7 to +7.5) mL/year (p = 0.004). FVC decline for all patients, including functionally stable/improved and functionally progressive subgroups regarding IPF and SSc-ILD, are presented in Figure 2.

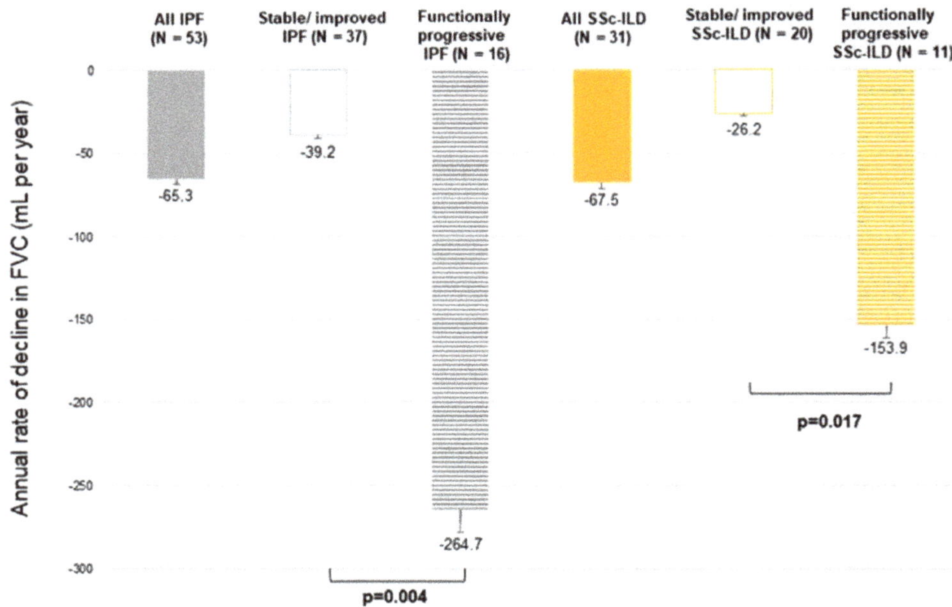

Figure 2. Annual rate of decline in FVC (mL per year) in IPF and SSc-ILD patients. FVC, forced vital capacity; IPF, idiopathic pulmonary fibrosis; SSc-ILD, systemic-sclerosis-associated interstitial lung disease.

The functionally progressive IPF and SSc-ILD subgroups' baseline characteristics and functional data are presented in Table 3. While in IPF, the functional progression was not associated with any difference in baseline characteristics or treatment, in SSc-ILD, cough and the presence of PH was significantly more common in the subgroup of patients showing functional decline over the one-year follow-up.

Table 3. Patient characteristics; HRCT pattern; treatment; and baseline lung function, ABG, and 6MWT functional parameters of the IPF and SSc-ILD subgroups.

Values	Functionally Stable/Improved IPF (n = 37)	Functionally Progressive IPF (n = 16)	Functionally Stable/Improved SSc-ILD (n = 20)	Functionally Progressive SSc-ILD (n = 11)
Age (years)	67.6 ± 9.1	71.8 ± 6.5	59.8 ± 13.1	59.7 ± 12.6
Sex (male:female)	20:17	8:8	1:19	1:10
Smoking history				
Ever smoker	23 (62.2)	10 (62.5)	3 (15.0)	4 (36.4)
Non-smoker	14 (37.8)	6 (37.5)	17 (85.0)	7 (63.6)
BMI (kg/m^2)	28.3 ± 4.4	26.8 ± 4.3	25.2 ± 4.4	23.7 ± 4.1
Overweight (25.0–29.9 kg/m^2)	15 (40.5%)	6 (37.5%)	5 (45.5%)	0
Signs and symptoms				
Dyspnea	37 (100)	15 (93.8)	7 (35.0)	4 (36.4)
Cough	19 (51.4)	7 (43.8)	**2 (10.0) ***	**7 (63.6) ***
Finger clubbing	6 (16.2)	6 (37.5)	0	0
Crackles	33 (89.2)	14 (87.5)	7 (35.0)	5 (45.5)
Raynaud phenomenon	0	0	16 (80.0)	7 (63.6)

Table 3. *Cont.*

Values	Functionally Stable/Improved IPF (*n* = 37)	Functionally Progressive IPF (*n* = 16)	Functionally Stable/Improved SSc-ILD (*n* = 20)	Functionally Progressive SSc-ILD (*n* = 11)
GAP score				
Stage I	35 (94.6)	15 (93.8)	20 (100.0)	11 (100.0)
Stage II	2 (5.4)	1 (6.2)	0	0
Stage III	0	0	0	0
Specific comorbidities				
PH	5 (13.5)	2 (13.3)	**1 (5.0)** #	**4 (36.4)** #
GERD	5 (13.5)	1 (6.2)	3 (15.0)	3 (27.3)
HRCT pattern				
UIP/pUIP	16/21 (100)	10/4 (87.5)	0/3 (15.0)	0
NSIP	0	0	16 (80.0)	10 (90.9)
Other	0	2 (12.5)	1 (5.0)	1 (9.1)
Therapy				
Nintedanib	27 (73.0)	12 (75.0)	0	0
Pirfenidone &	3 (8.1)	5 (31.2)	0	0
ISU	0	0	18 (90.0)	8 (72.7)
Biological treatment	0	0	7 (35.0)	2 (18.2)
None	10 (27.0)	4 (25.0)	1 (5.0)	2 (18.2)
Lung-function parameters				
FVC (mL)	3057.0 ± 840.8	2986.3 ± 850.7	2770.5 ± 681.2	2642.7 ± 631.4
FVC (% pred)	95.5 ± 12.8	98.4 ± 16.5	98.9 ± 13.7	98.4 ± 9.7
FEV_1 (mL)	2497.6 ± 707.6	2467.5 ± 692.0	2354.5 ± 609.4	2200.9 ± 510.8
FEV_1 (% pred)	96.8 ± 15.0	102.7 ± 18.5	100.0 ± 14.7	98.6 ± 11.3
FEV_1/FVC (%)	82.0 ± 8.7	82.9 ± 5.2	84.8 ± 4.5	83.6 ± 6.4
TLC (mL)	4683.5 ± 1459.1	4568.8 ± 1130.2	4343.0 ± 884.2	4199.1 ± 635.5
TLC (% pred)	79.7 ± 16.1	78.8 ± 9.5	89.7 ± 17.5	88.1 ± 10.7
Diffusion parameters				
DL_{CO} (mmol/min/kPa)	6.1 ± 1.8	5.3 ± 1.6	6.3 ± 1.5	6.6 ± 1.6
DL_{CO} (% pred)	76.6 ± 18.0	68.3 ± 15.5	82.5 ± 18.4	88.6 ± 14.5
KL_{CO} (mmol/min/kPa/L)	1.3 ± 0.4	1.2 ± 0.2	1.4 ± 0.3	1.5 ± 0.3
KL_{CO} (% pred)	89.2 ± 23.6	87.8 ± 26.3	70.3 ± 16.3	75.6 ± 12.9
ABGs				
pH	7.4 ±0.0	7.4 ±0.0	7.4 ±0.0	7.4 ±0.0
pCO_2 (mmHg)	38.5 ± 2.6	35.3 ± 10.0	37.7 ± 1.9	36.0 ± 2.8
pO_2 (mmHg)	69.7 ± 10.2	62.2 ± 11.8	77.0 ± 6.4	81.4 ± 11.7
6MWT				
Distance (m)	459.7 ± 102.5	442.5 ± 107.1	454.7 ± 68.8	438.4 ± 81.8
Initial SpO_2 (%)	95.6 ± 2.3	94.6 ± 3.8	95.6 ± 1.9	93.6 ± 4.4
Final SpO_2 (%)	89.4 ± 8.9	87.4 ± 9.3	93.6 ± 5.3	82.0 ± 14.0
Initial HR (1/min)	80.4 ± 12.9	82.3 ± 16.4	85.6 ± 15.6	82.4 ± 13.2
Final HR (1/min)	109.7 ± 19.7	114.7 ± 19.7	106.0 ± 16.7	107.2 ± 27.2
Initial Borg score (0–10)	0 (0–0)	0.5 (0–0)	0 (0–0)	0 (0–1)
Final Borg score (0–10)	1.5 (0–4)	2.8 (0–4)	2 (1–3)	1 (0–3)

Data are presented as *n* (%) or mean ± SD, median (IQR). 6MWT, 6-minute walk test; ABGs, arterialized capillary blood gases; BMI, body mass index; DL_{CO}, diffusing capacity of the lungs for carbon monoxide; FEV_1, forced expiratory volume in 1 s; FVC, forced vital capacity; GAP, Gender–Age–Physiology index; GERD, gastro-esophageal reflux disease; HR, heart rate; HRCT, high-resolution computed tomography; IPF, idiopathic pulmonary fibrosis; ISU, immunosuppressive therapy; KL_{CO}, transfer coefficient of the lung for carbon monoxide; NSIP, non-specific interstitial pneumonia; pCO_2; partial pressure of carbon dioxide; PH, pulmonary hypertension; pO2, partial pressure of oxygen; pUIP, probable usual interstitial pneumonia; SpO2, oxygen saturation; SSc-ILD, systemic-sclerosis-associated interstitial lung disease; TLC, total lung capacity; UIP, usual interstitial pneumonia;. GAP index: stage I = 0–3 points; stage II = 4–5 points; stage III = 6–8 points. Statistically significant values are highlighted in bold: * $p = 0.002$; # $p = 0.023$. & All treatment was included: a nintedanib-to-pirfenidone or pirfenidone-to-nintedanib change in therapy resulted in the higher number of treated vs. untreated IPF patients.

A multiple logistic regression did not show a significant baseline predictor of progression in the IPF group; however, in the SSc-ILD group, cough and PH were prognostic factors for functional progression (odds ratio: 36.2 (95% confidence interval: 1.8–711.9) and 36.4 (95% confidence interval: 1.1–1184.9), respectively). Most patients had a dry cough (SSc-ILD: $n = 7$; IPF: $n = 17$), and in the functionally progressive subgroups, a dry cough was predominant (SSc-ILD: 85.7% and IPF: 71.4%). The functional decline in individual patients according to treatment is presented in Figure 3.

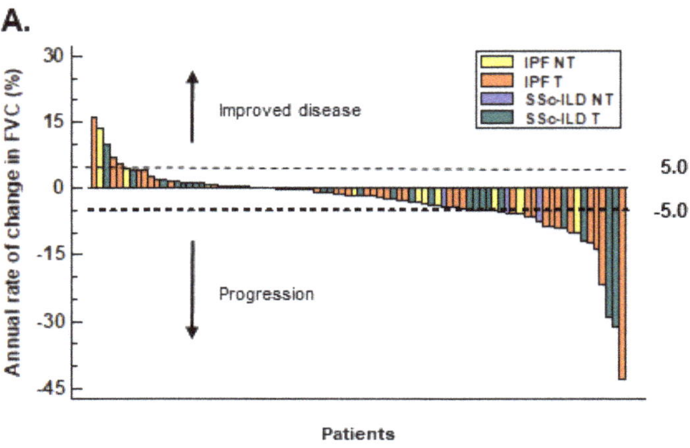

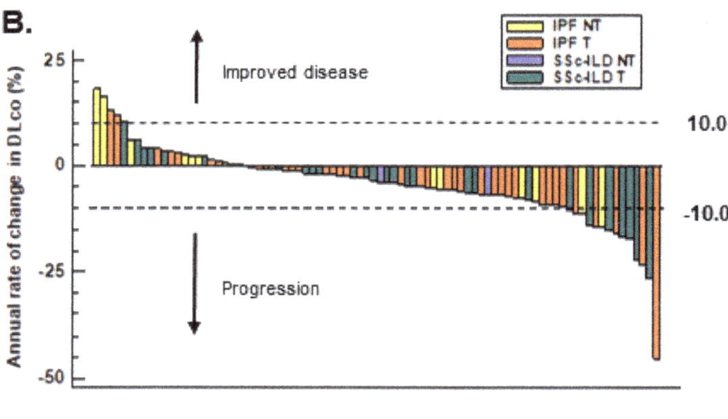

Figure 3. Annual changes in FVC (**A**) and DL_{CO} (**B**) % of the predicted value in all IPF and SSc-ILD patients according to specific treatment groups. DL_{CO}, diffusing capacity for carbon monoxide; FVC, forced vital capacity; IPF, idiopathic pulmonary fibrosis; NT, no treatment; SSc-ILD, systemic-sclerosis-associated interstitial lung disease; T, treatment.

4. Discussion

Our study confirmed that around one-third of SSc-ILD patients who had initial physiologic lung function parameters showed a decline in FVC and/or DL_{CO} over one year. While the disease course of SSc-ILD is heterogeneous, development of PPF might be a leading cause of death in this rare patient group. Although pulmonary functional decline is only one out of the three criteria defining PPF, PFT is an important non-invasive measurement that, in combination with symptoms, might promote clinicians to engage in early treatment interventions if PPF is confirmed [11].

According to our data, SSc patients with HRCT-confirmed ILD who are experiencing cough as the leading respiratory symptom might be at higher risk of developing PPF even with normal lung function results when presented at the respiratory specialist. Similarly, cough was also studied in a subgroup analysis of the SENSCIS trial and was found to be an important prognostic factor for functional decline and progression in SSc-ILD patients [17]. However, patients enrolled in the SENSCIS trial had less preserved lung function and patients with cough had an average decline of −95.6 mL/year, which was slightly higher than in our study. Importantly, nintedanib was less effective in SSc-ILD patients with a cough compared to patients without a cough. These two independent observations could mean that cough is an independent negative prognostic factor for functional decline. Cough was also found to be a marker of unfavorable therapeutic response in the Scleroderma Lung Study (SLS) II and was strongly correlated with lung-function decline [18–20]. It is important to note that the mean FVC decline in prospective average population cohort studies ranged in men between −47.2 and −78.4 mL/year, while in women it ranged between −14.1 and −65.6 mL/year, underlining functional loss in non-progressive patients within the expected range [21].

Additionally, our study confirmed the important prognostic role of PH for functional decline in SSc-ILD, which, to the best of our knowledge, has not been previously established in the literature. On the other hand, PH is a well-known severe complication of SSc, and several large investigations have confirmed its contribution to mortality [22–24]. However, our study was not powered to investigate PH as a risk of mortality due to the short observation period.

Surprisingly, we found that the 6MWT results did not differ between the SSc-ILD and IPF patients, although the SSc-ILD patients were significantly younger and therefore were expected to have better functional performance compared to older IPF patients. The underlying mechanisms might have included vascular manifestations such as PH and musculoskeletal conditions affecting exercise limitation [25–27].

IPF can be regarded as the prototype of rapidly progressive fibrotic lung disease. The common cellular mechanism in IPF and SSc-ILD is fibroblast activation and recruitment, as well as myofibroblast differentiation leading to interstitial extracellular matrix accumulation and lung fibrosis [9,15]. Our data confirmed that functional decline in our total IPF patient population was lower than in the nintedanib-treated population of the Safety and Efficacy of nintedanib at High Dose in IPF Patients (INPULSIS) trial (FVC decline of −65.3 vs. adjusted annual rate of change of −114.7 mL/year) [28]. Importantly, no difference was found in FVC decline between IPF patients receiving antifibrotics and the treatment-free subgroups. This highlighted that despite therapy, some patients might deteriorate even while receiving frontline antifibrotic therapy; therefore, further additional risk factors need to be identified. Notably, the original trials did not focus on patients with physiological lung function. The total IPF group (including untreated individuals) revealed a limited FVC decline; the stable/improved subgroup only showed a marginal loss in this parameter, while the progressive subgroup presented with a similar decline to that observed in patients on a placebo in clinical trials [28,29]. This real-world observation of IPF patients supported the beneficial effect of antifibrotic treatment, similar to our previous study on functionally advanced IPF cases [13].

Treatment of SSc-ILD is challenging. Our data confirmed better functional outcome in ISU and/or biological-therapy-treated SSc-ILD, emphasizing early and specific SSc treatment [8,30]. An SENSCIS subanalysis evaluated the effectiveness of nintedanib as an antifibrotic agent among SSc-ILD patients with or without baseline mycophenolate mofetil (MMF) treatment [7]. The inclusion criteria were: an FVC of at least 40% of the predicted value, a DL_{CO} of 30 to 89% of the predicted value, and a minimum of 10% fibrotic lung involvement on HRCT. At the end of the 52-week period, patients in the nintedanib + MMF group showed a significantly lower annual rate of decline in FVC than patients in the placebo + MMF group (−40.2 vs. −66.5 mL/year), and an even higher functional decline was present in the absence of ISU treatment (SSc-ILD on placebo: −119.3 mL/year). PPF SSc-ILD patients on the ISU therapy MMF with the addition of nintedanib had the highest advantage in lung-function preservation [7,31].

Treatment of PPF is still not well established; however, antifibrotic therapy might be an option in selected cases [11]. Nintedanib proved to be effective in reducing the annual rate of FVC decline in different fibrosing ILD patients in the Efficacy and Safety of Nintedanib in Patients with Progressive Fibrosing ILDs (INBUILD) trial and in the INPULSIS 1-2 IPF trials [28,32].

Optimal timing and dosing of ISU and/or biological therapy in SSc is of the utmost importance. Higher awareness of therapy initiation is needed in patients with normal lung function presenting with cough and in the presence of PH, and close pulmonary follow-up of these patients is suggested [33].

The limitations of our study were the retrospective single-center design and the low number of patients, which did not allow for a better stratification according to clinical features or treatment. Further prospective studies are needed to establish new markers of progression and to develop guidelines for the optimal timing of treatment introduction, with the adequate therapies being a combination of ISU and/or biological therapy and/or antifibrotics in the case of PPF in this special subgroup of SSc-ILD patients [34].

5. Conclusions

Progression in ILDs has an unfavorable effect on several clinical outcomes. Regular measurements of FVC are one of the most important parameters accepted for monitoring functional decline among IPF and SSc-ILD subjects. Patient-reported symptoms such as cough (especially dry cough) and the presence of PH as a lung-related comorbidity should be taken into consideration in connection with the disease progression in SSc-ILD patients. Defining high-risk patients for PPF is of utmost importance because timely and optimal introduction of ISU and/or biological and possibly antifibrotic agents might prevent disease deterioration. However, the timing and combination of treatments require further research in SSc-ILD and other PPF. Close monitoring and regular follow-ups are required in patients with normal lung function, especially in patients presenting with cough and PH.

Author Contributions: Conceptualization, T.N. and V.M.; methodology, software, and validation, T.N., B.C., L.P. and V.M.; formal analysis, V.M.; resources, V.M. data curation, T.N., N.M.T., E.P., A.N., E.B., Á.D.T., D.L.T., E.K., P.M.-H. and V.M.; writing—original draft preparation, T.N., B.C. and V.M.; writing—review and editing, Á.D.T., D.L.T., G.N., E.K., P.M.-H. and V.M.; visualization, T.N., A.B., K.V., N.E. and A.N.; supervision, N.E., K.V., E.B., A.B. and V.M.; T.N., N.M.T., E.P., L.P., N.E., K.V., E.B., A.B., Á.D.T., D.L.T., G.N., E.K., P.M.-H. and V.M. were treating physicians or radiologists and members of the multidisciplinary ILD team at Semmelweis University. All authors have read and agreed to the published version of the manuscript.

Funding: T.N. and L.P. were supported by the "Development of scientific workshops of medical, health sciences and pharmaceutical educations" (EFOP-3.6.3-VEKOP-1-00009). T.N. and A.N. were supported by the Hungarian Respiratory Society and Hungarian Respiratory Foundation. The EMPIRE registry was supported with funding from Boehringer Ingelheim (BI) and F. Hoffmann-La Roche (Roche). BI and Roche had no role in the study design, analysis, or interpretation of the results.

Institutional Review Board Statement: The study was conducted in accordance with the Declaration of Helsinki and approved by the Ethical Committees of TUKEB and Semmelweis University (Study No. 69/2015—24 February 2015; 181/2021—23 November 2021).

Informed Consent Statement: Informed consent was obtained from all IPF subjects involved in the study. In other cases, informed consent was not needed due to the retrospective design.

Data Availability Statement: The original contributions presented in the study are included in the article; further inquiries can be directed to the corresponding author.

Acknowledgments: Department of Pulmonology, Semmelweis University is a member of the European Reference Network for Respiratory Diseases (ERN-LUNG).

Conflicts of Interest: The authors declare no conflict of interest.

References

1. Bergamasco, A.; Hartmann, N.; Wallace, L.; Verpillat, P. Epidemiology of systemic sclerosis and systemic sclerosis-associated interstitial lung disease. *Clin. Epidemiol.* **2019**, *11*, 257–273. [CrossRef] [PubMed]
2. Denton, C.P.; Khanna, D. Systemic sclerosis. *Lancet* **2017**, *390*, 1685–1699. [CrossRef]
3. Panagopoulos, P.; Goules, A.; Hoffmann-Vold, A.M.; Matteson, E.L.; Tzioufas, A. Natural history and screening of interstitial lung disease in systemic autoimmune rheumatic disorders. *Ther. Adv. Musculoskelet. Dis.* **2021**, *13*, 1759720x211037519. [CrossRef]
4. Cottin, V.; Brown, K.K. Interstitial lung disease associated with systemic sclerosis (SSc-ILD). *Respir. Res.* **2019**, *20*, 13. [CrossRef]
5. Vonk, M.C.; Walker, U.A.; Volkmann, E.R.; Kreuter, M.; Johnson, S.R.; Allanore, Y. Natural variability in the disease course of SSc-ILD: Implications for treatment. *Eur. Respir. Rev.* **2021**, *30*, 200340. [CrossRef] [PubMed]
6. Goh, N.S.; Desai, S.R.; Veeraraghavan, S.; Hansell, D.M.; Copley, S.J.; Maher, T.M.; Corte, T.J.; Sander, C.R.; Ratoff, J.; Devaraj, A.; et al. Interstitial lung disease in systemic sclerosis: A simple staging system. *Am. J. Respir. Crit. Care Med.* **2008**, *177*, 1248–1254. [CrossRef] [PubMed]
7. Highland, K.B.; Distler, O.; Kuwana, M.; Allanore, Y.; Assassi, S.; Azuma, A.; Bourdin, A.; Denton, C.P.; Distler, J.H.W.; Hoffmann-Vold, A.M.; et al. Efficacy and safety of nintedanib in patients with systemic sclerosis-associated interstitial lung disease treated with mycophenolate: A subgroup analysis of the SENSCIS trial. *Lancet Respir. Med.* **2021**, *9*, 96–106. [CrossRef]
8. Nagy, A.; Nagy, T.; Kolonics-Farkas, A.M.; Eszes, N.; Vincze, K.; Barczi, E.; Tarnoki, A.D.; Tarnoki, D.L.; Nagy, G.; Kiss, E.; et al. Autoimmune Progressive Fibrosing Interstitial Lung Disease: Predictors of Fast Decline. *Front. Pharmacol.* **2021**, *12*, 778649. [CrossRef]
9. Richeldi, L.; Collard, H.R.; Jones, M.G. Idiopathic pulmonary fibrosis. *Lancet* **2017**, *389*, 1941–1952. [CrossRef]
10. Raghu, G.; Collard, H.R.; Egan, J.J.; Martinez, F.J.; Behr, J.; Brown, K.K.; Colby, T.V.; Cordier, J.F.; Flaherty, K.R.; Lasky, J.A.; et al. An official ATS/ERS/JRS/ALAT statement: Idiopathic pulmonary fibrosis: Evidence-based guidelines for diagnosis and management. *Am. J. Respir. Crit. Care Med.* **2011**, *183*, 788–824. [CrossRef]
11. Raghu, G.; Remy-Jardin, M.; Richeldi, L.; Thomson, C.C.; Inoue, Y.; Johkoh, T.; Kreuter, M.; Lynch, D.A.; Maher, T.M.; Martinez, F.J.; et al. Idiopathic Pulmonary Fibrosis (an Update) and Progressive Pulmonary Fibrosis in Adults: An Official ATS/ERS/JRS/ALAT Clinical Practice Guideline. *Am. J. Respir. Crit. Care Med.* **2022**, *205*, e18–e47. [CrossRef] [PubMed]
12. van den Hoogen, F.; Khanna, D.; Fransen, J.; Johnson, S.R.; Baron, M.; Tyndall, A.; Matucci-Cerinic, M.; Naden, R.P.; Medsger, T.A., Jr.; Carreira, P.E.; et al. 2013 classification criteria for systemic sclerosis: An American college of rheumatology/European league against rheumatism collaborative initiative. *Ann. Rheum. Dis.* **2013**, *72*, 1747–1755. [CrossRef]
13. Barczi, E.; Starobinski, L.; Kolonics-Farkas, A.; Eszes, N.; Bohacs, A.; Vasakova, M.; Hejduk, K.; Müller, V. Long-Term Effects and Adverse Events of Nintedanib Therapy in Idiopathic Pulmonary Fibrosis Patients with Functionally Advanced Disease. *Adv. Ther.* **2019**, *36*, 1221–1232. [CrossRef] [PubMed]
14. Ley, B.; Ryerson, C.J.; Vittinghoff, E.; Ryu, J.H.; Tomassetti, S.; Lee, J.S.; Poletti, V.; Buccioli, M.; Elicker, B.M.; Jones, K.D.; et al. A multidimensional index and staging system for idiopathic pulmonary fibrosis. *Ann. Intern. Med.* **2012**, *156*, 684–691. [CrossRef]
15. Khanna, D.; Tashkin, D.P.; Denton, C.P.; Renzoni, E.A.; Desai, S.R.; Varga, J. Etiology, Risk Factors, and Biomarkers in Systemic Sclerosis with Interstitial Lung Disease. *Am. J. Respir. Crit. Care Med* **2020**, *201*, 650–660. [CrossRef] [PubMed]
16. Winstone, T.A.; Assayag, D.; Wilcox, P.G.; Dunne, J.V.; Hague, C.J.; Leipsic, J.; Collard, H.R.; Ryerson, C.J. Predictors of mortality and progression in scleroderma-associated interstitial lung disease: A systematic review. *Chest* **2014**, *146*, 422–436. [CrossRef] [PubMed]
17. Volkmann, E.R.; Kreuter, M.; Hoffmann-Vold, A.M.; Wijsenbeek, M.; Smith, V.; Khanna, D.; Denton, C.P.; Wuyts, W.A.; Miede, C.; Alves, M.; et al. Dyspnoea and cough in patients with systemic sclerosis-associated interstitial lung disease in the SENSCIS trial. *Rheumatology* **2022**, keac091, *online ahead of print*. [CrossRef]
18. Cheng, J.Z.; Wilcox, P.G.; Glaspole, I.; Corte, T.J.; Murphy, D.; Hague, C.J.; Ryerson, C.J. Cough is less common and less severe in systemic sclerosis-associated interstitial lung disease compared to other fibrotic interstitial lung diseases. *Respirology* **2017**, *22*, 1592–1597. [CrossRef]
19. Tashkin, D.P.; Roth, M.D.; Clements, P.J.; Furst, D.E.; Khanna, D.; Kleerup, E.C.; Goldin, J.; Arriola, E.; Volkmann, E.R.; Kafaja, S.; et al. Mycophenolate mofetil versus oral cyclophosphamide in scleroderma-related interstitial lung disease (SLS II): A randomised controlled, double-blind, parallel group trial. *Lancet Respir. Med.* **2016**, *4*, 708–719. [CrossRef]
20. Tashkin, D.P.; Volkmann, E.R.; Tseng, C.H.; Roth, M.D.; Khanna, D.; Furst, D.E.; Clements, P.J.; Theodore, A.; Kafaja, S.; Kim, G.H.; et al. Improved Cough and Cough-Specific Quality of Life in Patients Treated for Scleroderma-Related Interstitial Lung Disease: Results of Scleroderma Lung Study II. *Chest* **2017**, *151*, 813–820. [CrossRef]
21. Thomas, E.T.; Guppy, M.; Straus, S.E.; Bell, K.J.L.; Glasziou, P. Rate of normal lung function decline in ageing adults: A systematic review of prospective cohort studies. *BMJ Open* **2019**, *9*, e028150. [CrossRef] [PubMed]

22. Fischer, A.; Swigris, J.J.; Bolster, M.B.; Chung, L.; Csuka, M.E.; Domsic, R.; Frech, T.; Hinchcliff, M.; Hsu, V.; Hummers, L.K.; et al. Pulmonary hypertension and interstitial lung disease within PHAROS: Impact of extent of fibrosis and pulmonary physiology on cardiac haemodynamic parameters. *Clin. Exp. Rheumatol.* **2014**, *32*, 109–114.
23. Launay, D.; Sobanski, V.; Hachulla, E.; Humbert, M. Pulmonary hypertension in systemic sclerosis: Different phenotypes. *Eur. Respir. Rev.* **2017**, *26*, 170056. [CrossRef] [PubMed]
24. Odler, B.; Foris, V.; Gungl, A.; Müller, V.; Hassoun, P.M.; Kwapiszewska, G.; Olschewski, H.; Kovacs, G. Biomarkers for Pulmonary Vascular Remodeling in Systemic Sclerosis: A Pathophysiological Approach. *Front. Physiol.* **2018**, *9*, 587. [CrossRef] [PubMed]
25. Bournia, V.K.; Kallianos, A.; Panopoulos, S.; Gialafos, E.; Velentza, L.; Vlachoyiannopoulos, P.G.; Sfikakis, P.P.; Trakada, G. Cardiopulmonary exercise testing and prognosis in patients with systemic sclerosis without baseline pulmonary hypertension: A prospective cohort study. *Rheumatol. Int.* **2022**, *42*, 303–309. [CrossRef]
26. Garin, M.C.; Highland, K.B.; Silver, R.M.; Strange, C. Limitations to the 6-minute walk test in interstitial lung disease and pulmonary hypertension in scleroderma. *J. Rheumatol.* **2009**, *36*, 330–336. [CrossRef]
27. Vandecasteele, E.; De Pauw, M.; De Keyser, F.; Decuman, S.; Deschepper, E.; Piette, Y.; Brusselle, G.; Smith, V. Six-minute walk test in systemic sclerosis: A systematic review and meta-analysis. *Int. J. Cardiol.* **2016**, *212*, 265–273. [CrossRef]
28. Richeldi, L.; du Bois, R.M.; Raghu, G.; Azuma, A.; Brown, K.K.; Costabel, U.; Cottin, V.; Flaherty, K.R.; Hansell, D.M.; Inoue, Y.; et al. Efficacy and safety of nintedanib in idiopathic pulmonary fibrosis. *N. Engl. J. Med.* **2014**, *370*, 2071–2082. [CrossRef]
29. King, T.E., Jr.; Bradford, W.Z.; Castro-Bernardini, S.; Fagan, E.A.; Glaspole, I.; Glassberg, M.K.; Gorina, E.; Hopkins, P.M.; Kardatzke, D.; Lancaster, L.; et al. A phase 3 trial of pirfenidone in patients with idiopathic pulmonary fibrosis. *N. Engl. J. Med.* **2014**, *370*, 2083–2092. [CrossRef]
30. Hoffmann-Vold, A.M.; Allanore, Y.; Alves, M.; Brunborg, C.; Airó, P.; Ananieva, L.P.; Czirják, L.; Guiducci, S.; Hachulla, E.; Li, M.; et al. Progressive interstitial lung disease in patients with systemic sclerosis-associated interstitial lung disease in the EUSTAR database. *Ann. Rheum. Dis.* **2021**, *80*, 219–227. [CrossRef]
31. Distler, O.; Highland, K.B.; Gahlemann, M.; Azuma, A.; Fischer, A.; Mayes, M.D.; Raghu, G.; Sauter, W.; Girard, M.; Alves, M.; et al. Nintedanib for Systemic Sclerosis-Associated Interstitial Lung Disease. *N. Engl. J. Med.* **2019**, *380*, 2518–2528. [CrossRef] [PubMed]
32. Flaherty, K.R.; Wells, A.U.; Cottin, V.; Devaraj, A.; Walsh, S.L.F.; Inoue, Y.; Richeldi, L.; Kolb, M.; Tetzlaff, K.; Stowasser, S.; et al. Nintedanib in Progressive Fibrosing Interstitial Lung Diseases. *N. Engl. J. Med.* **2019**, *381*, 1718–1727. [CrossRef] [PubMed]
33. Kowal-Bielecka, O.; Fransen, J.; Avouac, J.; Becker, M.; Kulak, A.; Allanore, Y.; Distler, O.; Clements, P.; Cutolo, M.; Czirjak, L.; et al. Update of EULAR recommendations for the treatment of systemic sclerosis. *Ann. Rheum. Dis.* **2017**, *76*, 1327–1339. [CrossRef] [PubMed]
34. Nagy, A.; Palmer, E.; Polivka, L.; Eszes, N.; Vincze, K.; Barczi, E.; Bohacs, A.; Tarnoki, A.D.; Tarnoki, D.L.; Nagy, G.; et al. Treatment and Systemic Sclerosis Interstitial Lung Disease Outcome: The Overweight Paradox. *Biomedicines* **2022**, *10*, 434. [CrossRef]

Article

Treatment and Systemic Sclerosis Interstitial Lung Disease Outcome: The Overweight Paradox

Alexandra Nagy [1,†], Erik Palmer [1,†], Lorinc Polivka [1], Noemi Eszes [1], Krisztina Vincze [1], Eniko Barczi [1], Aniko Bohacs [1], Adam Domonkos Tarnoki [2], David Laszlo Tarnoki [2], György Nagy [3,4], Emese Kiss [5,6], Pal Maurovich-Horvat [2] and Veronika Müller [1,*]

1 Department of Pulmonology, Semmelweis University, 1083 Budapest, Hungary; nagy.alexandra0718@gmail.com (A.N.); palmerperik@gmail.com (E.P.); polivka.lorinc@outlook.com (L.P.); noemi.eszes@gmail.com (N.E.); vinczekrisztina@gmail.com (K.V.); eniko.barczi@gmail.com (E.B.); bohacs.aniko@med.semmelweis-univ.hu (A.B.)
2 Medical Imaging Centre, Semmelweis University, 1082 Budapest, Hungary; tarnoki2@gmail.com (A.D.T.); tarnoki4@gmail.com (D.L.T.); maurovich-horvat.pal@med.semmelweis-univ.hu (P.M.-H.)
3 Department of Genetics, Cell- and Immunobiology, Semmelweis University, 1082 Budapest, Hungary; gyorgyngy@gmail.com
4 Department of Rheumatology and Clinical Immunology, Department of Internal Medicine and Oncology, Semmelweis University, 1083 Budapest, Hungary
5 Department of Clinical Immunology, Adult and Pediatric Rheumatology, National Institute of National Institute of Locomotor Diseases and Disabilities, 1023 Budapest, Hungary; drkiss.emese@gmail.com
6 3rd Department of Internal Medicine and Haematology, Semmelweis University, 1088 Budapest, Hungary
* Correspondence: muller.veronika@med.semmelweis-univ.hu
† These authors contributed equally to this work.

Abstract: (1) Background: Systemic sclerosis (SSc) is frequently associated with interstitial lung diseases (ILDs). The progressive form of SSc-ILD often limits patient survival. The aim of our study is to evaluate the clinical characteristics and predictors of lung function changes in SSc-ILD patients treated in a real-world setting. (2) Methods: All SSc-ILD cases previously confirmed by rheumatologists and a multidisciplinary ILD team between January 2017 and June 2019 were included (n = 54). The detailed medical history, clinical parameters and HRCT were analyzed. The longitudinal follow-up for pulmonary symptoms, functional parameters and treatment were performed for at least 2 years in no treatment, immunosuppression and biological treatment subgroups. (3) Results: In SSc-ILD patients (age 58.7 ± 13.3 years, 87.0% women), the main symptoms included dyspnea, cough, crackles and the Raynaud's phenomenon. The functional decline was most prominent in untreated patients, and a normal body mass index (BMI < 25 kg/m^2) was associated with a significant risk of deterioration. The majority of patients improved or were stable during follow-up. The progressive fibrosing-ILD criteria were met by 15 patients, the highest proportion being in the untreated subgroup. (4) Conclusions: SSc-ILD patients who are overweight are at a lower risk of the functional decline and progressive phenotype especially affecting untreated patients. The close monitoring of lung involvement and a regular BMI measurement are advised and early treatment interventions are encouraged.

Keywords: systemic sclerosis; interstitial lung disease; progressive fibrosing interstitial lung disease; lung function; body mass index; treatment

Citation: Nagy, A.; Palmer, E.; Polivka, L.; Eszes, N.; Vincze, K.; Barczi, E.; Bohacs, A.; Tarnoki, A.D.; Tarnoki, D.L.; Nagy, G.; et al. Treatment and Systemic Sclerosis Interstitial Lung Disease Outcome: The Overweight Paradox. *Biomedicines* **2022**, *10*, 434. https://doi.org/10.3390/biomedicines10020434

Academic Editor: Michal Tomcik

Received: 31 December 2021
Accepted: 11 February 2022
Published: 13 February 2022

1. Introduction

One of the systemic autoimmune rheumatic disorders—systemic sclerosis (SSc)—is often associated with interstitial lung disease (ILD). In SSc, an interaction of inflammation, vasculopathy, endothelial and fibroblast dysfunction leads to skin thickening, fingertip lesions, Raynaud's phenomenon and different internal organ manifestations. SSc has been divided according to its 3 phenotypes, which are limited cutaneous SSc (lcSSc, skin sclerosis that develops distal to the elbow and knee), diffuse cutaneous SSc (dcSSc, skin involvement that extends proximally to the elbow and knee and to the body surface with internal organs being affected early) and sine scleroderma SSc (ssSSc, whose internal organ manifestations and serological abnormalities are without skin involvement) [1]. Severe ILD is in general associated with dSSc and anti-SCL-70 antibodies, while it is relatively rare among individuals with anti-centromere antibodies [1,2]. Inflammatory cells accumulate in the lung parenchyma causing alveolar damage, which activates fibroblastic and myofibroblastic cells to produce extracellular matrices causing the scarring of the respiratory system, including diffuse parenchymal lung injury, frequently taking the form of pulmonary fibrosis, pulmonary hypertension and pleural involvement. Gastroesophageal reflux, malignancy, respiratory muscle weakness and heart failure are indirect pulmonary complications in SSc-ILD. Therefore, ILD is a common and usually early comorbidity of SSc, placing a significant burden on both the patient and the health care system, and can be the main cause of mortality [3–6].

The first symptoms of SSc-ILD are generally exertional dyspnea, fatigue and a dry cough, as an auscultation finding fine crepitation is often present [1]. The diagnosis of SSc-ILD is made by a multidisciplinary ILD-team (MDT), which is represented by respiratory, radiology, pathology specialists and often includes rheumatologists [4]. The presentation of SSc-ILD includes different types of ILDs and can also show different forms of progression. A slow decline in lung function (LF) and CO diffusion parameters can be present with periods of stability or improvement, but rapid progression and severe deterioration can also occur [3,7].

The definition of progressive fibrosing (PF)- ILD describes LF decline associated with either the worsening of respiratory symptoms (mainly dyspnea and cough) and/or a progression of fibrosis on high-resolution computed tomography (HRCT) [8–10]. Compared to previous years, nowadays we have a wide range of therapeutic options available. Immunosuppressive (ISU) drugs, such as cyclophosphamide (CYC) and mycophenolate mofetil (MMF), have shown to improve lung function and forced vital capacity (FVC) according to randomized clinical trials [9,11–14]. In progressive forms of SSc-ILD (PF-ILDs), the focus has shifted towards the tyrosine kinase inhibitor nintedanib, which was approved for progressive SSc-ILD treatment by the European Medicines Agency in 2020 [15,16]. While the co-treatment of MMF and nintedanib is not authorized in this subgroup, the combined therapy showed a more promising outcome, without increase in adverse events in the SENSCIS (A Trial to Compare Nintedanib with Placebo for Patients with Scleroderma Related Lung Fibrosis) trial [15]. High-dose corticosteroids are not recommended in SSc-ILD as first line treatment. In a smaller number of studies, the safety and beneficial adverse reaction profile of rituximab was demonstrated with a similar efficacy as CYC in SSc-ILD [17]. In some centers, bone marrow or lung transplantation is available in extremely carefully selected cases [14].

Measuring the body mass index (BMI) is a useful and simple way to evaluate excess weight. The role of weight loss and BMI in the disease prognosis is already well known in other respiratory diseases, but has not been studied in SSc-ILD [18–25].

Our goal is to assess the clinical course and risk factors for progression in SSc-ILD using longitudinal data on pulmonary evaluation.

2. Materials and Methods

2.1. Study Population

Our study is based on the retrospective data analysis of SSc-ILD outpatients classified under ICD10 J84 and M34 codes, who were presented at the Semmelweis University Pulmonology Department between January 2017 and June 2019. The diagnosis of SSc and treatment initiation was made by immunology and rheumatology specialists in immunological-rheumatological centers in central Hungary. The majority of patients had dcSSc (93%), while 4 patients had lcSSc, which had been known for more than 10 years in all cases. The dermatological care of patients with skin involvement took place in dermatology centers, and our study did not cover the exact assessment of these symptoms. All patients were evaluated by our MDT. A total of 54 SSc-ILD patients were identified and 42 of them had pulmonary follow-up in the given time interval and were included in the longitudinal evaluation. In accordance with the therapy, the patients were divided into 3 subgroups: no treatment (n = 12), immunosuppressive therapy (ISU, n = 21) and biological therapy (rituximab or tocilizumab) recipients (n = 9). The patient inclusion is summarized in Figure 1. The most frequently applied ISU therapies were CYC or MMF with or without low-dose glucocorticosteroids.

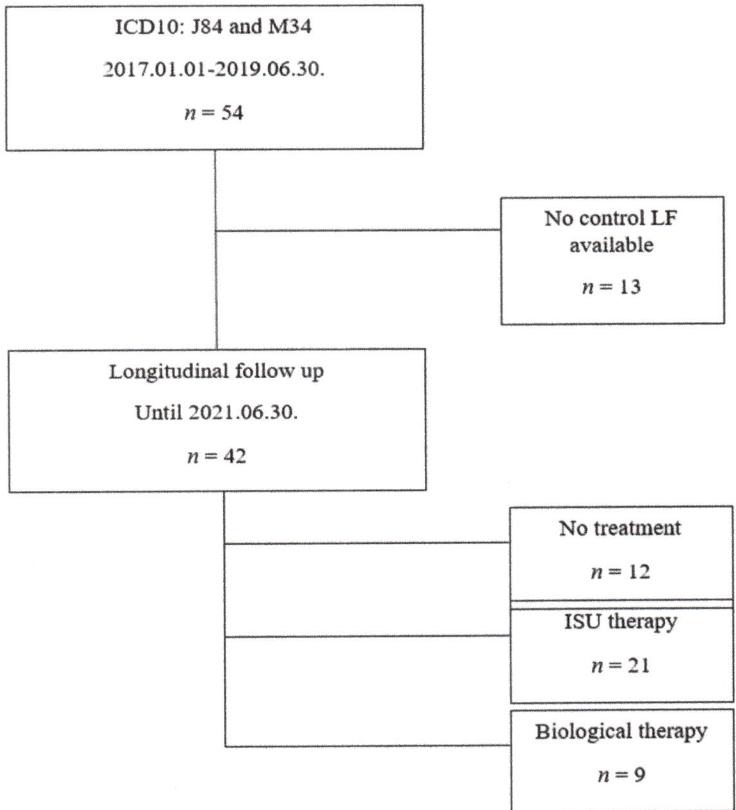

Figure 1. Study population.

SSc was previously diagnosed, based on the internationally accepted criteria of the American College of Rheumatology/European League Against Rheumatism Collaborative Initiative (EULAR-ACR), by rheumatology specialists working at rheumatology centers in central Hungary [5].

The SSc-ILD patients' data were collected retrospectively, based on the medical records. In each case, a detailed medical history was attained regarding the pulmonary symptoms, exposures and comorbidities. According to the WHO definition, the patients with a BMI between 25.0–29.9 kg/m^2 were considered as overweight. All the subjects underwent physical examination, chest X-ray at their pulmonary evaluation. Pulmonary function tests included the measurement of FVC, forced expiratory volume in 1 s (FEV1), FEV1/FVC and total lung capacity (TLC) according to the American Thoracic Society and European Respiratory Society (ARS/ETS) guidelines. For measuring the diffusing capacity for carbon monoxide (DLCO), we used the single breath CO method. In addition, the transfer coefficient of the lung for carbon monoxide (KLCO) was calculated (PDD-301/s, Piston, Budapest, Hungary). Gender–age–physiology (GAP) used in idiopathic pulmonary fibrosis (IPF) to estimate mortality was calculated as a potential risk assessment tool [26]. Patients were tested for physical activity using the 6 min walk test (6MWT). Arterialized capillary blood gases (ABGs) were evaluated at room temperature (Cobas b 221, Roche, Hungary). The Borg scale referring to dyspnea was used.

All patients had been confirmed as ILD by high resolution computed tomography (HRCT), performed in both inspiration and expiration using Philips Ingenuity Core 64 and Philips Brilliance 16 CT scanners. HRCT patterns, such as non-specific interstitial pneumonia (NSIP), usual interstitial pneumonia (UIP) and probable UIP (pUIP) and percentage involvement, were determined by MDT experts according to a visual scoring system [27]. The radiologic features of NSIP include ground-glass opacities (GGO), reticular opacities and traction bronchiectasis in the fibrotic subtype [28]. The UIP pattern comprises subpleural and basal predominance reticulation accompanied by honeycombing and traction bronchiectasis (which can be associated with presence of GGO) probable UIP (pUIP) are the same abnormalities without honeycombing [29,30].

The long-term care is comprised of pulmonological and rheumatological controls in correlation with the patients' disease requirements. The follow-up (and also treatment time) of at least 31 months (in the longest case 53 months) until June 2021 included radiological controls, therapy management, LF, DLCO and KLCO tests. In this analysis, PF-ILD was defined as a FVC relative annual decline $\geq$5% in addition to either a deterioration of clinical symptoms or a progression of fibrosis on HRCT, as mentioned previously [10,31].

2.2. Statistical Analysis

Statistical analysis was performed using Graph Pad software (GraphPad Prism 5.0 Software, Inc., La Jolla, CA, USA) and Microsoft Excel (Microsoft Corporation, Redmond, DC, USA). Continuous variables were expressed as the mean $\pm$ standard deviation. The Kolmogorov–Smirnow test was used for the normality test of data. Differences between the groups for continuous data were evaluated in normally distributed data with a Student's *t*-test. Otherwise the Mann–Whitney U test was used. Analysis of variance (ANOVA) and Tukey's post hoc analysis were used to examine the continuous variables in the therapeutic subgroups. The chi-squared test and two-tailed Fisher's exact test were applied for comparing the categorical variables. Predictors of the progression were analyzed using the odds ratio and plot analysis. The correlation between the BMI and FVC was analyzed by logarithmic transformation. All % values are expressed for the whole study population (all patients) or respective subgroups as indicated. A p-value < 0.05 was defined as statistically significant.

3. Results

The patient characteristics are summarized in Table 1. The mean age of the studied patients was 59 years, the majority of them being women. Non-smokers accounted for 74.1% of the case numbers. Ninety-five percent of patients had 0–3 points (Stage I) at GAP staging, while only 2 patients were in Stage II. The lowest BMI values were in the treatment-free subgroup, the highest in the ISU subgroup. The main symptoms included dyspnea, crackles and Raynaud's phenomenon, followed by cough, joint pain and digital clubbing. Significantly fewer patients reported cough in the ISU therapy subgroup compared to other subgroups. Gastrointestinal involvement was most often present in the treatment-free group. The most common HRCT radiological pattern is NSIP (predominantly less than 20% of lung involvement), followed by UIP or pUIP (approximately balanced involvement below and above 20%) (Table 1). In the ISU subgroup, significantly more patients had a NSIP pattern, while, from the patients receiving biological therapy, a pUIP/UIP pattern was predominant. LF data at baseline is summarized in Table 2.

Table 1. Patient characteristics.

Parameters	All Patients (n = 54) #	No Treatment (n = 12)	ISU Therapy (n = 21)	Biological Therapy (n = 9)
Age (year)	58.7 ± 13.3	66.1 ± 13.7	59.12 ± 12.4	62.7 ± 10.0
Sex (male: female) n (%)	7:47 (13.0:87.0)	1:11 (8.3:91.7)	3:18 (14.3:85.7)	1:8 (11.1:88.9)
GAP score n (%)				
Stage I (0–3 points)	48 (88.9)	12 (100)	20 (95.2)	8 (88.9)
Stage II (4–5 points)	6 (11.1)	0	1 (5)	1 (1.1)
Stage III (6–8 points)	0	0	0	0
Ever smoker n (%)	14 (25.9)	4 (33.3)	5 (23.8)	2 (22.2)
Non-smoker n (%)	40 (74.1)	8 (66.7)	16 (76.2)	7 (77.8)
BMI (kg/m^2)	24.8 ± 4.3	23.6 ± 3.1	25.0 ± 4.4	26.4 ± 4.5
Overweight n (%)	21 (38.9)	3 (25.0)	8 (38.1)	5 (55.6)
PF-ILD n (%)	15 (27.8)	5 (41.7)	7 (33.3)	3 (33.3)
Symptoms n (%)				
Dyspnea	26 (48.2)	6 (50.0)	8 (38.1)	5 (55.6)
Cough	15 (27.8)	4 (33.3)	1 (4.8) *	4 (44.4)
Chest pain	5 (9.3)	0	0	0
Joint pain	8 (14.8.)	1 (8.3)	5 (23.8)	1 (11.1)
Clubbing	1 (1.91)	0	0	1 (11.1)
Weight loss	3 (5.6%)	1 (8.3)	1 (4.9)	1 (11.1)
Crackles	15 (27.8)	5 (41.7)	6 (28.6)	2 (22.2)
Raynaud's phenomenon	38 (70.5)	8 (66.7)	13 (61.9)	5 (55.6)
GIT involvement	10 (18.5)	5 (41.7) ***	2 (9.5)	3 (33.3)
HRCT pattern n (%)				
NSIP n	34 (63.0)	11 (91.7)	15 (71.4)	3 (33.3) **
UIP/pUIP n	8 (14.8)	1 (8.3)	2 (9.5)	4 (44.4) ***
Other or no data n	10 (18.5)	0	4 (19.0)	2 (22.2)
Serological pattern n (%)				
ANA	23 (42.6)	8 (66.7)	12 (57.1)	3 (33.3)
ACA	1 (1.9)	1 (8.3)	0	0
RF	3 (5.6)	2 (1.7)	1 (4.8)	0
ACCP	1 (1.9)	0	1 (4.8)	0
Anti-RNA-polymerase	2 (3.7)	1 (8.3)	1 (4.8)	0
Anti-cytoplasmatic	5 (9.3)	1 (8.3)	2 (9.5)	2 (22.2)
Anti-chromatin	12 (22.2)	5 (41.7)	6 (28.6)	1 (11.1)
Anti-Smith	1 (1.9)	1 (8.3)	0	0
Anti-Jo-1	1 (1.9)	0	1 (4.8)	0
Anti-SSA	2 (3.7)	0	2 (9.5)	0
Anti-SSB	1 (1.9)	1 (8.3)	0	0
Anti-SCL-70	18 (33.3)	8 (66.7)	9 (42.9) **	1 (11.1)
Anti-RNP	4 (7.4)	3 (25)	1 (4.8)	0
Anti-dsDNA	3 (5.6)	1 (8.3)	1 (4.8)	1 (11.1)

ACA, anticentromere antibodies; ACCP, anti-cyclic citrullinated peptide antibodies; anti-dsDNA, antibodies to double-stranded deoxyribonucleic acid; anti-SSA, Ro autoantibodies; anti-SSB, anti-La antibodies; anti-SCL-70, anti-topoisomerase I antibodies; anti-RNP, antibodies to ribonucleoprotein; ANA, anti-nuclear antibodies; BMI, body mass index; GAP, gender–age–physiology index; GIT, gastrointestinal tract; HRCT, high-resolution computed tomography; ISU, immunosuppressive; NSIP, non-specific interstitial pneumonia; PF-ILD, progressive fibrosing interstitial lung disease; pUIP, probable usual interstitial pneumonia; RF, rheumatoid factor; UIP, usual interstitial pneumonia; # total number of patients were 54, but out of these only 42 had follow-up data; * $p < 0.05$ vs. no treatment and biological therapy subgroup ** $p < 0.05$ vs. no treatment subgroup, *** $p < 0.05$ vs. ISU subgroup.

Table 2. Lung function, ABG and 6MWT functional parameters.

Parameters	All Patients (n = 54)	No Treatment (n = 12)	ISU Therapy (n = 21)	Biological Therapy (n = 9)
Lung function				
FVC (L)	2.5 ± 0.8	2.5 ± 0.8	2.8 ± 0.8	2.2 ± 0.6
FVC (%predicted)	89.8 ± 23.2	97.6 ± 21.7	92.4 ± 26.6	82.2 ± 17.2
FEV1(L)	2.2 ± 0.6	2.1 ± 0.6	2.3 ± 0.7	2.0 ± 0.6
FEV1(%predicted)	90.2 ± 21.8	97.9 ± 21.6	92.2 ± 23.1	85.3 ± 21.5
FEV1/FVC (%)	84.7 ± 6.3	83.4 ± 6.3	84.9 ± 4.9	86.9 ± 12.4
TLC (L)	3.9 ± 1.1	4.0 ± 1.0	4.1 ± 1.4	3.8 ± 1.0
TLC (%predicted)	78.4 ± 21.0	81.4 ± 16.2	81.1 ± 23.0	79.2 ± 23.4
Diffusion parameters				
DLCO (mmol/min/kPa)	5.9 ± 2.0	6.2 ± 1.8	6.2 ± 2.1	5.1 ± 1.3
DLCO (%predicted)	75.2 ± 22.0	86.9 ± 24.5	77.4 ± 20.7	66.7 ± 14.7
KLCO (mmol/min/kPa/L)	1.4 ± 0.4	1.3 ± 0.3	1.4 ± 0.4	1.3 ± 0.3
KLCO (%predicted)	70.0 ± 18.1	66.0 ± 15.6	70.5 ± 18.9	67.0 ± 16.7
ABG				
pH	7.4 ± 0.0	7.4 ± 0.0	7.4 ± 0.0	7.4 ± 0.1
pCO_2 (mmHg)	38.0 ± 4.7	34.4 ± 3.1	43.4 ± 11.1	36.7 ± 3.6
pO_2 (mmHg)	74.2 ± 10.5	84.2 ± 13.1	71.2 ± 12.1	72.2 ± 14.6
6MWT				
Distance (m)	444.2 ± 119.8	365.3 ± 233.9	468.8 ± 108.0	342.0 ± 106.6
SpO_2 baseline (%)	94.9 ± 2.8	96.7 ± 3.2	97.3 ± 2.1	93.9 ± 4.4
SpO_2 post-exercise (%)	89.9 ± 10.0	82.7 ± 19.9	95.3 ± 2.4	87.6 ± 8.1
Desaturation (%)	4.9 ± 9.3	14.0 ± 16.6	2.5 ± 2.4	8.0 ± 6.3
Pulse baseline (1/min)	84.8 ± 14.6	78.7 ± 11.9	82.9 ± 9.6	86.4 ± 11.6
Pulse post-exercise (1/min)	108.5 ± 23.1	100.3 ± 29.5	108.4 ± 17.6	106.7 ± 26.4
Borg scale baseline (0–10)	0.2 ± 0.5	1.7 ± 2.9	0.1 ± 0.3	1.0 ± 1.4
Borg scale post-exercise (0–10)	1.8 ± 2.5	2.7 ± 3.8	1.6 ± 1.2	3.1 ± 2.5

6MWT, 6 min walk; ABG, arterialized capillary blood gases; DLCO, diffusing capacity for carbon monoxide; FVC, forced vital capacity; FEV1, forced expiratory volume in 1 s; ISU, immunosuppressive; KLCO, transfer coefficient of the lung for carbon monoxide; pCO_2; partial pressure of carbon dioxide; pO_2, partial pressure of oxygen ; SpO_2, oxygen saturation; TLC, total lung capacity.

Patients were characterized by mild restrictive functional decline. Annual FVC decline was more distinct in patients who did not receive therapy (−10.2 ± 13.0%), in comparison to patients who underwent ISU (−3.9 ± 5.1%) and biological treatment (−1.04 ± 7.8%). Patients receiving biological treatment showed the slightest degree of FVC deterioration, with 4 patients even showing an improvement. There was no difference in LF between the longitudinal subgroups, however annual FVC decline was less progressive in the treated subgroups. A decrease in DLCO is present in all three observed treatment subgroups, however the relative DLCO decline is greater with patients who received biological therapy (Figure 2). Out of the patients with longitudinal data, 15 patients met our PF-ILD criteria, while 27 did not show signs of deterioration during follow-up. The highest proportion of PF-ILDs was confirmed in the treatment-free subgroup, while over 3/4 of patients were stable or improved during the ISU and biological treatments.

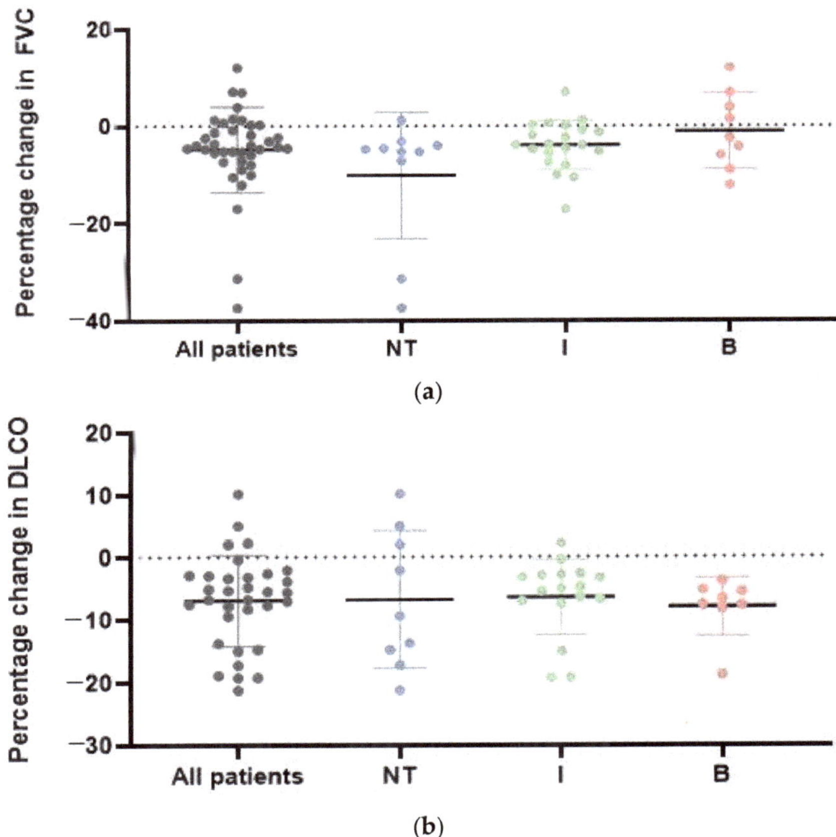

Figure 2. (**a**) Annual FVC changes in all SSc-ILD patients and in the specific treatment groups. Description of wha: is contained in the first panel; (**b**) annual DLCO changes in each specific treatment group; DLCO, diffusing capacity for carbon monoxide; FVC, forced vital capacity; I = ISU therapy; NT = No treatment: I = ISU therapy, B = Biological therapy.

The factors favoring functional stability are being overweight (according to the definition of World Health Organization (BMI > =25 kg/m^2)) and an absence of anti-SCL-70 positivity (Figure 3). Being overweight was mainly present in the biological treatment subgroup (55.6%) and affected the least patients in the treatment-free subgroup (25%). A significantly lower BMI is present in PF-ILD compared to stable SSc-ILD patients (22.9 kg/m^2 vs. 25.71 kg/m^2; $p = 0.03$) ,and a clear negative correlation between baseline BMI and annual FVC reduction ($r = -0.97$, $r^2 = 0.93$, $p < 0.001$) can be established (Figure 4).

A total of 10 patients had gastrointestinal involvement (mostly esophagus dysmotility) with a predominantly low or normal BMI ($p = 0.019$), compared to overweight patients. No significant association was found between gastrointestinal symptoms and PF-ILD.

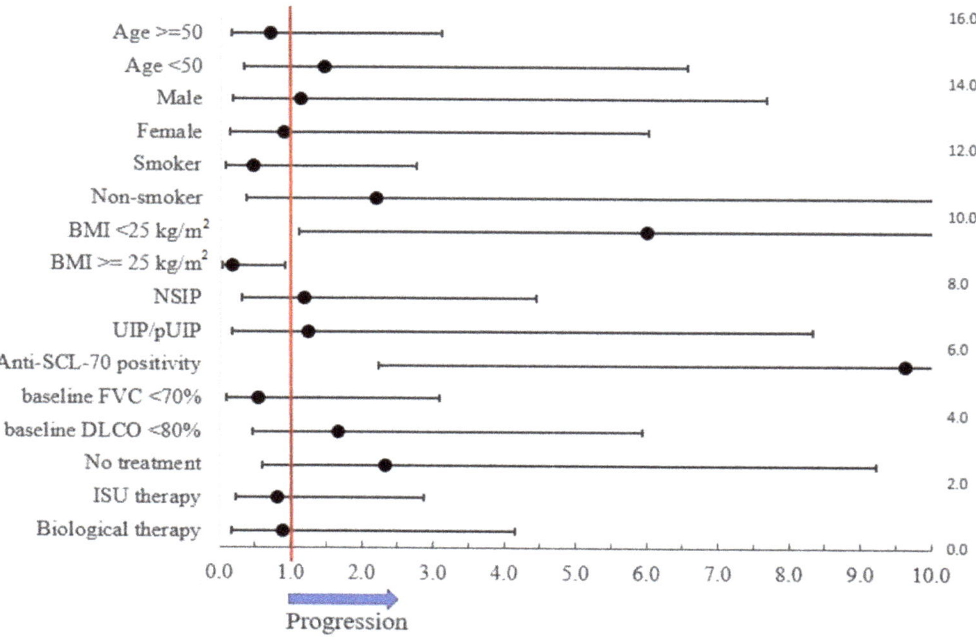

Figure 3. Risk factors of PF-ILD progression.

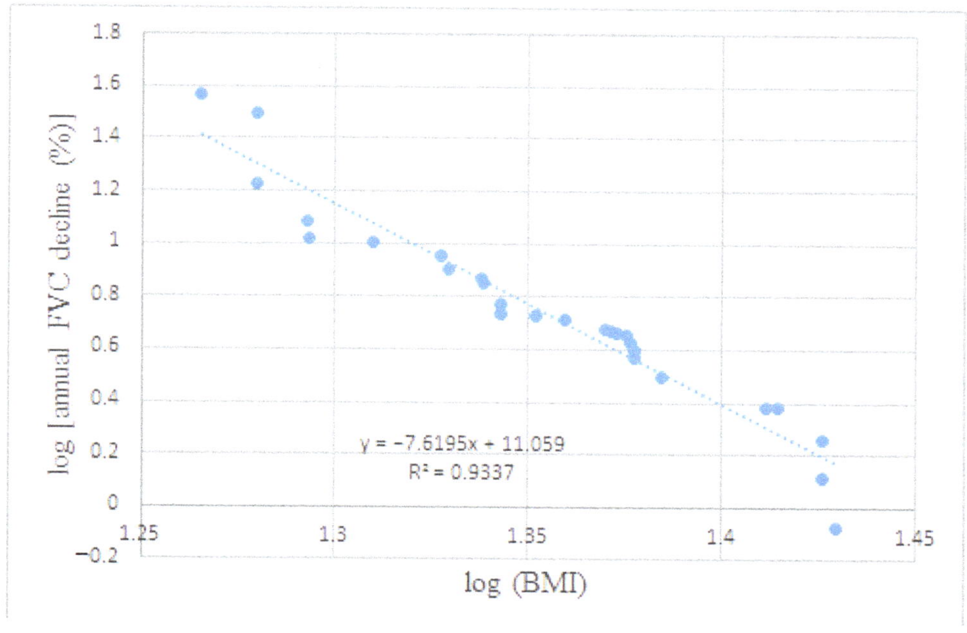

Figure 4. Negative correlation between BMI and annual FVC decline (%). BMI, body mass index; FVC, forced vital capacity.

4. Discussion

SSc-ILD has a variable disease course from slow decline with stable periods or improvement to rapidly progressive clinical course. The novelty of our study is that baseline overweight might be an important indicator for more favorable outcomes regarding PF-ILD.

In our study, 15 patients (35.7%) met our PF-ILD criteria, which is in correspondence with international data [1,3,18–20]. In 2014, Tiffany and colleagues analyzed 27 reviews on the topic of predictors of progression and mortality in SSc-ILD. In these studies, patient-specific, ILD-specific, and SSc-specific variables predicted the progression and mortality in SSc-ILD [32]. Based on this data, the major risk factors for progressive SSc-ILD include an older age, male sex, extent of lung involvement on initial HRCT, decreased DLCO and FVC at baseline. Our study could not quite verify these factors of PF-ILD, presumably due to the small sample size and the resulting large standard deviation. The presence of anti-Scl-70 antibodies and the absence of anti-centromere antibodies are also widely accepted negative predictors., and this association was also confirmed in our research [32,33].

In accordance with our study, the negative association between functional progression and BMI is a new and clinically important predictive marker in SSc-ILD. The question has to be asked: why is being overweight a protective factor in SSc-ILDs? The negative prognostic role of a lower BMI and weight loss is already known in IPF [18–22]. The post hoc analysis of INPULSIS, CAPACITY and other clinical trials confirm this condition, in which some association was found between the annual FVC decline and BMI at baseline [34,35]. However, contradictory data also exists [36,37], and some key studies, such as SLS (Scleroderma Lung Study) I and II, did also not include the BMI data [12,13]. The aggravating role of a lower BMI and weight loss are described in other respiratory diseases associated with chronic hypoxia, for example, in chronic obstructive pulmonary disease (COPD) [23]. While obesity in COPD appears to be a protective factor (obesity paradox) despite the comorbidities, there is only a limited amount of data about the comorbidities of extreme obesity due to the low number of obese patients in most large COPD studies [24]. The cause might be an electron transport chain dysfunction, which causes decreased muscle stamina in low BMI COPD patients. Considering the concept, a similar phenomenon may occur in SSc-ILD [25]. As it is well known, a lower BMI might be caused by gastrointestinal tract (GIT) involvement [38,39]. However, there was no significant difference in dysphagia and gastroesophageal reflux between the PF-ILD and non-PF-ILD forms, whereas more patients had GIT symptoms with normal BMI.

In our study, we investigated patient characteristics and progression according to therapy. The first line treatment of SSc-ILD is immunosuppression (mainly CYC and MMF) based on the results of previous clinical trials: SLS I and SLS II [11–13,40]. More recent biological therapies, such as rituximab and tocilizumab, have expanded the options of treatment [17,41,42]. Due to the risk of provoking renal crisis, the use of glucocorticosteroids in scleroderma is limited to low doses [14]. During the follow-up period, low dose steroid treatment was used only in the ISU subgroup. Although the weight-gaining side effect of glucocorticosteroids is well known and the mean BMI was the highest in the ISU subgroup, the overweight patients were more prevalent in the biological therapy subgroup. Additionally, the PF-ILD ratio was the same in the ISU and biological therapy subgroups (33.3%). The annual decrease in LF was more pronounced in the treatment-free subgroup and, simultaneously, the lowest BMI values were represented in these patients. The close monitoring of these patients (including regular body weight and BMI measurement) is the only way to initiate early treatment in order to intervene and slow down progression. It is important to emphasize that patients with preserved lung function and a normal BMI were less likely to be treated with ISU or biological therapy, and were the most prone to lung function loss. The choice of systemic treatment is decided by rheumatologists and immunologists. The systemic treatment of patients in our research was also carried out by immunologists or rheumatologist in an outpatient setting. However, in patients with mild symptoms, systemic treatment was not considered as necessary. Thus, small disease progression between visits was not noticeable, leading to late therapy initiation.

The correct introduction of systemic therapy is, to date, not mentioned in international recommendations [43–45]. More attention should be given to patients with good lung function and smaller extent lung involvement on HRCT, especially to patients with a normal BMI, as they are more prone to unnoticed functional decline. The Goh criteria set a low FVC treatment threshold (<70% predicted) that might result in severe functional loss in SSc-ILD patients before treatment initiation. In contrast, monitoring the functional decline and the earlier start of ISU or biological treatment might reduce the risk of deterioration. This concept of early treatment introduction was also pointed out in the EUSTAR database, although BMI was not included as a risk factor of functional decline in their evaluation [7]. Naturally, due to the comorbidities of being overweight, we do not recommend achieving obesity. The target BMI level may be in the upper normal BMI range, but this assumption requires further investigation.

The limitations of our study are the retrospective single center design and the low number of patients. Prospective studies should be encouraged to evaluate the correlation between a lower BMI, weight loss and SSc-ILD. Long term longitudinal follow-up, expert SSc-ILD evaluation by MDT and treatment defined subgroup analysis were the assets of our study.

5. Conclusions

BMI is an important prognostic factor in SSc-ILD progression and an initial low and normal BMI or weight loss should be critically followed by the clinical care team. The thorough monitoring of BMI in clinical practice is required and especially patients with normal BMI should be followed closely for deterioration. The timing of the introduction of ISU and biological therapy remain a major challenge for clinicians. A higher awareness and possibly lower threshold for therapy initiation is needed in patients with preserved lung function and a normal BMI, even in clinically or radiologically limited SSc-ILD as functional decline was the most prominent in untreated patients. Although it is important to note that patients with a normal BMI were presented more often with GIT symptoms, GIT involvement was not associated with PF-ILD. Regular LF and BMI measurements should be performed in all SSc patients starting from the initial diagnosis, and anti-SCL-70 positivity is helpful when considering therapy introduction. Future studies are needed to determine the optimal start and combination of therapy.

Author Contributions: Conceptualization, A.N., E.P. and V.M.; methodology, A.N., E.P. and L.P.; software, L.P.; validation, A.N., L.P. and V.M.; formal analysis, V.M.; resources, V.M.; data curation, A.N., E.P. and V.M.; writing—original draft preparation, A.N, E.P.; writing—review and editing, A.D.T., D.L.T., G.N., E.K., P.M.-H. and V.M.; visualization, A.N.; supervision, N.E., K.V., E.B., A.B. and V.M.; A.N., E.P., L.P., N.E., K.V., E.B., A.B., A.D.T., D.L.T., G.N., E.K., P.M.-H. and V.M. were treating physicians or radiologists, members of the multidisciplinary ILD team at Semmelweis University and contributed to the data, drafting and review of the manuscript. All authors have read and agreed to the published version of the manuscript.

Funding: This research received no external funding. Eniko Barczi was supported by the Hungarian Respiratory Society.

Institutional Review Board Statement: The study was conducted in accordance with the Declaration of Helsinki, and approved by the Ethical Committee of Semmelweis University (181/2021) for studies involving humans.

Informed Consent Statement: Informed consent was not needed due to the retrospective design.

Data Availability Statement: The original contributions presented in the study are included in the article; further inquiries can be directed to the corresponding author.

Conflicts of Interest: The authors declare no conflict of interest.

Abbreviations

6MWT	6 min walk
ABGs	arterialized capillary blood gases
ACAs	anticentromere antibodies
ANAs	anti-nuclear antibodies
Anti-dsDNA	antibodies to double-stranded deoxyribonucleic acid
Anti-RNP	antibodies to ribonucleoprotein
Anti-SCL-70	anti-topoisomerase I antibodies
Anti-SSA	Ro autoantibodies
Anti-SSB	anti-La antibodies
ATS	American Thoracic Society
BMI	body mass index
COPD	chronic obstructive pulmonary disease
CYC	cyclophosphamide
dcSSc	diffuse cutaneous SSc
DLCO	diffusing capacity for carbon monoxide
ERS	European Respiratory Society
EULAR-ACR	American College of Rheumatology/European League Against Rheumatism Collaborative Initiative
FEV1	forced expiratory volume in 1 s
FVC	forced vital capacity
GGOs	ground glass opacities
GIT	gastrointestinal tract
HRCT	high-resolution computed tomography
ILD	interstitial lung disease
IPF	idiopathic pulmonary fibrosis
ISU	immunosuppressive
KLCO	transfer coefficient of the lung for carbon monoxide
lcSSc	limited cutaneous SSc
LF	lung function
MDT	multidisciplinary ILD-team
MMF	mycophenolate mofetil
NSIP	non-specific interstitial pneumonia
pCO_2	partial pressure of carbon dioxide
PF-ILD	progressive fibrosing interstitial lung disease
pO_2	partial pressure of oxygen
pUIP	probable usual interstitial pneumonia
RF	rheumatoid factor
SENSCIS	A Trial to Compare Nintedanib with Placebo for Patients With Scleroderma-Related Lung Fibrosis
SLS	Scleroderma Lung Study
SpO_2	oxygen saturation
SSc	systemic sclerosis
ssSSc	sine scleroderma SSc
TLC	total lung capacity
UIP	usual interstitial pneumonia

References

1. Perelas, A.; Silver, R.M.; Arrossi, A.V.; Highland, K.B. Systemic sclerosis-associated interstitial lung disease. *Lancet Respir. Med.* **2020**, *8*, 304–320. [CrossRef]
2. Hu, P.Q.; Fertig, N.; Medsger, T.A., Jr.; Wright, T.M. Correlation of serum anti-DNA topoisomerase I antibody levels with disease severity and activity in systemic sclerosis. *Arthritis Care Res.* **2003**, *48*, 1363–1373. [CrossRef] [PubMed]
3. Panagopoulos, P.; Goules, A.; Hoffmann-Vold, A.M.; Matteson, E.L.; Tzioufas, A. Natural history and screening of interstitial lung disease in systemic autoimmune rheumatic disorders. *Ther. Adv. Musculoskelet. Dis.* **2021**, *13*, 1759720X211037519. [CrossRef] [PubMed]
4. Wijsenbeek, M.; Cottin, V. Spectrum of Fibrotic Lung Diseases. *N. Engl. J. Med.* **2020**, *383*, 958–968. [CrossRef]

5. van den Hoogen, F.; Khanna, D.; Fransen, J.; Johnson, S.R.; Baron, M.; Tyndall, A.; Matucci-Cerinic, M.; Naden, R.P.; Medsger, T.A., Jr.; Carreira, P.E.; et al. 2013 classification criteria for systemic sclerosis: An American College of Rheumatology/European League against Rheumatism collaborative initiative. *Arthritis Rheum.* **2013**, *65*, 2737–2747. [CrossRef]
6. Solomon, J.J.; Olson, A.L.; Fischer, A.; Bull, T.; Brown, K.K.; Raghu, G. Scleroderma lung disease. *Eur. Respir. Rev.* **2013**, *22*, 6–19. [CrossRef]
7. Hoffmann-Vold, A.M.; Allanore, Y.; Alves, M.; Brunborg, C.; Airó, P.; Ananieva, L.P.; Czirják, L.; Guiducci, S.; Hachulla, E.; Li, M.; et al. Progressive interstitial lung disease in patients with systemic sclerosis-associated interstitial lung disease in the EUSTAR database. *Ann. Rheum. Dis.* **2021**, *80*, 219–227. [CrossRef]
8. Brown, K.K.; Martinez, F.J.; Walsh, S.L.F.; Thannickal, V.J.; Prasse, A.; Schlenker-Herceg, R.; Goeldner, R.G.; Clerisme-Beaty, E.; Tetzlaff, K.; Cottin, V.; et al. The natural history of progressive fibrosing interstitial lung diseases. *Eur. Respir. J.* **2020**, *55*, 2000085. [CrossRef]
9. George, P.M.; Spagnolo, P.; Kreuter, M.; Altinisik, G.; Bonifazi, M.; Martinez, F.J.; Molyneaux, P.L.; Renzoni, E.A.; Richeldi, L.; Tomassetti, S.; et al. Progressive fibrosing interstitial lung disease: Clinical uncertainties, consensus recommendations, and research priorities. *Lancet Respir. Med.* **2020**, *8*, 925–934. [CrossRef]
10. Cottin, V.; Hirani, N.A.; Hotchkin, D.L.; Nambiar, A.M.; Ogura, T.; Otaola, M.; Skowasch, D.; Park, J.S.; Poonyagariyagorn, H.K.; Wuyts, W.; et al. Presentation, diagnosis and clinical course of the spectrum of progressive-fibrosing interstitial lung diseases. *Eur. Respir. Rev.* **2018**, *27*, 180076. [CrossRef]
11. Volkmann, E.R.; Tashkin, D.P.; Li, N.; Roth, M.D.; Khanna, D.; Hoffmann-Vold, A.M.; Kim, G.; Goldin, J.; Clements, P.J.; Furst, D.E.; et al. Mycophenolate Mofetil Versus Placebo for Systemic Sclerosis-Related Interstitial Lung Disease: An Analysis of Scleroderma Lung Studies I and II. *Arthritis Rheumatol.* **2017**, *69*, 1451–1460. [CrossRef] [PubMed]
12. Tashkin, D.P.; Roth, M.D.; Clements, P.J.; Furst, D.E.; Khanna, D.; Kleerup, E.C.; Goldin, J.; Arriola, E.; Volkmann, E.R.; Kafaja, S.; et al. Mycophenolate mofetil versus oral cyclophosphamide in scleroderma-related interstitial lung disease (SLS II): A randomised controlled, double-blind, parallel group trial. *Lancet Respir. Med.* **2016**, *4*, 708–719. [CrossRef]
13. Tashkin, D.P.; Elashoff, R.; Clements, P.J.; Goldin, J.; Roth, M.D.; Furst, D.E.; Arriola, E.; Silver, R.; Strange, C.; Bolster, M.; et al. Cyclophosphamide versus placebo in scleroderma lung disease. *N. Engl. J. Med.* **2006**, *354*, 2655–2666. [CrossRef] [PubMed]
14. Cappelli, S.; Guiducci, S.; Bellando Randone, S.; Matucci Cerinic, M. Immunosuppression for interstitial lung disease in systemic sclerosis. *Eur. Respir. Rev.* **2013**, *22*, 236–243. [CrossRef] [PubMed]
15. Distler, O.; Highland, K.B.; Gahlemann, M.; Azuma, A.; Fischer, A.; Mayes, M.D.; Raghu, G.; Sauter, W.; Girard, M.; Alves, M.; et al. Nintedanib for Systemic Sclerosis-Associated Interstitial Lung Disease. *N. Engl. J. Med.* **2019**, *380*, 2518–2528. [CrossRef]
16. Flaherty, K.R.; Wells, A.U.; Cottin, V.; Devaraj, A.; Walsh, S.L.F.; Inoue, Y.; Richeldi, L.; Kolb, M.; Tetzlaff, K.; Stowasser, S.; et al. Nintedanib in Progressive Fibrosing Interstitial Lung Diseases. *N. Engl. J. Med.* **2019**, *381*, 1718–1727. [CrossRef]
17. Sircar, G.; Goswami, R.P.; Sircar, D.; Ghosh, A.; Ghosh, P. Intravenous cyclophosphamide vs rituximab for the treatment of early diffuse scleroderma lung disease: Open label, randomized, controlled trial. *Rheumatology* **2018**, *57*, 2106–2113. [CrossRef]
18. Kishaba, T.; Nagano, H.; Nei, Y.; Yamashiro, S. Body mass index-percent forced vital capacity-respiratory hospitalization: New staging for idiopathic pulmonary fibrosis patients. *J. Thorac. Dis.* **2016**, *8*, 3596–3604. [CrossRef]
19. Alakhras, M.; Decker, P.A.; Nadrous, H.F.; Collazo-Clavell, M.; Ryu, J.H. Body mass index and mortality in patients with idiopathic pulmonary fibrosis. *Chest* **2007**, *131*, 1448–1453. [CrossRef]
20. Mogulkoc, N.; Sterclova, M.; Müller, V.; Kus, J.; Hajkova, M.; Jovanovic, D.; Tekavec-Trkanjec, J.; Janotova, M.; Vasakova, M. Does body mass index have prognostic significance for patients with idiopathic pulmonary fibrosis? *Eur. Respir. J.* **2018**, *52*, PA2210.
21. Nakatsuka, Y.; Handa, T.; Kokosi, M.; Tanizawa, K.; Puglisi, S.; Jacob, J.; Sokai, A.; Ikezoe, K.; Kanatani, K.T.; Kubo, T.; et al. The Clinical Significance of Body Weight Loss in Idiopathic Pulmonary Fibrosis Patients. *Respiration* **2018**, *96*, 338–347. [CrossRef] [PubMed]
22. Kulkarni, T.; Yuan, K.; Tran-Nguyen, T.K.; Kim, Y.I.; de Andrade, J.A.; Luckhardt, T.; Valentine, V.G.; Kass, D.J.; Duncan, S.R. Decrements of body mass index are associated with poor outcomes of idiopathic pulmonary fibrosis patients. *PLoS ONE* **2019**, *14*, e0221905. [CrossRef] [PubMed]
23. Wada, H.; Ikeda, A.; Maruyama, K.; Yamagishi, K.; Barnes, P.J.; Tanigawa, T.; Tamakoshi, A.; Iso, H. Low BMI and weight loss aggravate COPD mortality in men, findings from a large prospective cohort: The JACC study. *Sci. Rep.* **2021**, *11*, 1531. [CrossRef]
24. Divo, M.J.; Cabrera, C.; Casanova, C.; Marin, J.M.; Pinto-Plata, V.M.; de-Torres, J.P.; Zulueta, J.; Zagaceta, J.; Sanchez-Salcedo, P.; Berto, J.; et al. Comorbidity Distribution, Clinical Expression and Survival in COPD Patients with Different Body Mass Index. *Chronic. Obstr. Pulm. Dis.* **2014**, *1*, 229–238. [CrossRef]
25. Rabinovich, R.A.; Bastos, R.; Ardite, E.; Llinas, L.; Orozco-Levi, M.; Gea, J.; Vilaro, J.; Barbera, J.A.; Rodriguez-Roisin, R.; Fernandez-Checa, J.C.; et al. Mitochondrial dysfunction in COPD patients with low body mass index. *Eur. Respir. J.* **2007**, *29*, 643–650. [CrossRef] [PubMed]
26. Ley, B.; Ryerson, C.J.; Vittinghoff, E.; Ryu, J.H.; Tomassetti, S.; Lee, J.S.; Poletti, V.; Buccioli, M.; Elicker, B.M.; Jones, K.D.; et al. A multidimensional index and staging system for idiopathic pulmonary fibrosis. *Ann. Intern. Med* **2012**, *156*, 684–691. [CrossRef]
27. Wangkaew, S.; Euathrongchit, J.; Wattanawittawas, P.; Kasitanon, N. Correlation of delta high-resolution computed tomography (HRCT) score with delta clinical variables in early systemic sclerosis (SSc) patients. *Quant. Imaging Med. Surg.* **2016**, *6*, 381–390. [CrossRef]

28. Kligerman, S.J.; Groshong, S.; Brown, K.K.; Lynch, D.A. Nonspecific interstitial pneumonia: Radiologic, clinical, and pathologic considerations. *Radiographics* **2009**, *29*, 73–87. [CrossRef]
29. Raghu, G.; Collard, H.R.; Egan, J.J.; Martinez, F.J.; Behr, J.; Brown, K.K.; Colby, T.V.; Cordier, J.F.; Flaherty, K.R.; Lasky, J.A.; et al. An official ATS/ERS/JRS/ALAT statement: Idiopathic pulmonary fibrosis: Evidence-based guidelines for diagnosis and management. *Am. J. Respir. Crit. Care Med.* **2011**, *183*, 788–824. [CrossRef]
30. Raghu, G.; Remy-Jardin, M.; Myers, J.L.; Richeldi, L.; Ryerson, C.J.; Lederer, D.J.; Behr, J.; Cottin, V.; Danoff, S.K.; Morell, F.; et al. Diagnosis of Idiopathic Pulmonary Fibrosis. An Official ATS/ERS/JRS/ALAT Clinical Practice Guideline. *Am. J. Respir. Crit. Care Med.* **2018**, *198*, e44–e68. [CrossRef]
31. Nagy, A.; Nagy, T.; Kolonics-Farkas, A.M.; Eszes, N.; Vincze, K.; Barczi, E.; Tarnoki, A.D.; Tarnoki, D.L.; Nagy, G.; Kiss, E.; et al. Autoimmune Progressive Fibrosing Interstitial Lung Disease: Predictors of Fast Decline. *Front. Pharmacol.* **2021**, *12*, 778649. [CrossRef] [PubMed]
32. Winstone, T.A.; Assayag, D.; Wilcox, P.G.; Dunne, J.V.; Hague, C.J.; Leipsic, J.; Collard, H.R.; Ryerson, C.J. Predictors of mortality and progression in scleroderma-associated interstitial lung disease: A systematic review. *Chest* **2014**, *146*, 422–436. [CrossRef] [PubMed]
33. Khanna, D.; Tashkin, D.P.; Denton, C.P.; Renzoni, E.A.; Desai, S.R.; Varga, J. Etiology, Risk Factors, and Biomarkers in Systemic Sclerosis with Interstitial Lung Disease. *Am. J. Respir. Crit. Care Med.* **2020**, *201*, 650–660. [CrossRef] [PubMed]
34. Jouneau, S.; Crestani, B.; Thibault, R.; Lederlin, M.; Vernhet, L.; Valenzuela, C.; Wijsenbeek, M.; Kreuter, M.; Stansen, W.; Quaresma, M.; et al. Analysis of body mass index, weight loss and progression of idiopathic pulmonary fibrosis. *Respir. Res.* **2020**, *21*, 312. [CrossRef]
35. Jouneau, S.; Crestani, B.; Thibault, R.; Lederlin, M.; Vernhet, L.; Yang, M.; Morgenthien, E.; Kirchgaessler, K.U.; Cottin, V. Post hoc Analysis of Clinical Outcomes in Placebo- and Pirfenidone-Treated Patients with IPF Stratified by BMI and Weight Loss. *Respiration* **2022**, *101*, 142–154. [CrossRef]
36. Nishiyama, O.; Yamazaki, R.; Sano, H.; Iwanaga, T.; Higashimoto, Y.; Kume, H.; Tohda, Y. Fat-free mass index predicts survival in patients with idiopathic pulmonary fibrosis. *Respirology* **2017**, *22*, 480–485. [CrossRef]
37. Doubkova, M.; Svancara, J.; Svoboda, M.; Sterclova, M.; Bartos, V.; Plackova, M.; Lacina, L.; Zurkova, M.; Binkova, I.; Bittenglova, R.; et al. EMPIRE Registry, Czech Part: Impact of demographics, pulmonary function and HRCT on survival and clinical course in idiopathic pulmonary fibrosis. *Clin. Respir. J.* **2018**, *12*, 1526–1535. [CrossRef]
38. Caporali, R.; Caccialanza, R.; Bonino, C.; Klersy, C.; Cereda, E.; Xoxi, B.; Crippa, A.; Rava, M.L.; Orlandi, M.; Bonardi, C.; et al. Disease-related malnutrition in outpatients with systemic sclerosis. *Clin. Nutr.* **2012**, *31*, 666–671. [CrossRef]
39. Hvas, C.L.; Harrison, E.; Eriksen, M.K.; Herrick, A.L.; McLaughlin, J.T.; Lal, S. Nutritional status and predictors of weight loss in patients with systemic sclerosis. *Clin. Nutr ESPEN* **2020**, *40*, 164–170. [CrossRef]
40. Naidu, G.; Sharma, S.K.; Adarsh, M.B.; Dhir, V.; Sinha, A.; Dhooria, S.; Jain, S. Effect of mycophenolate mofetil (MMF) on systemic sclerosis-related interstitial lung disease with mildly impaired lung function: A double-blind, placebo-controlled, randomized trial. *Rheumatol. Int.* **2020**, *40*, 207–216. [CrossRef]
41. Khanna, D.; Denton, C.P.; Lin, C.J.F.; van Laar, J.M.; Frech, T.M.; Anderson, M.E.; Baron, M.; Chung, L.; Fierlbeck, G.; Lakshminarayanan, S.; et al. Safety and efficacy of subcutaneous tocilizumab in systemic sclerosis: Results from the open-label period of a phase II randomised controlled trial (faSScinate). *Ann. Rheum. Dis.* **2018**, *77*, 212–220. [CrossRef] [PubMed]
42. Khanna, D.; Lin, C.J.F.; Furst, D.E.; Goldin, J.; Kim, G.; Kuwana, M.; Allanore, Y.; Matucci-Cerinic, M.; Distler, O.; Shima, Y.; et al. Tocilizumab in systemic sclerosis: A randomised, double-blind, placebo-controlled, phase 3 trial. *Lancet Respir. Med.* **2020**, *8*, 963–974. [CrossRef]
43. Allanore, Y.; Simms, R.; Distler, O.; Trojanowska, M.; Pope, J.; Denton, C.P.; Varga, J. Systemic sclerosis. *Nat. Rev. Dis. Primers* **2015**, *1*, 15002. [CrossRef] [PubMed]
44. Volkmann, E.R.; Varga, J. Emerging targets of disease-modifying therapy for systemic sclerosis. *Nat. Rev. Rheumatol.* **2019**, *15*, 208–224. [CrossRef] [PubMed]
45. Kowal-Bielecka, O.; Fransen, J.; Avouac, J.; Becker, M.; Kulak, A.; Allanore, Y.; Distler, O.; Clements, P.; Cutolo, M.; Czirjak, L.; et al. Update of EULAR recommendations for the treatment of systemic sclerosis. *Ann. Rheum. Dis.* **2017**, *76*, 1327–1339. [CrossRef]

Article

Bisphosphonates for the Treatment of Calcinosis Cutis—A Retrospective Single-Center Study

Lilian Rauch [1], Rüdiger Hein [1], Tilo Biedermann [1], Kilian Eyerich [1,2] and Felix Lauffer [1,*]

[1] Department of Dermatology and Allergy, Technical University of Munich, 80802 Munich, Germany; lilianrauch@t-online.de (L.R.); ruediger.hein@mri.tum.de (R.H.); tilo.biedermann@tum.de (T.B.); kilian.eyerich@tum.de (K.E.)

[2] Center for Molecular Medicine, Division of Dermatology and Venereology, Department of Medicine Solna, Karolinska Institute, 17177 Stockholm, Sweden

* Correspondence: felix.lauffer@mri.tum.de; Tel.: +49-89-4140-3170

Abstract: (1) Background: Calcinosis cutis is a frequent symptom of autoimmune connective tissue diseases leading to pain, transcutaneous expulsion of calcified material and bacterial superinfection. There is a high need for new therapeutic options as no standardized treatment algorithm is established. While case reports indicate beneficial effects of bisphosphonates, standardized evaluation of treatment effects is missing. (2) Methods: In this retrospective analysis we evaluate the effects of intravenous pamidronate, a second-generation bisphosphonate, in seven patients with calcinosis cutis using consecutive clinical pictures, radiological examinations and patient's subjective evaluation. (3) Results: 5/6 patients reported a reduction of pain, improvement of general condition and cessation of calcinosis progression. Regression of skin lesions was detectable in clinical pictures of 2/6 patients, while 1/6 patients had stable disease. Radiological examination revealed improvement or stable disease in 3/5 patients. Fever was the most common side effect. One out of seven patients developed osteonecrosis of the jaw. (4) Conclusions: Bisphosphonates appear to have beneficial effects in a subgroup of calcinosis cutis patients. While patient's subjective evaluation was mainly positive, objective assessments showed improvement in approximately half of the cases. With regard to potential severe side effects, a careful risk-benefit evaluation is necessary before treatment initiation.

Keywords: calcinosis cutis; bisphosphonates; pamidronate; autoimmune connective tissue diseases; systemic sclerosis; dermatomyositis; treatment

Citation: Rauch, L.; Hein, R.; Biedermann, T.; Eyerich, K.; Lauffer, F. Bisphosphonates for the Treatment of Calcinosis Cutis—A Retrospective Single-Center Study. *Biomedicines* **2021**, *9*, 1698. https://doi.org/10.3390/biomedicines9111698

Academic Editor: Jun Lu

Received: 28 September 2021
Accepted: 12 November 2021
Published: 16 November 2021

1. Introduction

Calcinosis cutis (CC) is a frequent symptom of autoimmune connective tissue diseases, which severely impacts patient's quality of life. It is defined as a deposition of insoluble calcium in the subcutaneous tissue, while serum calcium and/or phosphate levels are not elevated in most cases [1]. CC forms subcutaneous nodules, which are usually localized at mechanically stressed body sites, such as fingers or above joints. Often it causes inflammation and devastating perforations of the skin, leading to pain and baring the risk of bacterial superinfection. CC can occur in all autoimmune connective tissue diseases, but is most frequently seen in patients suffering from dermatomyositis and systemic sclerosis [2]. The pathogenesis is so far poorly understood. Chronic hypoxia, local trauma and inflammation are potential factors favouring the de novo development and maintenance of CC. Known associations to CC are female sex, chronic disease course, osteoporosis, digital ulcers and anti-centromere autoantibodies [3]. The treatment of CC is challenging as no standardized treatment guidelines have been established. Most recommendations are based on case reports or small case series, while randomized controlled trials are missing. Published treatment attempts include intralesional corticosteroid injections, topical or intravenous thiosulfate, warfarin, minocycline, colchicine, calcium channel blockers, intravenous immunoglobulins, surgical excision and ablative LASER therapy [4]. However, variability of individual response to therapy is high and additional therapeutic concepts are needed.

Pamidronate is a second-generation disodium bisphosphonate, which is approved for the treatment of hypercalcemia of malignancy, osteolytic bone metastases of breast cancer and multiple myeloma, as well as Paget disease of the bone. Second-generation bisphosphonates inhibit farnesyl pyrophosphate synthase, a key enzyme of the mevalonate pathway. The inhibition leads to a reduction of posttranslational modifications of guanosine triphosphate binding proteins, which are essential for the activity of osteoclasts [5]. Based on this mode of action, bisphosphonates prevent atrophy of bone tissue and lead to a decrease of calcium serum levels. Stabilization and/or improvement of CC was observed in two case reports [6,7]. However, a case series of six patients showed no beneficial effects of etidronate on clinical and radiological assessments of CC in six patients with dermatomyositis and systemic sclerosis [8].

In this retrospective study we evaluate the effects of intravenous pamidronate in seven patients with dermatomyositis, systemic sclerosis or mixed connective tissue diseases. We assessed clinical outcomes based on photo documentation and radiological examination, as well as subjective therapy-related benefits perceived by the patients using questionnaires.

2. Materials and Methods

Seven patients diagnosed with "calcinosis cutis" who received treatment with pamidronate at our department between 2010 and 2020 were identified. This comprises all cases of severe CC in our department during this period of time. Of those, 2 patients died before the beginning of the study. Clinical information was assessed by clinical records, radiological reports and photography taken during hospital stays. Clinical and radiological improvement were always evaluated in the last photograph or radiological examination, which was available. The median time between the first pamidronate infusion and the last photograph for clinical assessment was 41.5 (9–66) months. The median time between the first pamidronate infusion and the last radiologic assessment was 19 (11–37) months. Five patients filled out a questionnaire (see Supplementary Materials) about personal perceptions of bisphosphonate therapy effects. The median time between the last pamidronate infusion and the completion of the questionnaire (follow up) was 30 (15–98) months. The full questionnaire can be found in Supplementary Material S1. Data was visualized using GraphPad Prism 7 software (San Diego, CA, USA). The study protocol was approved by the Ethics Committee of the School of Medicine, Technical University of Munich (321/20) and written informed consent was obtained from all patients.

3. Results

3.1. Patient's Characteristics

Between 2010 and 2020 six female and one male patient received 70–75 mg of intravenous pamidronate every 12 weeks for three consecutive days for the treatment of CC at our department (Table 1). Underlying autoimmune diseases were dermatomyositis (n = 2), mixed connective tissue diseases (n = 3) and systemic sclerosis (n = 2). The time between the onset of CC and the initiation of the pamidronate treatment was 7 $\pm$ 4.2 years. Six out of seven patients received immunosuppressive co-medication before and during pamidronate treatment, mostly with prednisolone, methotrexate or intravenous immunoglobulins. The number of bisphosphonate cycles per patient was 8.6 $\pm$ 5.7 and 75 mg was the dosage administered in 6/7 cases. Fever was the most common side effect present in 3/7 patients. Additionally, one patient developed low blood pressure and one patient limb pain and shivering. The most severe adverse event was a necrosis of the jaw, which occurred in 1/7 patients. Four patients did not report any adverse effects.

Table 1. Patients' characteristics.

Participant ID	1	2	3	4	5	6	7
sex	male	female	female	female	female	female	female
diagnosis	dermato-myositis	dermato-myositis	mixed connective tissue disease	mixed connective tissue disease	systemic sclerosis	mixed connective tissue disease	systemic sclerosis
age at disease onset (years)	26	46	48	10	58	47	38
age at onset of calcinosis cutis (years)	26	unknown	50	14	59	49	unknown
time between beginning of calcinosis cutis and initiation of bisphospho-nate treatment (years)	8	unknown	8	13	4	2	unknown
number of bisphospho-nate cycles	7	12	3	6	12	18	2
dosage (mg)	70	75	75	75	75	75	75
previous treatments	intravenous immuno-globulins cyclophos-phamid methyl-prednisolone azathioprine	methotrexate prednisolone azathioprine	intravenous immuno-globuline methotrexate prednisolone azathioprine chloroquine PUVA, iloprost	methotrexate prednisolone cyclosporine chloroquine	bosentan pentoxi-fylline	hydroxy-chloroquine azathioprin mycophenolate mofetil intravenous immuno-globulins prednisolone iloprost	prednisolone
treatment during bisphospho-nate therapy	methyl-prednisolone	methotrexate prednisolone	intravenous immuno-globulines methotrexate prednisolone	methotrexate prednisolone	pentoxi-fylline	-	prednisolone
adverse events	-	fever low blood pressure	-	fever shivering limb pain	fever osteonecrosis of the jaw	-	-

3.2. Patients Evaluate Bisphosphonate Treatment Positively

To assess patient satisfaction with bisphosphonate therapy, we asked five patients to estimate the effects on different aspects of CC (Figure 1) and analyzed the well-documented physician's records of one deceased patient.

Three out of six patients noticed a softening of the CC skin lesions. Furthermore, 5/6 patients reported a reduction of pain, cessation of CC progression and improvement of their general condition. Three patients suffered from immobility of one or more joints before starting the bisphosphonate therapy. Of those, only one observed an improvement of mobility under treatment. Overall, 5/6 patients evaluated bisphosphonate therapy positively.

Objective assessment reveals heterogeneous response to bisphosphonate therapy. While subjective perception is often biased by expected benefits, we evaluated clinical images and radiological reports before and after bisphosphonate therapy. Data were available for six patients. Here, we identified two patients with clear regression of CC skin lesions visible in clinical pictures after the initiation of bisphosphonate therapy (Figure 2).

One patient showed stable disease according to physician's records. Three patients, however, developed additional or showed a progression of pre-existing CC lesions (Figure 3).

Radiological reports were available for 5/7 patients. The size of the CC was measured by conventional X-ray, computer tomography or magnetic resonance imaging. Here, improvement was reported in 3/5 patients, whereas 1/5 patients showed stable disease and 1/5 patients radiological progression. Concomitant treatment with prednisolone or methotrexate was not associated with a better outcome of CC. Thus, bisphosphonate treatment resulted in clinical and/or radiological improvement in approximately half of the patients, while 3/6 patients did not profit according to objective measurements (Figure 4).

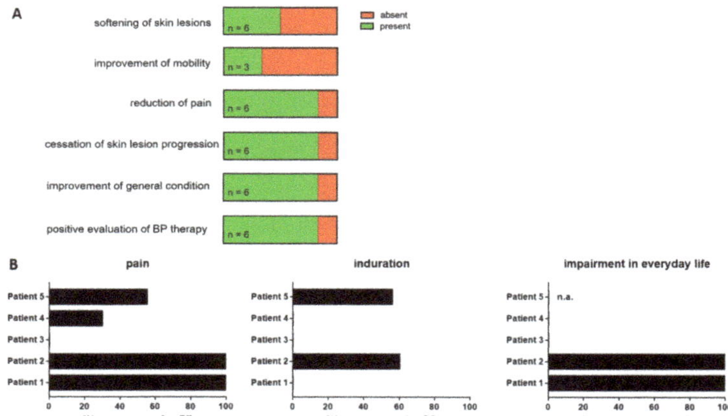

Figure 1. Patient's subjective evaluation of bisphosphonate therapy effects. (**A**) Absence or presence of different treatment outcomes after initiation of bisphosphonate therapy as assessed by a postal questionnaire. (**B**) Percent improvement of pain, induration of calcinosis cutis skin lesions and impairment in everyday life after initiation of bisphosphonate therapy for each individual patient taking part in the questionnaire evaluation.

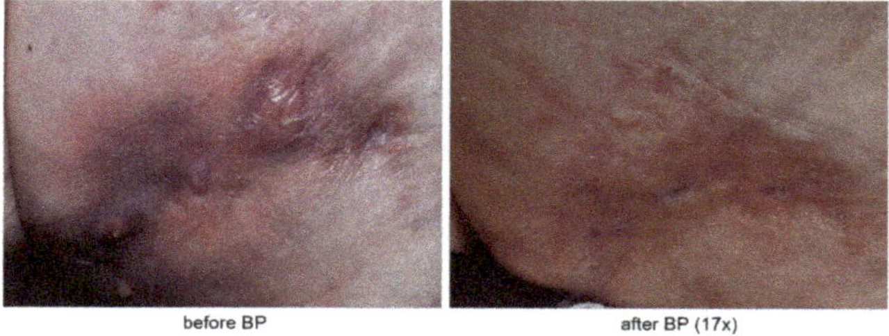

Figure 2. Representative clinical pictures of calcinosis cutis before and after bisphosphonate (BP) therapy (17 cycles).

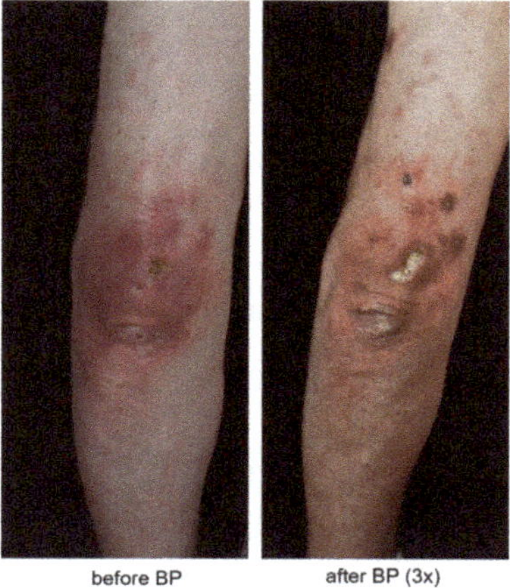

Figure 3. Representative clinical pictures of calcinosis cutis before and after bisphosphonate (BP) therapy.

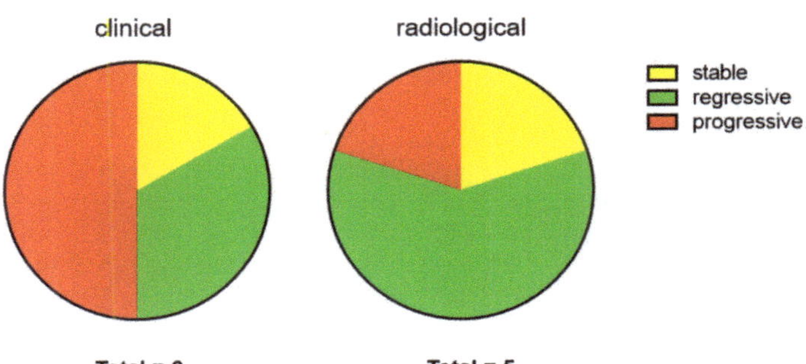

Figure 4. Objective evaluation of bisphosphonate therapy effects by dermatologists (clinical) and radiological examinations.

4. Discussion

In this retrospective study we observed diverse clinical response to bisphosphonate therapy for the treatment of CC. While the vast majority of patients evaluated therapeutic effects positively, an objective improvement could only be observed in half of the patients. The most common side effect was fever. However, one patient developed osteonecrosis of the jaw, a known side effect of bisphosphonate treatment, which led to the discontinuation of therapy.

CC is a common symptom of autoimmune connective tissue diseases and one of the biggest therapeutic challenges. While negative effects on patient's quality of life are immense, evidence for successful therapeutic concepts is rare. Dealing with different diseases, disease stages and heterogenous patient cohorts hampers the generation of homogenized and comparable patient cohorts and thereby the creation of evidence-based treatment guidelines [9]. Therefore, retrospective analysis of treatment outcome is a valuable tool to gain more insight into the therapeutic potential of different compounds.

Previous reports about bisphosphonates for CC are contradictory. Apart from positive results in single patient reports, the so far largest case series with six patients showed no beneficial effects of etidronate on CC [8]. Publication bias might account for an overestimation of therapeutical effects, as negative results are rarely published [10]. Our analysis includes all patients treated at our department within the last ten years. We found objective improvement or stabilization of CC in 4/7 patients. The diverse clinical outcome might be based on variance in disease duration, different mechanisms of CC formation or so far unknown individual factors contributing to CC development and maintenance.

The way that pamidronate influences CC is still speculative. Bisphosphonates decrease systemic calcium levels by inhibiting osteoclast activity. Thus, one potential mechanism of how pamidronate influences CC is an osmotic adjustment favouring resorption of calcium deposits from the subcutaneous tissue into blood vessels by lowering serum calcium levels. However, in contrast to metastatic calcification, CC in autoimmune diseases often develops independently from systemic shifts of calcium homeostasis [2]. Therefore, bisphosphonates might also have additional effects on CC lesions. Like many other mechanisms of autoimmune diseases, CC is thought to proceed in different stages. In the early stage, hypoxia or repeated local trauma triggers chronic inflammation at acral sites of the body. Destruction of connective tissue, tissue necrosis, immunological response to tissue degeneration and decreased activity of calcification inhibitors favour the accumulation of calcium depositions [11]. Often calcium deposits are accompanied by granulomatous inflammation comprising macrophages and foreign body giant cells. There is increasing evidence that bisphosphonate might also have anti-inflammatory effects on monocytes and macrophages [12,13]. Therefore, regression of CC under pamidronate therapy might be based on interrupting chronic inflammation. Given this potential mode of action, an analysis of the immunological signature accompanying CC might allow future stratification of patients. Patients with a high number of macrophages and monocytes due to foreign body reaction might be prone to better respond to bisphosphonate treatment than patients without inflammatory activity around CC lesions. Therefore, future studies should include histological and immunological evaluation of CC biopsies before the initiation of new therapeutic strategies.

Our study has limitations: First, we performed a retrospective study and not a randomized, placebo-controlled clinical trial. Therefore, there is no control group of untreated patients or patients receiving a placebo. Second, as we summarized data from clinical routines, the time between the first pamidronate infusion and radiological assessment varies between different patients. Third, there is a potential recall bias as patients filled out the questionnaire years after the last pamidronate cycle. However, as CC is a very difficult-to-treat symptom of connective tissue diseases with a high unmet medical need for new therapeutic options, this comprehensive retrospective analysis including subjective and objective outcome parameters is of high value for clinicians treating patients with CC.

In summary our study demonstrates that a subgroup of CC patients might profit from bisphosphonate treatment. However, as parameters predicting therapeutic response are still missing and severe adverse events can occur, treatment with bisphosphonates has to be evaluated carefully. Assessment of treatment outcomes should be based on objective findings rather than on a patient's subjective perception, which is biased by individual expectations. Future studies better investigating the exact mechanisms of CC formation in autoimmune diseases will allow better identification of patient subgroups with a high probability to respond well to bisphosphonate treatment.

Supplementary Materials: The following are available online at https://www.mdpi.com/article/10.3390/biomedicines9111698/s1. Questionnaire for the research project: Disodium pamidronate for the treatment of calcinosis cutis–a retrospective single-center study.

Author Contributions: L.R. collected clinical data from patient's records, clinical pictures, radiological examinations and performed the questionnaire study. R.H., K.E. and T.B. critically reviewed the manuscript. F.L. supervised the study, created the figures and wrote the manuscript. All authors have read and agreed to the published version of the manuscript.

Funding: This research received no external funding.

Institutional Review Board Statement: The study was conducted according to the guidelines of the Declaration of Helsinki, and approved by the Ethics Committee of the School of Medicine, Technical University of Munich (Approval code: 321/20, Approval date: 22 June 2020).

Informed Consent Statement: Informed consent was obtained from all subjects involved in the study. The manuscript does not contain patient's identifying information.

Data Availability Statement: The authors confirm that the data supporting the findings of this study are available within the article and its Supplementary Materials.

Conflicts of Interest: The authors declare no conflict of interest.

References

1. Reiter, N.; El-Shabrawi, L.; Leinweber, B.; Berghold, A.; Aberer, E. Calcinosis cutis: Part I. Diagnostic pathway. *J. Am. Acad. Dermatol.* **2011**, *65*, 1–12. [CrossRef] [PubMed]
2. Gutierrez, A., Jr.; Wetter, D.A. Calcinosis cutis in autoimmune connective tissue diseases. *Dermatol. Ther.* **2012**, *25*, 195–206. [CrossRef] [PubMed]
3. Valenzuela, A.; Baron, M.; Canadian Scleroderma Research Group; Herrick, A.L.; Proudman, S.; Stevens, W.; Australian Scleroderma Interest Group; Rodriguez-Reyna, T.S.; Vacca, A.; Medsger, T.A., Jr.; et al. Calcinosis is associated with digital ulcers and osteoporosis in patients with systemic sclerosis: A Scleroderma Clinical Trials Consortium study. *Semin. Arthritis Rheum.* **2016**, *46*, 344–349. [CrossRef] [PubMed]
4. Valenzuela, A.; Song, P.; Chung, L. Calcinosis in scleroderma. *Curr. Opin. Rheumatol.* **2018**, *30*, 554–561. [CrossRef] [PubMed]
5. Drake, M.T.; Clarke, B.L.; Khosla, S. Bisphosphonates: Mechanism of action and role in clinical practice. *Mayo Clin. Proc.* **2008**, *83*, 1032–1045. [CrossRef] [PubMed]
6. Fujii, N.; Hamano, T.; Isaka, Y.; Ito, T.; Imai, E. Risedronate: A possible treatment for extraosseous calcification. *Clin. Calcium* **2005**, *15* (Suppl. 1), 75–78. [PubMed]

7. Rabens, S.F.; Bethune, J.E. Disodium etidronate therapy for dystrophic cutaneous calcification. *Arch. Dermatol.* **1975**, *111*, 357–361. [CrossRef] [PubMed]
8. Metzger, A.L.; Singer, F.R.; Bluestone, R.; Pearson, C.M. Failure of disodium etidronate in calcinosis due to dermatomyositis and scleroderma. *N. Engl. J. Med.* **1974**, *291*, 1294–1296. [CrossRef] [PubMed]
9. Reiter, N.; El-Shabrawi, L.; Leinweber, B.; Berghold, A.; Aberer, E. Calcinosis cutis: Part II. Treatment options. *J. Am. Acad. Dermatol.* **2011**, *65*, 15–22. [CrossRef] [PubMed]
10. Mlinaric, A.; Horvat, M.; Supak Smolcic, V. Dealing with the positive publication bias: Why you should really publish your negative results. *Biochem. Med.* **2017**, *27*, 030201. [CrossRef] [PubMed]
11. Jimenez-Gallo, D.; Ossorio-Garcia, L.; Linares-Barrios, M. Calcinosis Cutis and Calciphylaxis. *Actas Dermosifiliogr.* **2015**, *106*, 785–794. [CrossRef] [PubMed]
12. Rogers, T.L.; Holen, I. Tumour macrophages as potential targets of bisphosphonates. *J. Transl. Med.* **2011**, *9*, 177. [CrossRef] [PubMed]
13. Roelofs, A.J.; Thompson, K.; Ebetino, F.H.; Rogers, M.J.; Coxon, F.P. Bisphosphonates: Molecular mechanisms of action and effects on bone cells, monocytes and macrophages. *Curr. Pharm. Des.* **2010**, *16*, 2950–2960. [CrossRef] [PubMed]

MDPI
St. Alban-Anlage 66
4052 Basel
Switzerland
Tel. +41 61 683 77 34
Fax +41 61 302 89 18
www.mdpi.com

Biomedicines Editorial Office
E-mail: biomedicines@mdpi.com
www.mdpi.com/journal/biomedicines

www.ingramcontent.com/pod-product-compliance
Lightning Source LLC
LaVergne TN
LVHW071033170726
843515LV00006B/1270

* 9 7 8 3 0 3 6 5 6 1 2 1 9 *